Introduction to Visual SLAM

Xiang Gao · Tao Zhang

Introduction to Visual SLAM

From Theory to Practice

中国工信出版集团
電子工業出版社
PUBLISHING HOUSE OF ELECTRONICS INDUSTRY
http://www.phei.com.cn
Springer

Xiang Gao
Tsinghua University
Beijing, China

Tao Zhang
Tsinghua University
Beijing, China

ISBN 978-981-16-4938-7 ISBN 978-981-16-4939-4 (eBook)
https://doi.org/10.1007/978-981-16-4939-4

Jointly published with Publishing House of Electronics Industry
The print edition is not for sale in China (Mainland). Customers from China (Mainland) please order the print book from: Publishing House of Electronics Industry.

Translation from the Chinese Simplified language edition: 视觉SLAM十四讲: 从理论到实践 (第2版) by Xiang Gao, and Tao Zhang, © Publishing House of Electronics Industry 2019. Published by Publishing House of Electronics Industry. All Rights Reserved.

This Springer imprint is published by the registered company Springer Nature Singapore Pte Ltd.
The registered company address is: 152 Beach Road, #21-01/04 Gateway East, Singapore 189721, Singapore

To my beloved Lilian and Shenghan

Preface

What is This Book Talking About?

This book introduces visual SLAM, and it is probably the first Chinese book solely focused on this specific topic. With a lot of help from the commit, it was translated into English in 2020.

So, what is SLAM?

SLAM stands for **S**imultaneous **L**ocalization **a**nd **M**apping. It usually refers to a robot or a moving rigid body, equipped with a specific **sensor**, estimates its **motion** and builds a **model** (certain kinds of description) of the surrounding environment, without a *priori* information [1]. If the sensor referred to here is mainly a camera, it is called **Visual SLAM**.

Visual SLAM is the subject of this book. We deliberately put a long definition into one single sentence so that the readers can have a clear concept. First of all, SLAM aims at solving the *localization* and *map building* issues at the same time. In other words, it is a problem of how to estimate the location of a sensor itself, while estimating the model of the environment. So how to achieve it? SLAM requires a good understanding of sensor information. A sensor can observe the external world in a particular form, but the specific approaches for utilizing such observations are usually different. And, why is this problem worth spending an entire book to discuss? Because it is difficult, especially if we want to do SLAM in **real-time** and **without any prior knowledge**. When we talk about visual SLAM, we need to estimate the trajectory and map based on a set of continuous images (which form a video sequence).

This seems to be quite intuitive. When we human beings enter an unfamiliar environment, aren't we doing exactly the same thing? So, the question is whether we can write programs and make computers do so.

At the birth of computer vision, people imagined that one-day computers could act like humans, watching and observing the world and understanding the surrounding environment. The ability to explore unknown areas is a beautiful and romantic dream, attracting numerous researchers striving on this problem day and night [2]. We thought that this would not be that difficult, but the progress turned out to be not as smooth as expected. Flowers, trees, insects, birds, and animals are recorded so differently in computers: they are just numerical matrices consisted of numbers. To make computers understand the contents of images is as difficult as making us humans understand those blocks of numbers. We didn't even know how we understand images, nor do we know how to make computers do so. However, after decades of struggling, we finally started to see signs of success—through Artificial Intelligence (AI) and Machine Learning (ML) technologies, which gradually enable computers to recognize objects, faces, voices, texts, although in a way (probabilistic modeling) that is still so different from us.

On the other hand, after nearly three decades of development in SLAM, our cameras begin to capture their movements and know their positions. However, there is still a massive gap between the capability of computers and humans. Researchers have successfully built a variety of real-time SLAM systems. Some of them can efficiently track their locations, and others can even do the three-dimensional reconstruction in real-time.

This is really difficult, but we have made remarkable progress. What's more exciting is that, in recent years, we have seen the emergence of a large number of SLAM-related applications. The sensor location could be very useful in many areas: indoor sweeping machines and mobile robots, self-driving cars, Unmanned Aerial Vehicles (UAVs), Virtual Reality (VR), and Augmented Reality (AR). SLAM is so important. Without it, the sweeping machine cannot maneuver in a room autonomously but wandering blindly instead; domestic robots cannot follow instructions to accurately reach a specific room; Virtual reality devices will always be limited within a seat. If none of these innovations could be seen in real life, what a pity it would be. Today's researchers and developers are increasingly aware of the importance of SLAM technology. SLAM has over 30 years of research history, and it has been a hot topic in both robotics and computer vision communities. Since the twenty-first century, visual SLAM technology has undergone a significant change and breakthrough in both theory and practice and is gradually moving from laboratories into the real-world. At the same time, we regretfully find that, at least in the Chinese language, SLAM-related papers and books are still very scarce, making many beginners of this area unable to get started smoothly. Although SLAM's theoretical framework has basically become mature, implementing a complete SLAM system is still very challenging and requires a high level of technical expertise. Researchers new to the area have to spend a long time learning a significant amount of scattered knowledge and often have to go through several detours to get close to the real core.

This book systematically explains the visual SLAM technology. We hope that it will (at least partially) fill the current gap. We will detail SLAM's theoretical background, system architecture, and the various mainstream modules. At the same time, we emphasize the practice: all the essential algorithms introduced in this book will be provided with runnable code that can be tested by yourself so that readers can reach a more in-depth understanding. Visual SLAM, after all, is a technology for real applications. Although the mathematical theory can be beautiful, if you cannot convert it into code, it will be like a castle in the air, bringing little practical impact.

We believe that practice brings real knowledge (and true love). After getting your hands dirty with the algorithms, you can truly understand SLAM and claim that you have fallen in love with SLAM research.

Since its inception in 1986 [3], SLAM has been a hot research topic in robotics. It is very difficult to provide a complete introduction to all the algorithms and their variants in the SLAM history, and we consider it unnecessary as well. This book will first introduce the background knowledge, such as the 3D geometry, computer vision, state estimation theory, and Lie Group/Lie algebra. We will show the trunk of the SLAM tree and omit those complicated and oddly-shaped leaves. We think this is effective. If the reader can master the trunk's essence, they have already gained the ability to explore the frontier research details. So we aim to help SLAM beginners quickly grow into qualified researchers and developers. On the other hand, even if you are already an experienced SLAM researcher, this book may reveal areas that you are unfamiliar with and provide you with new insights.

There have already been a few SLAM-related books around, such as *Probabilistic Robotics* [4], *Multiple View Geometry in Computer Vision* [2], and *State Estimation for Robotics: A Matrix-Lie-Group Approach* [5]. They provide rich content, comprehensive discussions, and rigorous derivations, and therefore are the most popular textbooks among SLAM researchers. However, there are two critical issues: Firstly, the purpose of these books is often to introduce the fundamental mathematical theory, with SLAM being only one of its applications. Therefore, they cannot be considered as specifically visual SLAM focused. Secondly, they place great emphasis on mathematical theory but are relatively weak in programming. This makes readers still fumbling when trying to apply the knowledge they learn from the books. Our belief is one can only claim a real understanding of a problem only after coding, debugging, and tweaking algorithms and parameters with his own hands.

This book will introduce the history, theory, algorithms, and research status in SLAM and explain a complete SLAM system by decomposing it into several modules: *visual odometry*, *backend optimization*, *map building*, and *loop closure detection*. We will accompany the readers step by step to implement each core algorithm, discuss why they are effective, under what situations they are ill-conditioned, and guide them by running the code on your own machines. You will be exposed to the critical mathematical theory and programming knowledge and will use various libraries including *Eigen*, *OpenCV*, *PCL*, *g2o*, and *Ceres*, and learn their usage in Linux.

Well, enough talking, wish you a pleasant journey!

How to Use This Book?

This book is entitled as *Introduction to Visual SLAM: From Theory to Practice*. We will organize the contents into lectures like studying in a classroom. Each lecture focuses on one specific topic, organized in a logical order. Each chapter will include both a *theoretical* part and a *practical* part, with the theoretical usually coming first. We will introduce the mathematics essential to understand the algorithms, and most of the time in a narrative way, rather than in a *definition, theorem, inference* approach adopted by most mathematical textbooks. We think this will be much easier to understand, but of course, with the price of being less rigorous sometimes. In practical parts, we will provide code, discuss the various components' meaning, and demonstrate some experimental results. So, when you see chapters with the word practice in the title, you should turn on your computer and start to program with us, joyfully.

The book can be divided into two parts: The first part will be mainly focused on fundamental math knowledge, which contains

1. Preface (the one you are reading now), introducing the book's contents and structure.
2. Lecture 1: an overview of a SLAM system. It describes each module of a typical SLAM system and explains what to do and how to do it. The practice section introduces basic C++ programming in a Linux environment and the use of an IDE.
3. Lecture 2: rigid body motion in 3D space. You will learn about rotation matrices, quaternions, Euler angles and practice them with the *Eigen* library.
4. Lecture 3: Lie group and Lie algebra. It doesn't matter if you have never heard of them. You will learn the basics of the Lie group and manipulate them with *Sophus*.
5. Lecture 4: pinhole camera model and image expression in computer. You will use *OpenCV* to retrieve the camera's intrinsic and extrinsic parameters and generate a point cloud using the depth information through Point Cloud Library (*PCL*) .
6. Lecture 5: nonlinear optimization, including state estimation, least squares, and gradient descent methods, e.g., Gauss-Newton and Levenburg-Marquardt method. You will solve a curve-fitting problem using the *Ceres* and *g2o* library. From lecture 6, we will be discussing SLAM algorithms, starting with visual odometry (VO) and followed by the map building problems:
7. Lecture 6: feature-based visual odometry, which is currently the mainstream in VO. Contents include feature extraction and matching, epipolar geometry calculation, Perspective-n-Point (PnP) algorithm, Iterative Closest Point (ICP) algorithm, and Bundle Adjustment (BA), etc. You will run these algorithms either by calling *OpenCV* functions or constructing your own optimization problem in *Ceres* and *g2o*.
8. Lecture 7: direct (or intensity-based) method for VO. You will learn the optical flow principle and the direct method. The practice part is about writing single-layer and multi-layer optical flow and direct method to implement a two-view VO.
9. Lecture 8: backend optimization. We will discuss Bundle Adjustment in detail and show the relationship between its sparse structure and the corresponding graph model. You will use *Ceres* and *g2o* separately to solve the same BA problem.

10. Lecture 9: pose graph in the backend optimization. Pose graph is a more compact representation for BA, which converts all map points into constraints between keyframes. You will use *g2o* to optimize a pose graph.
11. Lecture 10: loop closure detection, mainly Bag-of-Word (BoW) based method. You will use DBoW3 to train a dictionary from images and detect loops in videos.
12. Lecture 11: map building. We will discuss how to estimate the depth of pixels in monocular SLAM (and show why they are unreliable). Compared with monocular depth estimation, building a dense map with RGB-D cameras is much easier. You will write programs for epipolar line search and patch matching to estimate depth from monocular images and then build a point cloud map and octagonal treemap from RGB-D data.
13. Lecture 12: a practice chapter for stereo VO. You will build a visual odometry framework by yourself by integrating the previously learned knowledge and solve problems such as frame and map point management, keyframe selection, and optimization control.
14. Lecture 13: current open-source SLAM projects and future development direction. We believe that after reading the previous chapters, you can understand other people's approaches easily and be capable of achieving new ideas of your own.

Finally, if you don't understand what we are talking about at all, congratulations! This book is right for you!

Source Code

All source code in this book is hosted on Github: https://github.com/gaoxiang12/slambook2

Note the *slambook2* refers to the second version in which we added a lot of extra experiments.

Check out the English version by: git checkout -b en origin-en

It is strongly recommended that readers download them for viewing at any time. The code is divided into chapters. For example, the contents of the *7th* lecture will be placed in folder *ch7*. Some of the small libraries used in the book can be found in the "3rdparty" folder as compressed packages. For large and medium-sized libraries like *OpenCV*, we will introduce their installation methods when they first appear. If you have any questions regarding the code, click the *issue* button on GitHub to submit. If there is indeed a problem with the code, we will correct them in time. If you are not accustomed to using Git, you can also click the *Download* button on the right side to download a zipped file to your local drive.

Targeted Readers

This book is for students and researchers interested in SLAM. Reading this book needs specific prerequisites, and we assume that you have the following knowledge:

- Calculus, Linear Algebra, Probability Theory. These are the fundamental mathematical knowledge that most readers should have learned during undergraduate study. You should at least understand what a matrix and a vector are, and what it means by doing differentiation and integration. For more advanced mathematical knowledge required, we will introduce in this book as we proceed.
- Basic C++ Programming. As we will be using C++ as our major programming language, it is recommended that the readers are at least familiar with its basic concepts and syntax. For example, you should know what a class is, how to use the C++ standard library, how to use template classes, etc. We will try our best to avoid using tricks, but we really cannot avert them in certain situations. We will also adopt some of the C++11 standards, but don't worry. They will be explained if necessary.
- Linux Basics. Our development environment is Linux instead of Windows, and we will only provide source code for Linux. We believe that mastering Linux is an essential skill for SLAM researchers, and please don't ask for Windows-related issues. After going through this book's contents, we think you will agree with us.[1] In Linux, the configuration of related libraries is so convenient, and you will gradually appreciate the benefit of mastering it. If you have never used a Linux system, it will be beneficial to find some Linux learning materials and spend some time reading them (the first few chapters of an introductory book should be sufficient). We do not ask readers to have superb Linux operating skills, but we do hope readers know how to find a terminal and enter a code directory. There are some self-test questions on Linux at the end of this chapter. If you have answers to them, you should be able to quickly understand the code in this book.

Readers interested in SLAM but do not have the knowledge mentioned above may find it difficult to proceed with this book. If you do not understand the basics of C++, you can read some introductory books such as C++ *Primer Plus*. If you do not have the relevant math knowledge, we also suggest reading some relevant math textbooks first. Nevertheless, most readers who have completed undergraduate study should already have the necessary mathematical backgrounds. Regarding the code, we recommend that you spend time typing them by yourself and tweaking the parameters to see how they affect outputs. This will be very helpful.

This book can be used as a textbook for SLAM-related courses or as self-study materials.

[1] Linux is not that popular in China as our computer science education starts very lately around the 1990s.

Style

This book covers both mathematical theory and programming implementation. Therefore, for the convenience of reading, we will be using different layouts to distinguish the contents.

1. Mathematical formulas will be listed separately, and important formulas will be assigned with an equation number on the right end of the line, for example,

$$\mathbf{y} = \mathbf{A}\mathbf{x}. \tag{1}$$

   Italics are used for scalars like a. Bold symbols are used for vectors and matrices like $\mathbf{a}$, $\mathbf{A}$. Hollow bold represents special sets, e.g., the real number set $\mathbb{R}$ and the integer set $\mathbb{Z}$. Gothic is used for Lie Algebra, e.g., $\mathfrak{se}(3)$.
2. Source code will be framed into boxes, using a smaller font size, with line numbers on the left. If a code block is long, the box may continue to the next page:

Listing 1 Code example:

```
#include <iostream>
using namespace std;

int main (int argc, char** argv) {
  cout << "Hello" << endl;
  return 0;
}
```

3. When the code block is too long or contains repeated parts with previously listed code, it is not appropriate to be listed entirely. We will only give the important parts and mark them with *part*. Therefore, we strongly recommend that readers download all the source code on GitHub and complete the exercises to better understand the book.
4. Due to typographical reasons, the book's code may be slightly different from the code in GitHub. In that case, please use the code on GitHub.
5. For each of the libraries we use, it will be explained in detail when first appearing but not repeated in the follow-up. Therefore, it is recommended that readers read this book in order.
6. A *goal of study* part will be presented at the beginning of each lecture. A summary and some exercises will be given at the end. The cited references are listed at the end of the book.
7. The chapters with an asterisk mark in front are optional readings, and readers can read them according to their interests. Skipping them will not hinder the understanding of subsequent chapters.
8. Important contents will be marked in **bold** or *italic*, as we are already accustomed to.

9. Most of the experiments we designed are demonstrative. Understanding them does not mean that you are already familiar with the entire library. Otherwise, this book will be an *OpenCV* or *PCL* document. So we recommend that you spend time on yourselves in further exploring the important libraries frequently used in the book.
10. The book's exercises and optional readings may require you to search for additional materials, so you need to learn to use search engines.

Exercises (Self-test Questions)

1. Suppose we have a linear equation $\mathbf{Ax} = \mathbf{b}$. If $\mathbf{A}$ and $\mathbf{b}$ are known, how to solve the $\mathbf{x}$? What are the requirements for $\mathbf{A}$ and $\mathbf{b}$ if we want a unique $\mathbf{x}$? (Hint: check the rank of $\mathbf{A}$ and $\mathbf{b}$).
2. What is a Gaussian distribution? What does it look like in a one-dimensional case? How about in a high-dimensional case?
3. What is the **class** in C++? Do you know STL? Have you ever used them?
4. How do you write a C++ program? (It's completely fine if your answer is "using Visual C++ 6.0" [2]).
5. Do you know the C++11 standard? Which new features have you heard of or used? Are you familiar with any other standard?
6. Do you know Linux? Have you used at least one of the popular distributions (not including Android), such as Ubuntu?
7. What is the directory structure of Linux? What basic commands do you know? (e.g., *ls*, *cat*, etc.)
8. How to install the software in Ubuntu (without using the Software Center)? What directories are software usually installed under? If you only know the fuzzy name of a software (for example, you want to install a library with the word "eigen" in its name), how to search it?
9. *Spend an hour learning *vim*. You will be using it sooner or later. You can *vimtutor* into a terminal and read through its contents. We do not require you to operate it very skillfully, as long as you can use it to edit the code in the process of learning this book. Do not waste time on its plugins for now. Do not try to turn vim into an IDE. We will only use it for text editing in this book.

Beijing, China

Xiang Gao
Tao Zhang

[2] As I know, many of our undergraduate students are still using this version of VC++ in the university.

Acknowledgments

In the process of writing this book, a large number of documents and papers have been referenced. Most of the theoretical knowledge of mathematics is the result of previous research, not my original creation. A small part of the experimental design also comes from various open-source code demonstration programs, but most of them are written by myself. In addition, there are some pictures taken from published journals or conference papers, which have been cited in the text. Unexplained images are either original or fetched from the Internet. I don't want to infringe anyone's picture copyright. If readers find any problems, please contact me to modify it.

As I'm not a native English speaker, the translation work is based on Google translation and some afterward modifications. If you think the quality of translation can be improved and willing to do this, please contact me or send an issue on Github. Any help will be welcome!

My friends, Dr. Yi Liu and Qinrei Yan, helped me a lot in the Chinese edition of this book. And I also thank them very much about this. Thanks for the following friend's help in the translation time: Nicolas Rosa, Carrie (Yan Ran), Collen Jones, Hong Ma. And also, thanks for your attention and support!

Please contact me through GitHub or email: gao.xiang.thu@gmail.com.

Contents

Part I
Fundamental Knowledge

Chapter 1
Introduction to SLAM

Check for updates

> **Goal of Study**
> 1. Understand which modules a visual SLAM framework consists of, and what task each module carries out.
> 2. Set up the programming environment, and prepare for experiments.
> 3. Understand how to compile and run a program under Linux. If there is a problem, how to debug it.
> 4. Learn the basic usage of CMake.

This lecture summarizes the structure of a visual SLAM system as an outline of subsequent chapters. The practice part introduces the fundamentals of environment setup and program development. We will make a small "Hello SLAM" program at the end.

1.1 Meet "Little Carrot"

Suppose we assembled a robot called *Little Carrot*, as shown in the following Fig. 1.1:

Although it looks a bit like the Android robot, it has nothing to do with the Android system. We put a laptop into its trunk (so that we can debug programs at any time). So, what do we expect it to do?

We hope *Little Carrot* has the ability of autonomous moving. Many domestic robots are placed statically on desktops. They can chat with people or play music, but a tablet computer nowadays can also deliver the same tasks. As a robot, we hope Little Carrot can move freely in a room. Wherever we say hello to it, it can come to us right away.

First of all, such a robot needs wheels and motors to move, so we installed wheels on it (gait control for humanoid robots is very complicated, which we will not be considered here). Now with the wheels, the robot is able to move, but without an

X. Gao and T. Zhang, *Introduction to Visual SLAM*,
https://doi.org/10.1007/978-981-16-4939-4_1

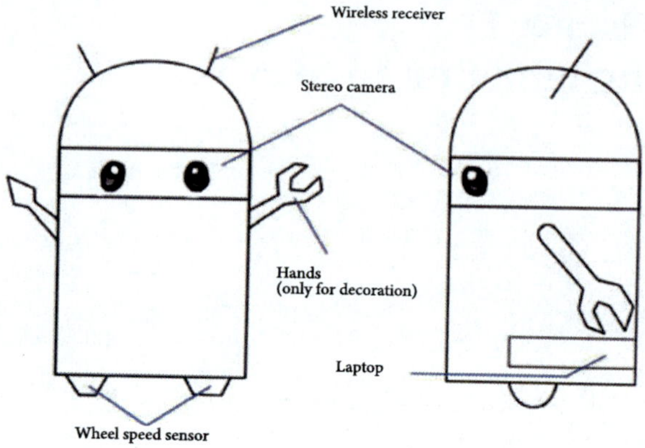

Fig. 1.1 The sketch of robot *Little Carrot*

effective navigation system, *Little Carrot* does not know where a target of action is, and it can do nothing but wander around blindly. Even worse, it may hit a wall and cause damage. To avoid this, we installed cameras on its head, with the intuition that such a robot should look similar to a human. Indeed, with eyes, brains, and limbs, humans can walk freely and explore any environment, so we (somehow naively) think that our robot should be able to achieve it too. Well, to make Little Carrot able to explore a room, we find it at least needs to know two things:

1. Where am I?—It's about *localization.*
2. What is the surrounding environment like?—It's about *map building.*

Localization and *map building* can be seen as the perception in both inward and outward directions. As a completely autonomous robot, *Little Carrot* needs to understand its own state (i.e., the location) and the external environment (i.e., the map). Of course, there are many different approaches to solve these two problems. For example, we can lay guiding rails on the floor of the room, or paste a lot of artificial markers such as QR code pictures on the wall, or mount radio localization devices on the table. If you are outdoor, you can also install a GNSS receiver (like the one in a cell phone or a car) on the head of Little Carrot. With these devices, can we claim that the localization problem has been resolved? Let's categorize these sensors (see Fig. 1.2) into two classes.

The first class is non-intrusive sensors, which are entirely self-contained inside a robot, such as wheel encoders, cameras, laser scanners, etc. They do not assume a cooperative environment around the robot. The other class is intrusive sensors depending on a prepared environment, such as the above-mentioned guiding rails, QR codes, etc. Intrusive sensors can usually locate a robot directly, solving the localization problem in a simple and effective manner. However, since they require changes in the environment, the usage scope is often limited to a certain degree. For example, if there is no GPS signal or guiding rails cannot be laid, what should we do in those cases?

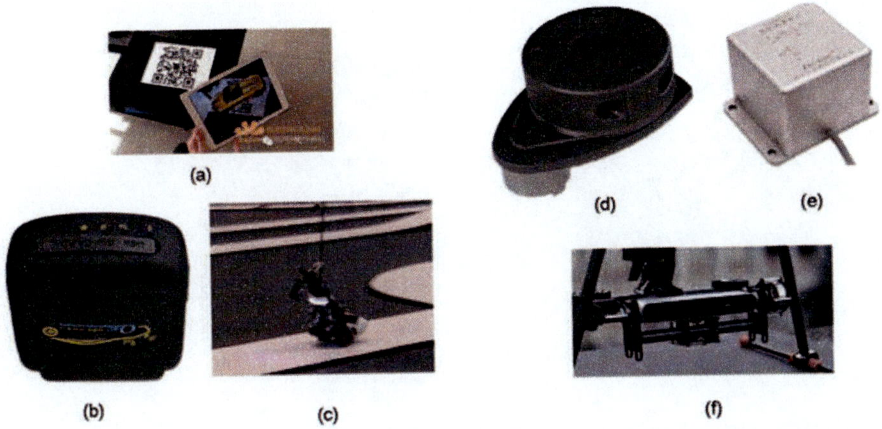

Fig. 1.2 Different kinds of sensors: **a** QR code; **b** GNSS receiver; **c** Guiding rails; **d** Laser range finder; **e** Inertial measurement unit; **f** Stereo camera

We can see that the intrusive sensors place certain constraints on the external environment. A localization system based on them can only function properly when those constraints are met in the real world. Otherwise, the localization approach cannot be carried out anymore, like GPS localization system normally doesn't work well in indoor environments. Therefore, although this type of sensor is reliable and straightforward, it does not work as a general solution. In contrast, non-intrusive sensors, such as laser scanners, cameras, wheel encoders, Inertial Measurement Units (IMUs), etc., can only observe indirect physical quantities rather than the direct location. For example, a wheel encoder measures the wheel rotation angle. An IMU measures the angular velocity and the acceleration. A camera or a laser scanner observes the external environment in a specific form like point-clouds and images. We have to apply algorithms to infer positions from these indirect observations. While this sounds like a roundabout tactic, the more obvious benefit is that it does not make any demands on the environment, making it possible for this localization framework to be applied to an unknown environment. Therefore, they are called self-localization in many research areas.

Looking back at the SLAM definitions discussed earlier, we emphasized an unknown environment in SLAM problems. In theory, we should not presume which environment the *Little Carrot* will be used (but in reality, we will have a rough range, such as indoor or outdoor), which means that we can not assume that external sensors like GPS can work smoothly. Therefore, the use of portable non-intrusive sensors to achieve SLAM is our main focus. In particular, when talking about visual SLAM, we generally refer to *cameras*' usage to solve the localization and map building problems.

Visual SLAM is the main subject of this book, so we are particularly interested in what the Little Carrot's eyes can do. The cameras used in SLAM are different from the commonly seen single-lens reflex (SLR) cameras. It is often much cheaper and does

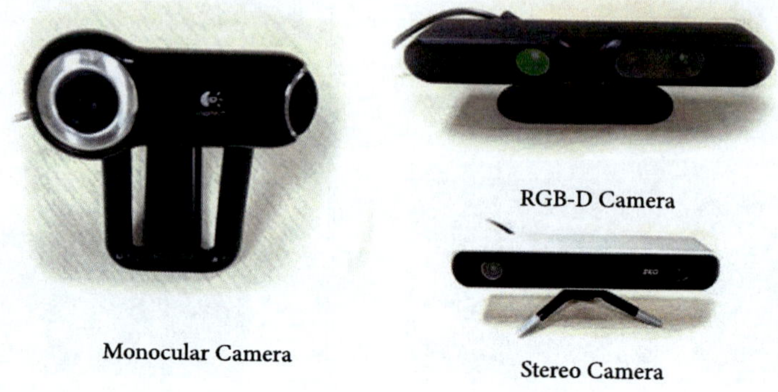

Fig. 1.3 Different kinds of cameras: monocular, RGB-D and stereo

not carry an expensive lens. It shoots at the surrounding environment at a specific rate, forming a continuous video stream. An ordinary camera can capture images at 30 frames per second, while high-speed cameras can do faster. The camera can be roughly divided into three categories: monocular, stereo, and RGB-D, as shown by the following Fig. 1.3. Intuitively, a monocular camera has only one camera. A stereo camera has two. The principle of an RGB-D camera is more complicated. In addition to collecting color images, it can also measure the distance of the scene from the camera for each pixel. RGB-D devices usually carry multiple cameras and may adopt a variety of different working principles. In the fifth lecture, we will detail their working principles, and readers just need an intuitive impression for now. Some new special and emerging camera types can also be applied to SLAM, such as panorama camera [6], event camera [7]. Although they are occasionally seen in SLAM applications, so far, they have not become mainstream. From the appearance, we can infer that Little Carrot seems to carry a stereo camera.

Now, let's take a look at the pros and cons of using different types of cameras for SLAM.

1.1.1 Monocular Camera

The SLAM system that uses only one camera is called Monocular SLAM. This sensor structure is particularly simple, and the cost is particularly low. Therefore the monocular SLAM has been very attractive to researchers. You must have seen the output data of a monocular camera: photo. Yes, what can we do with a single photo?

A photo is essentially a *projection* of a scene onto a camera's imaging plane. It reflects a three-dimensional world in a two-dimensional form. Evidently, there is one dimension lost during this projection process: the so-called *depth* (or distance). In a monocular case, we can not obtain the distance between objects in the scene and the

Fig. 1.4 We cannot tell if the people are real humans or just small toys from a single image

camera by using a single image. Later, we will see that this distance is critical for SLAM. Because we humans have seen many photos, we formed a natural sense of distances for most scenes, helping us determine the distance relationship among the objects in the image. For example, we can recognize objects in the image and correlate them with their approximate size obtained from daily experience. The close objects will occlude the distant objects; the sun, the moon, and other celestial objects are infinitely far away; an object will have shadow if it is under sunlight. This common-sense can help us determine the distance of objects, but there are also certain cases that confuse us, and we can no longer distinguish the distance and proper size of an object. The following Fig. 1.4 shows an example. We can not determine whether the figures are real people or small toys purely based on the image itself. Unless we change our view angle, explore the three-dimensional structure of the scene. It may be a big but far away object, but it may also be a close but small object. They may appear to be the same size in an image due to the perspective projection effect.

Since the image taken by a monocular camera is just a 2D projection of the 3D space, if we want to recover the 3D structure, we have to change the camera's view angle. Monocular SLAM adopts the same principle. We move the camera and estimate its motion, as well as the distances and sizes of the objects in the scene, namely the structure of the scene. So how do we estimate these movements and structures? From the everyday experience, we know that if a camera moves to the right, the objects in the image will move to the left, which gives us an inspiration of inferring motion. On the other hand, we also know that closer objects move faster, while distant objects move slower. Thus, when the camera moves, the movement of these objects on the image forms pixel *disparity*. Through calculating the disparity, we can quantitatively determine which objects are far away and which objects are close.

However, even if we know which objects are near and far, they are still only relative values. For example, when we are watching a movie, we can tell which objects in the movie scene are bigger than the others, but we can not determine the real size of those objects—are the buildings real high-rise buildings or just models on a table? Is it a real monster that destroys a building, or just an actor wearing special clothing? Intuitively, if the camera's movement and the scene size are doubled simultaneously, monocular cameras see the same. Likewise, multiplying this size by any factor, we will still get the same picture. This demonstrates that the trajectory and map obtained from monocular SLAM estimation will differ from the actual trajectory and map with an unknown factor, which is called the *scale*.[1] Since monocular SLAM can not determine this real scale purely based on images, this is also called the scale ambiguity.

In monocular SLAM, depth can only be calculated with translational movement, and the real scale cannot be determined. These two things could cause significant trouble when applying monocular SLAM into real-world applications. The fundamental cause is that depth can not be determined from a single image. So, in order to obtain real-scaled depth, we start to use stereo and RGB-D cameras.

1.1.2 Stereo Cameras and RGB-D Cameras

The purpose of using stereo and RGB-D cameras is to measure the distance between objects and the camera, to overcome the shortcomings of monocular cameras that distances are unknown. Once distances are known, the 3D structure of a scene can be recovered from a single frame and eliminates scale ambiguity. Although both stereo and RGB-D cameras can measure the distance, their principles are not the same. A stereo camera consists of two synchronized monocular cameras, displaced with a known distance, namely the baseline. Because the physical distance of the baseline is known, we can calculate each pixel's 3D position in a very similar way to our human eyes. We can estimate the objects' distances based on the differences between the images from the left and right eye, and we can try to do the same on computers (see Fig. 1.5). We can also extend stereo camera to multi-camera systems if needed, but basically, there is no much difference.

Stereo cameras usually require a significant amount of computational power to (unreliably) estimate depth for each pixel, which is really clumsy compared to human beings. The depth range measured by a stereo camera is related to the baseline length. The longer a baseline is, the farther it can measure. So stereo cameras mounted on autonomous vehicles are usually quite big. Depth estimation for stereo cameras is achieved by comparing images from the left and right cameras and not relying on

---

[1] We will explain the mathematical reason in the visual odometry chapter.

Fig. 1.5 Distance is calculated from the disparity of two stereo image pair

other sensing equipment. Thus stereo cameras can be applied both indoor and outdoor. The disadvantage of stereo cameras or multi-camera systems is that the configuration and calibration process is complicated. Their depth range and accuracy are limited by baseline length and camera resolution. Moreover, stereo matching and disparity calculation also consume many computational resources and usually require GPU or FPGA to accelerate to generate real-time depth maps. Therefore, in most state-of-the-art algorithms, the computational cost is still one of the major problems of stereo cameras.

Depth camera (also known as RGB-D camera, RGB-D will be used in this book) is a new type of camera rising since 2010. Similar to laser scanners, RGB-D cameras adopt infrared structure of light or Time-of-Flight (ToF) principles and measure the distance between objects and the camera by actively emitting light to the object and receive the returned light. This part is not solved by software as a stereo camera, but by physical sensors to save many computational resources compared to stereo cameras (see Fig. 1.6). Common RGB-D cameras include Kinect/Kinect V2, Xtion Pro Live, RealSense, etc. However, most of the RGB-D cameras still suffer from issues including narrow measurement range, noisy data, small field of view, susceptibility to sunlight interference, and unable to measure transparent material. For SLAM purposes, RGB-D cameras are mainly used in indoor environments and are not suitable for outdoor applications.[2]

We have discussed the common types of cameras, and we believe you should have gained an intuitive understanding of them. Now, imagine a camera is moving in a scene, we will get a series of continuously changing images.[3] The goal of visual SLAM is to localize and build a map using these images. This is not a simple task like you think. It is not a single algorithm that continuously outputs positions and map information as long as we feed it with input data. SLAM requires an algorithm framework, and after decades of hard work by researchers, the framework has been matured in recent years.

---

[2] Note that this book is written in 2016. The statements may be outdated in future.

[3] You can try to use your phone to record a video clip.

Fig. 1.6 RGBD cameras measure the distance and can build a point cloud with a single image frame

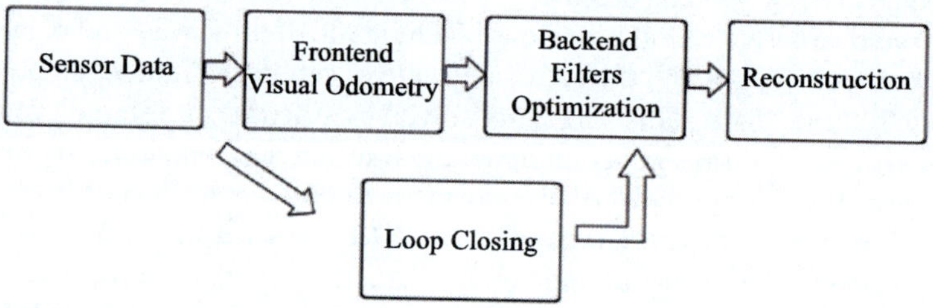

Fig. 1.7 The classic visual SLAM framework

1.2 Classical Visual SLAM Framework

Let's take a look at the classic visual SLAM framework, shown in the following Fig. 1.7:

A typical visual SLAM workflow includes the following steps:

1. Sensor data acquisition. In visual SLAM, this mainly refers to for acquisition and preprocessing of camera images. For a mobile robot, this will also include the acquisition and synchronization with motor encoders, IMU sensors, etc.
2. Visual Odometry (VO). VO's task is to estimate the camera movement between adjacent frames (ego-motion) and generate a rough local map. VO is also known as the *frontend*.
3. Backend filtering/optimization. The backend receives camera poses at different time stamps from VO and results from loop closing, and then applies optimization

to generate a fully optimized trajectory and map. Because it is connected after the VO, it is also known as the *backend*.

4. Loop Closing. Loop closing determines whether the robot has returned to its previous position in order to reduce the accumulated drift. If a loop is detected, it will provide information to the backend for further optimization.
5. Reconstruction. It constructs a task-specific map based on the estimated camera trajectory.

The classic visual SLAM framework is the result of more than a decade's research endeavor. The framework itself and the algorithms have been basically finalized and provided as essential functions in several public vision and robotics libraries. Relying on these algorithms, we can build visual SLAM systems performing real-time localization and mapping in static environments. Therefore, we can reach a rough conclusion that if the working environment is limited to fixed and rigid with stable lighting conditions and no human interference, the visual SLAM problem is basically solved [8].

The readers may not have fully understood the concepts of the modules mentioned above yet, so we will detail each module's functionality in the following sections. However, a deeper understanding of their working principles requires certain mathematical knowledge, which will be expanded in this book's second part. For now, an intuitive and qualitative understanding of each module is good enough.

1.2.1 Visual Odometry

The visual odometry is concerned with the movement of a camera between adjacent image frames, and the simplest case is, of course, the motion between two successive images. For example, when we see the images in Fig. 1.8, we will naturally tell that the right image should be the result of the left image after a rotation to the left with a certain angle (it will be easier if we have a video input). Let's consider this question: how do we know the motion is "turning left"? Humans have long been accustomed to using our eyes to explore the world and estimating our own positions, but this intuition is often difficult to explain, especially in natural language. When we see these images, we will naturally think that, ok, the bar is close to us, the walls and the blackboard are farther away. When the camera turns to the left, the bar's closer part starts to appear, and the cabinet on the right side starts to move out of our sight. With this information, we conclude that the camera should be rotating to the left.

But if we go a step further: can we determine how much the camera has rotated or translated in units of degrees or centimeters? It is still difficult for us to give a quantitative answer. Because our intuition is not good at calculating numbers. But for a computer, movements have to be described with such numbers. So we will ask: how should a computer determine a camera's motion only based on images?

As mentioned earlier, in the field of computer vision, a task that seems natural to a human can be very challenging for a computer. Images are nothing but numerical

Fig. 1.8 Camera motion can be inferred from two consecutive image frames. Images are from the NYUD dataset [9]

matrices in computers. A computer has no idea what these matrices *mean* (this is the problem that machine learning is also trying to solve). In visual SLAM, we can only see blocks of pixels, knowing that they are the results of projections by spatial points onto the camera's imaging plane. In order to quantify a camera's movement, we must first understand the geometric relationship between a camera and the spatial points.

Some background knowledge is needed to clarify this geometric relationship and the realization of VO methods. Here we only want to convey an intuitive concept. For now, you just need to take away that VO can estimate camera motions from images of adjacent frames and restore the 3D structures of the scene. It is named odometry because it is similar to a real wheel odometry, which only calculates the ego-motion at neighboring moments and does not estimate a global map or an absolute pose. In this regard, VO is like a species with only a short memory.

Now, assuming that we have the visual odometry, we are able to estimate camera movements between every two successive frames. If we connect the adjacent movements, this naturally constitutes the robot trajectory movement and addresses the localization problem. On the other hand, we can calculate the 3D position for each pixel according to the camera position at each time step, and they will form a map. Up to here, it seems with a VO, the SLAM problem is already solved. Or, is it?

Visual odometry is indeed a key technology for solving the visual SLAM problem. We will be spending a significant part explaining it in detail. However, using only a VO to estimate trajectories will inevitably cause accumulative drift. This is because the visual odometry (in the simplest case) only estimates the movement between two frames. We know that each estimate is accompanied by a certain error. Because of the way odometry works, errors from previous moments will be carried forward to the next moments, resulting in inaccurate estimation after some time (see Fig. 1.9). For example, the robot first turns left 90° and then turns right 90°. Due to error, we estimate the first 90° as 89°, which is possible to happen in real-world applications. Then we will be embarrassed to find that after the right turn, the robot's estimated position will not return to the origin. What's worse, even the following estimates are perfectly calculated, they will always be carrying this 1° error compared to the true trajectory.

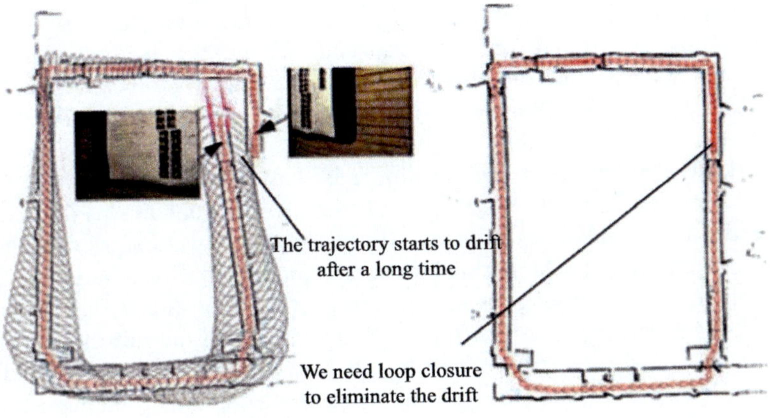

Fig. 1.9 Drift will be accumulated if we only have a relative motion estimation [10]

The accumulated drift will make us unable to build a consistent map. A straight corridor may oblique, and a 90° angle may be crooked—this is really an unbearable matter! To solve the drifting problem, we also need other two components: the *backend optimization*[4] and *loop closing*. Loop closing is responsible for detecting whether the robot returns to its previous position, while the backend optimization corrects the shape of the entire trajectory based on this information.

1.2.2 Backend Optimization

Generally speaking, backend optimization mainly refers to the process of dealing with the *noise* in SLAM systems. We hope that all the sensor data is accurate, but in reality, even the most expensive sensors still have a certain amount of noise. Cheap sensors usually have larger measurement errors, while that of expensive ones may be small. Moreover, many sensors' performance is affected by changes in the magnetic field, temperature, etc. Therefore, in addition to solving the problem of estimating camera movements from images, we also care about how much noise this estimation contains, how it is carried forward from the last time step to the next, and how confident we have in the current estimation. The backend optimization solves the problem of estimating the state of the entire system from noisy input data and calculate their uncertainty. The state here includes both the robot's own trajectory and the environment map.

In contrast, the visual odometry part is usually referred to as the *frontend*. In a SLAM framework, the frontend provides data to be optimized by the backend and the initial values. Because the backend is responsible for the overall optimization,

[4] It is usually known as the *backend*. Since it is often implemented by optimization, so we use the term backend optimization.

we only care about the data itself instead of where it comes from. In other words, we only have numbers and matrices in the backend without those beautiful images. In visual SLAM, the frontend is more relevant to *computer vision* topics, such as image feature extraction and matching, while the backend is relevant to the *state estimation* research area.

Historically, the backend optimization part has been equivalent to "SLAM research" for a long time. In the early days, the SLAM problem was described as a state estimation problem, exactly what the backend optimization tries to solve. In the earliest papers on SLAM, researchers at that time called it "estimation of spatial uncertainty" [3, 11]. Although it sounds a little obscure, it does reflect the nature of the SLAM problem: the estimation of the uncertainty of the self-movement and the surrounding environment. In order to solve the SLAM problem, we need state estimation theory to express the uncertainty of localization and map construction and then use filters or nonlinear optimization to estimate the mean and uncertainty (covariance) of the states. The details of state estimation and nonlinear optimization will be explained in Chaps. 5, 8, and 9.

1.2.3 Loop Closing

Loop Closing, also known as *loop closure detection*, mainly addresses the drifting problem of position estimation in SLAM. So how to solve it? Assuming a robot has returned to origin after a movement period, the estimated does not return to the origin due to drift. How to correct it? Imagine that if there is some way to let the robot know that it has returned to the origin, we can pull the estimated locations to the origin to eliminate drifts, which is what the loop closing exactly does.

Loop closing has a close relationship with both localization and map building. In fact, the main purpose of building a map is to enable a robot to know the places it has been to. To achieve loop closing, we need to let the robot have the ability to identify the scenes it has visited before. There are different alternatives to achieve this goal. For example, as we mentioned earlier, we can set a marker where the robot starts, such as a QR code. If the sign was seen again, we know that the robot has returned to the origin. However, the marker is essentially an intrusive sensor that sets additional constraints to the application environment. We prefer the robot can use its non-intrusive sensors, e.g., the image itself, to complete this task. A possible approach would be to detect similarities between images. This is inspired by us humans. When we see two similar images, it is easy to identify that they are taken from the same place. If the loop closing is successful, the accumulative error can be significantly reduced. Therefore, visual loop detection is essentially an algorithm for calculating similarities of images. Note that the loop closing problem also exists in laser-based SLAM. The rich information contained in images can remarkably reduce the difficulty of making a correct loop detection.

After a loop is detected, we will tell the backend optimization algorithm like, "OK, A and B are the same point". Based on this new information, the trajectory

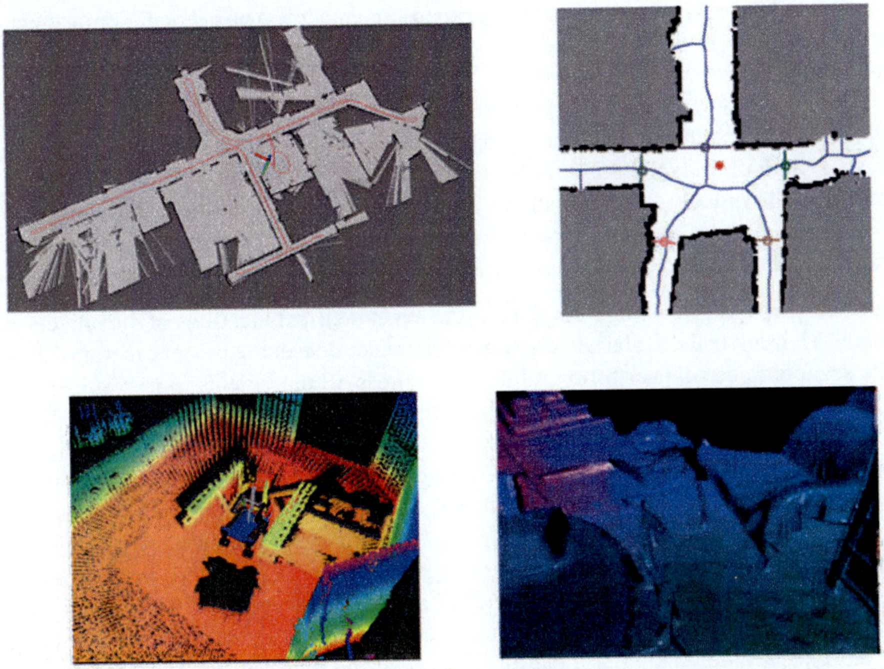

Fig. 1.10 Different kinds of maps: 2D grid map, 2D topological map, 3D point clouds and 3D meshes [12]

and the map will be adjusted to match the loop detection result. In this way, if we have sufficient and reliable loop detection, we can eliminate cumulative errors and get globally consistent trajectories and maps.

1.2.4 Mapping

Mapping means the process of building a map, whatever kind it is. A map (see Fig. 1.10) is a description of the environment, but the way of description is not fixed and depends on the actual application.

Let's take the domestic cleaning robots as an example. Since they basically move on the ground, a two-dimensional map with marks for open areas and obstacles, built by a single-line laser scanner, would be sufficient for navigation for them. And for a camera, we need at least a three-dimensional map for its 6 degrees-of-freedom movement. Sometimes, we want a smooth and beautiful reconstruction result, not just a set of points but also a texture of triangular faces. And at other times, we do not care about the map, just need to know things like "point A and point B are connected, while point B and point C are not", which is a topological way to understand the environment. Sometimes maps may not even be needed. For instance, a level-2 autonomous driving car can make a lane-following driving only knowing its relative motion with the lanes.

For maps, we have various ideas and demands. Compared to the previously mentioned VO, loop closure detection, and backend optimization, map building does not have a particular algorithm. A collection of spatial points can be called a map. A beautiful 3D model is also a map, so is a picture of a city, a village, railways, and rivers. The form of the map depends on the application of SLAM. In general, they can be divided into to categories: *metrical maps* and *topological maps*.

Metric Maps Metrical maps emphasize the exact metrical locations of the objects in maps. They are usually classified as either sparse or dense. Sparse metric maps store the scene into a compact form and do not express all the objects. For example, we can construct a sparse map by selecting representative landmarks such as the lanes and traffic signs and ignore other parts. In contrast, dense metrical maps focus on modeling all the things that are seen. A sparse map would be enough for localization, while for navigation, a dense map is usually needed (otherwise, we may hit a wall between two landmarks). A dense map usually consists of a number of small grids at a certain resolution. It can be small occupancy grids for 2D metric maps or small voxel grids for 3D maps. For example, in a 2D occupancy grid map, a grid may have three states: occupied, not occupied, and unknown, to express whether there is an object. When a spatial location is queried, the map can provide information about whether the location is passable. This type of maps can be used for various navigation algorithms, such as A*, D*,[5] etc., and thus attract the attention of robotics researchers. But we can also see that all the grid status are store in the map, thus being storage expensive. There are also some open issues in building a metric map. For example, in large-scale metrical maps, a small steering error may cause the walls of two rooms to overlap, making the map ineffective.

Topological Maps Compared to the accurate metrical maps, topological maps emphasize the relationships among map elements. A topological map is a graph composed of nodes and edges, only considering the connectivity between nodes. For instance, we only care about that point A and point B are connected, regardless of how we could travel from point A to point B. It relaxes the requirements on precise locations of a map by removing map details and is, therefore, a more compact expression. However, topological maps are not good at representing maps with complex structures. Questions such as how to split a map to form nodes and edges and how to use a topological map for navigation and path planning are still open problems to be studied.

[5] See https://en.wikipedia.org/wiki/A*_search_algorithm.

1.3 Mathematical Formulation of SLAM Problems

Through the previous introduction, readers should have gained an intuitive understanding of the modules in a SLAM system and each module's main functionality. However, we cannot write runnable programs only based on intuitive impressions. We want to raise it to a rational and rigorous level, using mathematical symbols to formulate a SLAM process. We will be using variables and formulas, but please rest assured that we will try our best to keep it clear enough.

Assuming that our Little Carrot is moving in an unknown environment, carrying some sensors. How can this be described in mathematical language? First, since sensors usually collect data at different time points, we are only concerned with the locations and the map at these moments. This turns a continuous process into discrete time steps, say $1, \ldots, k$, at which data sampling happens. We use $\mathbf{x}$ to indicate positions of Little Carrot. So the positions at different time steps can be written as $\mathbf{x}_1, \ldots, \mathbf{x}_k$, which constitute the trajectory of Little Carrot. In terms of the map, we assume that the map is made up of several *landmarks*, and at each time step, the sensors can see a part of the landmarks and record their observations. Assume there is a total of N landmarks in the map, and we will use $\mathbf{y}_1, \ldots, \mathbf{y}_N$ to denote them.

With such a setting, the process that *Little Carrot moves in the environment with sensors* basically has two parts:

1. What is its *motion*? We want to describe how $\mathbf{x}$ changes from time step $k-1$ to k.
2. What are the sensor *observations*? Assuming that the Little Carrot detects a specific landmark, let's say $\mathbf{y}_j$ at position $\mathbf{x}_k$, we need to describe this event in mathematical language.

Let's first take a look at motion. Typically, we may send some motion commands to the robots like "turn 15 degrees to left". These commands will be finally carried out by the controller, probably in many different ways. Sometimes we control the position of robots, but acceleration or angular velocity would always be reasonable alternates. However, no matter what the controller is, we can use a universal and abstract mathematical model to describe it:

$$\mathbf{x}_k = f\left(\mathbf{x}_{k-1}, \mathbf{u}_k, \mathbf{w}_k\right), \tag{1.1}$$

where $\mathbf{u}_k$ is the input commands, and $\mathbf{w}_k$ is noise. Note that we use a general $f(\cdot)$ to describe the process, instead of specifying the exact form of f. This allows the function to represent any motion input, rather than being limited to a particular one. We call it the *motion equation*.

The presence of noise turns this model into a stochastic model. In other words, even if we give an order as "move forward one meter", it does not mean that our robot really drives one meter in front. If all the instructions are accurate, there is no need to estimate anything. In fact, the robot may only forward by, say, 0.9 m, and at another moment, it moves by 1.1 m. Thus, the noise during each movement is random. If we ignore this noise, the position determined only by the input command maybe a hundred miles away from the real position after several minutes.

Corresponding to the motion equation, we also have an *observation equation.* The observation equation describes the process that the Little Carrot sees a landmark point $\mathbf{y}_j$ at $\mathbf{x}_k$ and generates an observation data $\mathbf{z}_{k,j}$. Likewise, we will describe this relationship with an abstract function $h(\cdot)$:

$$\mathbf{z}_{k,j} = h\left(\mathbf{y}_j, \mathbf{x}_k, \mathbf{v}_{k,j}\right), \tag{1.2}$$

where $\mathbf{v}_{k,j}$ is the noise in this observation. Since there are various observation sensors, the observed data $\mathbf{z}$ and the observation equation h may also have many different forms.

Readers may say that the function f, h here does not seem to specify what is going on in the motion and observation. Also, what do the variables $\mathbf{x}$, $\mathbf{y}$, $\mathbf{z}$ mean here? Depending on the actual motion and the type of sensor, there are several kinds of parameterization methods. What is parameterization? For example, suppose our robot moves in a plane, then its pose[6] is described by two $x-y$ coordinates and an angle, i.e., $\mathbf{x}_k = [x_1, x_2, \theta]_k^{\mathrm{T}}$, where x_1, x_2 are positions on two axes and θ is the angle. At the same time, the input command is the position and angle change between the time interval: $\mathbf{u}_k = [\Delta x_1, \Delta x_2, \Delta\theta]_k^{\mathrm{T}}$, so the motion equation can be parameterized as:

$$\begin{bmatrix} x_1 \\ x_2 \\ \theta \end{bmatrix}_k = \begin{bmatrix} x_1 \\ x_2 \\ \theta \end{bmatrix}_{k-1} + \begin{bmatrix} \Delta x_1 \\ \Delta x_2 \\ \Delta\theta \end{bmatrix}_k + \mathbf{w}_k, \tag{1.3}$$

where $\mathbf{w}_k$ is the noise again. This is a simple linear relationship. However, not all input commands are position and angular changes. For example, the input of "throttle" or "joystick" is the speed or acceleration, so there are other forms of more complex motion equations. At that time, we would say the kinetic analysis is required.

Regarding the observation equation, for example, the robot carries a two-dimensional laser sensor. We know that a laser observes a 2D landmark by measuring two quantities: the distance r between the landmark point and the robot, and the angle ϕ. Let's say the landmark is at $\mathbf{y}_j = [y_1, y_2]_j^{\mathrm{T}}$, the pose is $\mathbf{x}_k = [x_1, x_2]_k^{\mathrm{T}}$, and the observed data is $\mathbf{z}_{k,j} = [r_{k,j}, \phi_{k,j}]^{\mathrm{T}}$, then the observation equation is written as:

$$\begin{bmatrix} r_{k,j} \\ \phi_{k,j} \end{bmatrix} = \begin{bmatrix} \sqrt{\left(y_{1,j} - x_{1,k}\right)^2 + \left(y_{2,j} - x_{2,k}\right)^2} \\ \arctan\left(\frac{y_{2,j} - x_{2,k}}{y_{1,j} - x_{1,k}}\right) \end{bmatrix} + \mathbf{v}_{k,j}. \tag{1.4}$$

When considering visual SLAM, the sensor is a camera, then the observation equation is a process like "getting the pixels in the image of the landmarks." This process involves a description of the camera model, which will be covered in detail in Chap. 4.

---

[6] In this book, we use the word "pose" to mean "position" plus "rotation".

Obviously, it can be seen that the two equations have different parameterized forms for different sensors. If we maintain versatility and take them into a common abstract form, then the SLAM process can be summarized into two basic equations:

$$\begin{cases} \mathbf{x}_k = f\left(\mathbf{x}_{k-1}, \mathbf{u}_k, \mathbf{w}_k\right), & k = 1, \ldots, K \\ \mathbf{z}_{k,j} = h\left(\mathbf{y}_j, \mathbf{x}_k, \mathbf{v}_{k,j}\right), & (k, j) \in \mathcal{O} \end{cases}, \tag{1.5}$$

where $\mathcal{O}$ is a set that contains the information at which pose the landmark was observed (usually not all the landmarks can be seen at every moment—we are likely to see only a small part of the landmarks at a single moment). These two equations together describe a basic SLAM problem: how to solve the estimate $\mathbf{x}$ (localization) and $\mathbf{y}$ (mapping) problem with the noisy control input $\mathbf{u}$ and the sensor reading $\mathbf{z}$ data? Now, as we see, we have modeled the SLAM problem as a **state estimation problem**: How to estimate the internal, hidden state variables through the noisy measurement data?

The solution to the state estimation problem is related to the two equations' specific form and the noise probability distribution. Whether the motion and observation equations are *linear* and whether the noise is assumed to be *Gaussian*, it is divided into **linear/nonlinear** and **Gaussian/non-Gaussian** systems. The Linear Gaussian (LG system) is the simplest, and its unbiased optimal estimation can be given by the Kalman Filter (KF). In the complex nonlinear non-Gaussian (NLNG system), we basically rely on two methods: Extended Kalman Filter (EKF) and nonlinear optimization. Until the early 21st century, the EKF-based filter method still dominated SLAM. We will linearize the system at the working point and solve it in two steps: predict-update (see Chap. 8). The earliest real-time visual SLAM system was developed based on EKF [1]. Subsequently, in order to overcome the shortcomings of EKF (such as the linearization error and noise Gaussian distribution assumptions), people began to use other filters such as particle filters, and even use nonlinear optimization methods. Today, the mainstream of visual SLAM uses state-of-the-art optimization techniques represented by *graph optimization* [13]. We believe that optimization methods are clearly superior to filters, and as long as computing resources allow, optimization methods are often preferred (see Chaps. 8 and 9).

Now we believe the reader has a general understanding of SLAM's mathematical model, but we still need to clarify some issues. First, we should explain what the *pose* $\mathbf{x}$ is. The word *pose* we just said is somewhat vague. Perhaps the reader can understand that a robot moving in the plane can be parameterized by two coordinates plus an angle. However, more robots are more like things in three-dimensional space. We know that the motion of the three-dimensional space consists of three axes, so the movement of the robot is described by the translation on the three axes and the rotation around the three axes, totally having six *degrees of freedom* (DoF). But wait, does that mean that we can describe it with a vector in $\mathbb{R}^6$? We will find that things are not that simple. For a 6 DoF **pose**, how to express it? How to optimize it? How to describe its mathematical properties? This will be the main content of the Chaps. 2 and 3. Next, we will explain how the *observation equation* is parameterized in the visual SLAM. In other words, how the landmark points in space are projected onto a photo? This requires an explanation of the camera's projection model and distortion, which we will cover in Chap. 4. Finally, when we know this information, how to solve the above state estimation problem? This requires knowledge of nonlinear optimization and is the content of Chap. 5.

These contents form part of the mathematical knowledge of this book. After laying the groundwork for them, we can discuss more detailed knowledge of visual odometry, backend optimization, and more. It can be seen that the content of this lecture constitutes a summary of the book. If you don't understand the above concepts well, you may want to go back and read them again. Now let's start the introduction of programming!

1.4 Practice: Basics

1.4.1 Installing Linux

Finally, we come to the exciting practice session! Are you ready? In order to complete the practice of this book, we need to prepare a computer. You can use a laptop or desktop, preferably your personal computer, because we need to install an operating system on it for experiments.

Our program is based on C++ programs on Linux. During the experiment, we will use several open-source libraries. Most libraries are only supported in Linux, while configuration on Windows is relatively (or quite) troublesome. Therefore, we have to assume that you already have a basic knowledge of Linux (see the previous lecture exercises), including using basic commands to understand how to install the software. Of course, you don't have to know how to develop C++ programs under Linux, which is what we want to talk about below.

Let's start by installing the experimental environment required for this book. As a book for beginners, we recommend Ubuntu as a development environment. Ubuntu and its variances have enjoyed a good reputation as a novice user in all major Linux distributions. Ubuntu is an open-source operating system. Its system and software can be downloaded freely on the official website (http://ubuntu.com), which provides detailed installation instructions. At the same time, Tsinghua University, China Science and Technology University, and other major universities in China have also provided Ubuntu software mirrors, making the software installation very convenient (probably there are also mirror websites in your country).

The first version of this book uses Ubuntu 14.04 as the default development environment. In the second edition, we updated the default version to the newer **Ubuntu 18.04** (Fig. 1.11) for later research. If you want to change the styles, then Ubuntu Kylin, Debian, Deepin, and Linux Mint are also good choices. I promise that all the code in the book has been well tested under Ubuntu 18.04, but if you choose a different distribution, I am not sure if you will encounter some minor problems. You may need to spend some time solving small issues (but you can also take them as

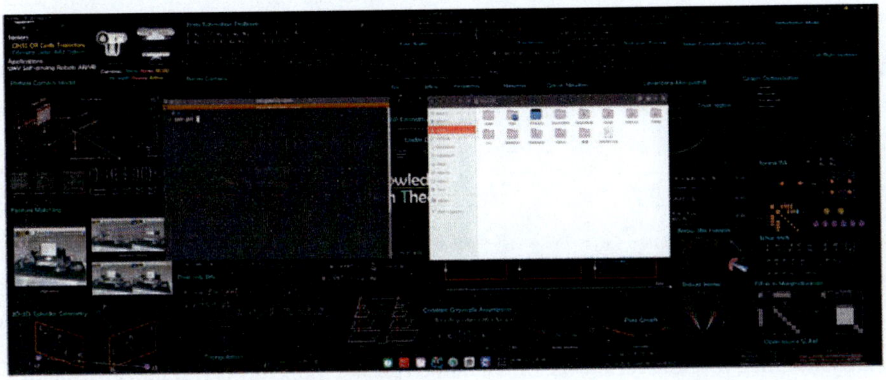

Fig. 1.11 Ubuntu 18.04 in virtual machine

opportunities to exercise yourself). In general, Ubuntu's support for various libraries is relatively complete and rich. Although we don't limit which Linux distribution you use, in the explanation, **we will use Ubuntu 18.04 as an example** and mainly use Ubuntu commands (such as apt-get). So, in other versions of Ubuntu, there will be no obvious differences below. In general, the migration of programs between Linux is not very difficult. But if you want to use the programs in this book under Windows or OS X, you need to have some porting experience.

Now, I assume there's an Ubuntu 18.04 installed on your PC. Regarding the installation of Ubuntu, you can find a lot of tutorials on the Internet. Just do it. I'll skip here. The easiest way is to use a virtual machine (see Fig. 1.11), but it takes a lot of memory (our experience is more than 4GB) and CPU to be smooth. You can also install dual systems, which will be faster, but a blank USB flash drive is required as the boot disk. In addition, virtual machine software support for external hardware is often not good enough. If you want to use real sensors (cameras, Kinects, etc.), it is recommended that you use dual systems to install Linux.

Now, let's say you have successfully installed Ubuntu, either a virtual machine or a dual system. If you are not familiar with Ubuntu, try its various software and experience its interface and interaction mode.[7] But I have to remind you, especially to the novice friends: don't spend too much time on Ubuntu's user interface. Linux has a lot of chances to waste your time. You may find some niche software, some games, and even spend a lot of time looking for wallpapers. But remember, you are working with Linux. Especially in this book, you are using Linux to learn SLAM, so try to spend your time learning SLAM.

Ok, let's choose a directory and put the code for the SLAM program in this book. For example, you can put the code under *slambook2* in the home directory (/home/username). We will refer to this directory *code root* in the future. At the same time, you can choose another directory to copy the git code of this book, which is convenient for comparison when doing experiments. The code for this book is divided

[7] Most people think Ubuntu is cool for the first time.

into chapters. For example, this chapter's code will be under slambook2/ch2, and the next one will be under slambook2/ch3. So, now please go into the slambook2/ch2 directory (you should create a new folder and enter the folder name).

1.4.2 Hello SLAM

Like many computer books, let's write a *HelloSLAM* program. But before we do this, let's talk about what a program is.

In Linux, a program is a file with execute permissions. It can be a script or a binary file, but we don't limit its suffix name (unlike Windows, you need to specify it as a .exe file). The commonly used binaries such as *cd* and *ls* are executable files located in the */bin* directory. For an executable program elsewhere, as long as it has executable permissions, it will run when we enter the program name in the terminal. When programming in C++, we first write a text file:

Listing 1.1 slambook2/ch2/helloSLAM.cpp

```
#include <iostream>
using namespace std;

int main(int argc, char **argv) {
  cout << "Hello SLAM!" << endl;
  return 0;
}
```

Then use a program called a **compiler** to turn this text file into an executable program. Obviously, this is a very simple program that just prints a hello slam text. You should be able to understand it effortlessly, so there is no explanation here—if this is not the case, please take a look at the basics of C++. This program just outputs a string to the screen. You can use the text editor like *gedit* (or *vim*, if you have learned vim in the previous tutorial) and enter the code and save it in the path listed above. Now, we compile it into an executable using the compiler g++ (g++ is a C++ compiler). Enter:

Listing 1.2 Terminal input:

```
g++ helloSLAM.cpp
```

If it goes well, this command should have no output. If the "command not found" error message appears on the screen, you may not have g++ installed yet. Please use the following command to install it:

Listing 1.3 Terminal input:

```
sudo apt-get install g++
```

If there are other errors, please check again if the program you just entered is correct.

This command compiles the text file "helloSLAM.cpp" into an executable program. We check the current directory and find that there is an additional a.out file.

It has executed permissions (the color shall be different in default settings). We can enter ./a.out to run the program[8]:

Listing 1.4 Terminal input:

```
% ./a.out
Hello SLAM!
```

As we thought, this program outputs "Hello SLAM!", telling us that it is running correctly.

Please review what we did before. In this example, we used the editor to enter the source code for "helloSLAM.cpp", then called the $g{+}{+}$ compiler to compile it and get the executable. By default, $g{+}{+}$ compiles the source file into a program of the name *a.out* (it is a bit weird but acceptable). If we like, we can also specify the file name of this output. This is an extremely simple example. We actually **use a lot of hidden default parameters, almost omitting all intermediate steps**, in order to give the reader a simple impression (although you may not have realized it). Below we will use CMake to compile this program.

1.4.3 Use CMake

Theoretically, any C++ program can be compiled with g++. But when the program size is getting bigger and bigger, a project may have many folders and source files, and the compiled commands will be longer and longer. Usually, a small C++ project may contain more than a dozen classes, and there are complex dependencies between these classes. Some of them are compiled into executables, and some are compiled into libraries. If we only rely on the g++ command, we need to enter many commands, and the whole compilation process will become very long. Therefore, for C++ projects, using some engineering management tools is more efficient. In history, engineers used *makefile* to compile automatically, but the CMake to be discussed below is more convenient than it. And CMake is widely used in engineering, we will see that most of the libraries mentioned later use CMake to manage the source code.

In a CMake project, we will use the *cmake* command to generate a *makefile*, and then use the make command to compile the entire project based on the makefile contents. The reader may not know what a makefile is, but it doesn't matter. We will learn by example. Still taking the above "helloSLAM.cpp" as an example, this time we are not using g++ directly, but using CMake to build a project and then compiling it. Create a new "CMakeLists.txt" file in "slambook2/ch2/" with the following contents:

Listing 1.5 slambook2/ch2/CMakeLists.txt

```
cmake_minimum_required( VERSION 2.8 )
project( HelloSLAM )
add_executable( helloSLAM helloSLAM.cpp )
```

[8] Don't type the first %.

The "CMakeLists.txt" file is used to tell CMake what we want to do with the files in this directory. The contents of the "CMakeLists.txt" file need to follow the CMake syntax. This example demonstrates the most basic project: specifying a project name and an executable program. According to the comments, the reader should understand what each sentence does.

Now, in the current directory (slambook2/ch2/), call *cmake* to compile the project[9]:

Listing 1.6 Terminal input

```
cmake .
```

cmake will output some compilation information, and then generate some intermediate files in the current directory, the most important of which is the makefile[10]. Since *makefile* is automatically generated, we don't have to modify it. Now, compile the project with the make command.

Listing 1.7 Terminal input

```
% make
Scanning dependencies of target helloSLAM
[100%] Building CXX object CMakeFiles/helloSLAM.dir/helloSLAM.cpp.o
Linking CXX executable helloSLAM
[100%] Built target helloSLAM
```

The compiler will show a process percent during compilation. We then get the declared executable *helloSLAM* in our "CMakeLists.txt" if the compilation is successful. Just type:

Listing 1.8 Terminal Input

```
% ./helloSLAM
Hello SLAM!
```

to run it. Because we didn't modify the source code, we got the same result as before. Please think about the difference between this practice and the previous use of the g++ compiler. This time we used the *cmake-make* process. The *cmake* process handles the relationship between the project files, and the *make* process actually calls g++ to compile the program. By calling this *cmake-make* process, we have good management for the project: **from inputting a string of g++ commands to maintaining several relatively intuitive "CMakeLists.txt" files**, which will drastically reduce the difficulty of maintaining the entire project. For example, if you want to add another executable file, just add a line "add_executable" in CMakeLists.txt, and the subsequent steps are unchanged. CMake will help us resolve code dependencies without having to type in a bunch of g++ commands.

The only thing that is dissatisfied with this process is that the intermediate files generated by CMake are still in our code files. When we want to release the code, we don't want to publish these intermediate files together. At this time, we still need to delete them one by one, which is very inconvenient. A better approach is to have these

---

[9] Note that there's a dot at the end of the command, please don't forget it, which means using CMake in the current directory.

[10] The *makefile* is an automated compilation script. Please just take it as an automatically generated compiler instructions without taking care of its content.

intermediate files in an intermediate directory. After the compilation is successful, we will delete the intermediate directory. Therefore, the more common practice of compiling CMake projects is as follows:

Listing 1.9 Terminal input

```
mkdir build
cd build
cmake ..
make
```

We created a new intermediate folder "*build*", and then entered the build folder, use the "*cmake ..*" command to compile the previous folder, which is the folder where the code is located. In this way, the intermediate files generated by CMake will be in the "*build*" folder, separate from the source code. When publishing the source code, we just delete the *build* folder. Please try to compile the code in ch2 in this way, and then call the generated executable (please remember to delete the intermediate file generated in the last section).

1.4.4 Use Libraries

In a C++ project, not all code is compiled into executables. Only executable files with the main function will generate executable programs. For other codes, we just want to package them into a packet for other programs to call. This packet is called *library*.

A library is often just a collection of many algorithms and programs, and we will be exposed to other libraries in later exercises. For example, the OpenCV library provides many computer vision-related algorithms, while the *Eigen* library provides calculations of matrix algebra. Therefore, we need to learn how to use CMake to generate libraries and use the library's functions. Now let's demonstrate how to write a library yourself. Write the following "libHelloSLAM.cpp" file:

Listing 1.10 slambook2/ch2/libHelloSLAM.cpp

```cpp
#include <iostream>
using namespace std;

// Just a function printing hello message
void printHello() {
  cout << "Hello SLAM" << endl;
}
```

This library provides a "printHello" function that will output a message. But it doesn't have the main function, which means there are no executables in this library. We add the following to "CMakeLists.txt":

Listing 1.11 slambook2/ch2/CMakeLists.txt

```
add_library( hello libHelloSLAM.cpp )
```

This line tells CMake that we want to compile this file into a library called "hello". Then, as above, compile the entire project using *cmake*:

Listing 1.12 Terminal input

```
cd build
cmake ..
make
```

At this point, a "libhello.a" file is generated in the build folder, which is the library we declared.

In Linux, library files are divided into static libraries and shared libraries. Static libraries have a ".a" extension, and shared libraries end with ".so". All libraries are collections of functions that are packaged. The difference is that a static library will generate a copy each time it is called, and the shared library has only one copy, which saves space. If you want to generate a shared library instead of a static library, just use the following statement:

Listing 1.13 slambook2/ch2/CMakeLists.txt

```
add_library( hello_shared SHARED libHelloSLAM.cpp )
```

Then we will get a libhello_shared.so.

The library file is a compressed package with compiled binary functions. However, if there is only a ".a" or ".so" library file, then we don't know what the function is and how to call it. In order for others (or ourselves) to use this library, we need to provide a *header file* to indicate what is in the library. Therefore, for the library users, *you can call this library as long as you get the header and library files*. Write the header file for "libhello" below.

Listing 1.14 slambook2/ch2/libHelloSLAM.h

```
#ifndef LIBHELLOSLAM_H_
#define LIBHELLOSLAM_H_

// Declares a function in header file
void printHello();

#endif
```

In this way, according to this file and the library file we just compiled, you can use the *printHello* function. Finally, we write an executable program to call this simple function:

Listing 1.15 slambook2/ch2/useHello.cpp

```
#include "libHelloSLAM.h"

// Call printHello() in libHelloSLAM.h
int main(int argc, char **argv) {
  printHello();
  return 0;
}
```

Then, declare an executable in "CMakeLists.txt" and *link* it to the library:

Listing 1.16 slambook2/ch2/CMakeLists.txt

```
add_executable( useHello useHello.cpp )
target_link_libraries( useHello hello_shared )
```

Through these two lines of statements, the "useHello" program can successfully use the code in the hello_shared library. This small example demonstrates how to generate and call a library. Please note that we can also call other libraries' functions in the same way and integrate them into our own programs.

In addition to the features already demonstrated, *cmake* has many more syntax and options. Of course, we can not list all of them here. In fact, cmake is very similar to a normal programming language, with variables and conditional control statements, so you can learn cmake just like learning programming. The exercises contain some reading materials for cmake, which can be read by interested readers. Now, a brief review of what we did before:

1. First, the program code consists of a header file and a source file.
2. The source file with the main function is compiled into an executable program, and the other is compiled into a library file.
3. If the executable wants to call a function in the library file, it needs to refer to the library's header file to understand the format of the function. Also, we need to link the executable to the library file.

These steps should be simple and clear, but you may encounter some problems in the actual operation. For example, what happens if the executable references a library function but we forget to link the library? Try removing the link command in "CMakeLists.txt" and see what happens. Can you understand the error message reported by *cmake*?

1.4.5 Use IDE

Finally, let's talk about how to use the Integrated Development Environments (IDEs). The previous programming can be done with a simple text editor. However, you may need to jump between files to query the declaration and implementation of a function. This can be a little annoying when there are too many files. The IDE provides developers with a lot of convenient functions such as jump, completion, breakpoint debugging, etc. Therefore, we recommend that the reader choose an IDE for development.

There are many kinds of IDEs under Linux. Although there are still some gaps with the best IDE (I mean Visual Studio in Windows), there are several C++ IDEs in Linux, such as Eclipse, Qt Creator, Code::Blocks, CLion, Visual Studio Code, and so on. Again, we don't force readers to use a particular IDE but only give our advice.[11] We are using KDevelop and CLion (see Figs. 1.12 and 1.15).[12] KDevelop is a free software located in Ubuntu's software repository, meaning you can install it with apt-get; CLion is a paid software, but you can use the student mailbox for

[11] Yes, vim is cool. But I still suggest use vim in IDE, not vim as IDE.

[12] However, the recent Visual Studio Code is getting better and better. It's free. It's very popular among developers. You may have a try.

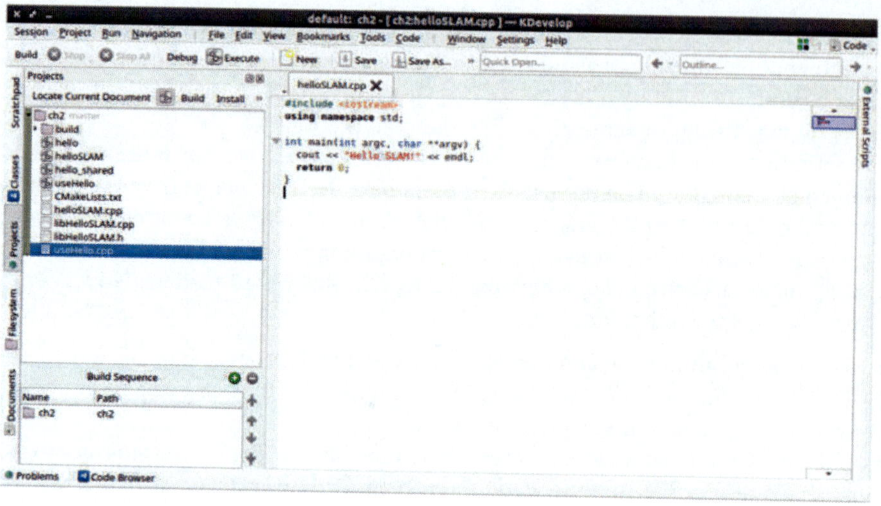

Fig. 1.12 Kdevelop in Ubuntu

free for one year. Both are good C++ development environments, the advantages are listed below:

1. Support CMake projects natively.
2. Support C++ better (including the 11 and later standards). There are highlighting, jumping, and finishing functions. Can automatically format the code.
3. Makes it easy to see individual files and directory trees.
4. Has one-click compilation, breakpoint debugging, and other functions.

Below we take a little bit of space to introduce KDevelop and CLion.

1.4.5.1 Use KDevelop

KDevelop natively supports the CMake project. To do this, after creating "CMakeLists.txt" in the terminal, open the "CMakeLists.txt" with "Project →Open/Import Project" in KDevelop. The software will ask you a few questions, and by default, create a build folder to help you call the cmake and make commands. These can be done automatically by pressing the shortcut key F8. The following section of Fig. 1.12 shows the compilation information.

We hand over the task of adapting to the IDE to the reader. If you are transferring from Windows, you will find its interface similar to Visual C++ or Visual Studio. Please use KDevelop to open the previous project and compile it to see what information it outputs. I believe you will feel more convenient than opening the terminal.

Next, let's show to debug in the IDE. Most of the students who program under Windows will have experience of breakpoint debugging under Visual Studio. However, in Linux, the default debugging tool gdb only provides a text interface, which is not convenient for novices. Some IDEs provide breakpoint debugging (the bot-

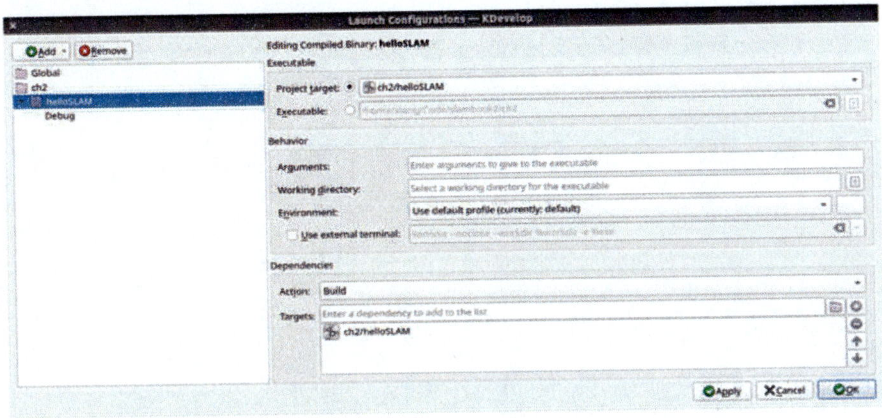

Fig. 1.13 Config launches. We can choose a launch target and set parameters here

tom layer is still gdb), and KDevelop is one of them. To use KDevelop's breakpoint debugging feature, you need to do the following:

1. Set the project to Debug compilation mode in "CMakeLists.txt", and don't use optimization options (not used by default).
2. Tell KDevelop which program you want to run. If there are parameters, also configure its parameters and working directory.
3. Enter the breakpoint debugging interface, you can single step, see the value of the intermediate variable.

The first step is to set the compilation mode by adding the following command to "CMakeLists.txt":

Listing 1.17 slambook2/ch2/CMakeLists.txt

```
Set( CMAKE_BUILD_TYPE "Debug" )
```

CMake has some compilation-related built-in variables that give you more precise control over the compilation process. There is usually a *debug* mode for debugging and a *release* mode for publishing. In debug mode, the program runs slower, but breakpoint debugging is possible, and you can see the values of the variables. In contrast, release mode is faster, but there is probably no debugging information. We set the program to Debug mode and place the breakpoint. Next, tell KDevelop which program you want to launch.

In the second step, open "Run →Configure Launches" and click on "Add New → Application" on the left. In this step, our task is to tell KDevelop which program to launch. As shown in Fig. 1.13, you can either select a CMake project target (that is, the executable we built with the add_executable directive) or point to a binary file. The second approach is recommended, and in our experience, this is less of a problem.

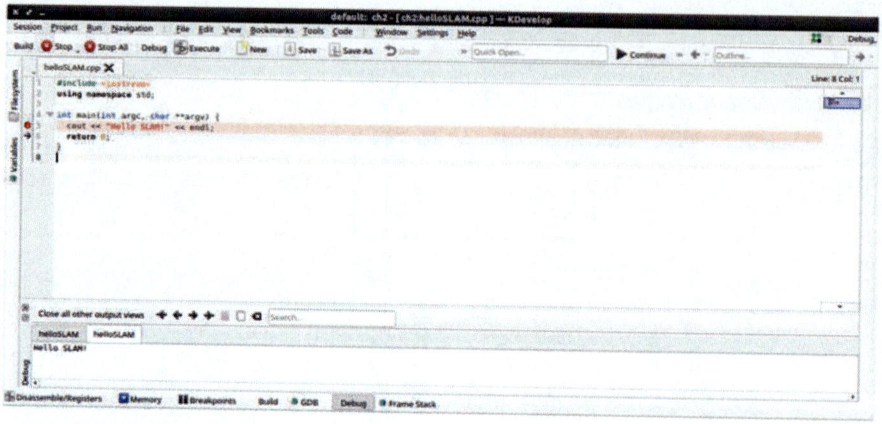

Fig. 1.14 Debug interface

In the second column, you can set the program's parameters and working directory. Sometimes programs have runtime parameters that are passed in as arguments to the main function. If not, leave it blank, as is the working directory. After configuring these two items, click the "OK" button to save the configuration results.

In these steps, we have configured an application launcher. We can click the "Execute" button to start the program directly or click the "Debug" button to debug it for each application. Readers can try to click the "Execute" button to see the results of the output. To debug this program, click on the left side of the "printHello" line and add a breakpoint. Then, click on the "Debug" button, and the program will wait at the breakpoint, as shown by Fig. 1.14.

When debugging, KDevelop will switch to debug mode, and the interface will change a bit. At the breakpoint, you can control the operation of the program with a single-step operation (F10 key), single-step follow up (F11 key), and single-step jump (F12 key) function. At the same time, you can click the interface on the left to view the local variable's value. Or select the "Stop" button to end debugging. After debugging, KDevelop will return to the normal development interface.

Now you should be familiar with the entire process of breakpoint debugging. In the future, if an error occurs during the running phase of the program, causing the program to crash, you can use breakpoint debugging to determine the location of the error, and then modify it.[13]

1.4.5.2 Use CLion

CLion is more complete than KDevelop, but it requires a user account, and the memory/CPU requirements for the host will be higher. In CLion, you can also open

[13] Instead of directly sending us an email asking how to deal with the problems.

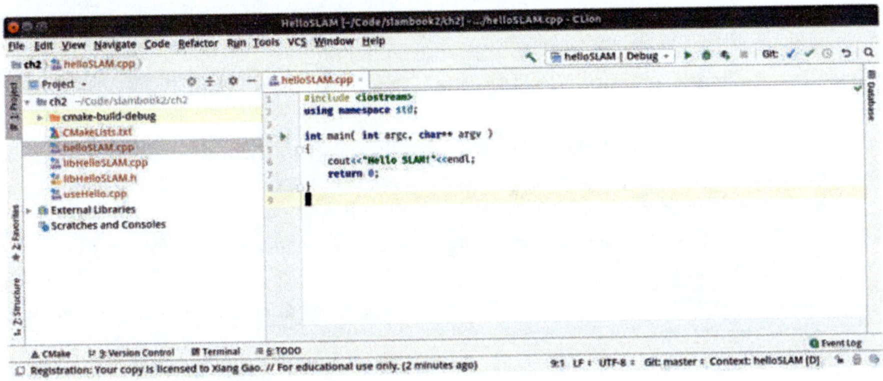

Fig. 1.15 CLion interface

a "CMakeLists.txt" or specify a directory. CLion will complete the *cmake-make* process for you. Its running interface is shown in Fig. 1.15.

Similarly, after opening CLion, you can select the programs you want to run or debug in the upper right corner of the interface and adjust their startup parameters and working directory. Click the small beetle button in this column to start the breakpoint debugging mode. CLion also has several convenient features, such as automatically creating classes, changing functions, and automatically adjusting the coding style. Please try it.

OK, if you are already familiar with the IDE's usage, then the second chapter will stop here. You may already feel that I have talked too much, so in the following practice section, we will not introduce things like creating a new build folder, calling the *cmake* and *make* commands to compile the program. I believe that readers should master these simple steps. Similarly, since most of the third-party libraries used in this book are cmake projects, you will continue to be familiar with the compilation process. Next, we will start the formal chapter and introduce some related mathematics.

Exercises

1.* Read SLAM's review literature, such as [8, 14–17] and so on. What are the similarities and differences between these papers on SLAM and this book?
2. What are the parameters of the g++ command? If I want to change the generated program file name, how should I call g++?
3. Use the build folder to compile your CMake project, then try it in KDevelop.
4. Deliberately add some syntax errors to the code to see what information the build will generate. Can you read the error message of g++?

5. If you forgot to link the library to the executable, will the compiler report an error? What kind of mistakes are reported?
6.* Read "CMake practice" (or other materials) to learn about the grammars of CMake.
7.* Improve the hello SLAM problem, make it a small library, and install it on your local hard drive. Then, create a new project, use find_package to find the library, and call it.
8.* Read other CMake instructional materials, such as https://github.com/TheErk/CMake-tutorial.
9. Find the official website of KDevelop and see what other features it has. Are you using it?
10. If you learned vim in the last lecture, please try KDevelop's/CLion's vim editing function.

Chapter 2
3D Rigid Body Motion

Goal of Study

1. Study the rigid body geometry in three-dimensional space: rotation matrix, transformation matrix, quaternion, and Euler angle.
2. Learn the usage of the *Eigen* library's matrix and geometry module.

In the last lecture, we explained the framework and content of visual SLAM. This lecture will introduce one of the fundamental problems of visual SLAM: How to describe a rigid body's motion in three-dimensional space? Intuitively, we certainly know that this consists of one rotation plus one translation. The translation part does not really have any problems, but the rotation part is questionable. We will introduce the meaning of rotation matrices, quaternions, Euler angles and how they are computed and transformed. In the practice section, we will introduce one of the widely used linear algebra libraries: *Eigen*. It provides a C++ matrix calculation, and its geometry module also provides the necessary data structures and operations, like quaternions. *Eigen* is heavily optimized, but there are still some special issues to be discussed about its usage. We will leave it to the practice part.

2.1 Rotation Matrix

2.1.1 Points, Vectors, and Coordinate Systems

The space of our daily life is three-dimensional, so we are born to be used to 3D movements. The 3D space consists of three axes, so the position of one spatial point can be specified by three coordinates. However, we should now consider a rigid body, which has its *position* and *orientation*. The camera can also be viewed as a rigid body in three dimensions, so what we care about in VSLAM are the problem of the camera's position and orientation. Combined, we can say, "the camera is at

X. Gao and T. Zhang, *Introduction to Visual SLAM*,
https://doi.org/10.1007/978-981-16-4939-4_2

the (0, 0, 0) point, facing the front". But this natural language is troublesome, and we prefer to describe it in a mathematical language.

We start from the most basic content: *points* and *vectors*. Points are the basic elements in space, no length, no volume. Connecting the two points forms a vector. A vector can be thought of as an arrow pointing from one point to another. Here we need to remind the reader that please do not confuse the vector with its coordinates. A vector is one thing in space, such as **a**. Here **a** does not need to be associated with several real numbers. We can naturally talk about the plus or minus operation of two vectors, without relating to any real numbers. Only when we specify a coordinate system in this 3D space can we talk about the vector's coordinates in this system, finding several real numbers corresponding to this vector.

With the knowledge of linear algebra, the coordinates of a point in 3D space can be described by $\mathbb{R}^3$. How to describe it? Suppose that in this linear space, we find a set of base[1] $(\mathbf{e}_1, \mathbf{e}_2, \mathbf{e}_3)$, then, an arbitrary vector **a** has a *coordinate* under this base:

$$\mathbf{a} = [\mathbf{e}_1, \mathbf{e}_2, \mathbf{e}_3] \begin{bmatrix} a_1 \\ a_2 \\ a_3 \end{bmatrix} = a_1\mathbf{e}_1 + a_2\mathbf{e}_2 + a_3\mathbf{e}_3. \tag{2.1}$$

Here $(a_1, a_2, a_3)^T$ is called **a**'s coordinates.[2] The coordinates' specific values are related to the vector itself and the selection of the bases. In $\mathbb{R}^3$, the coordinate system usually consists of 3 orthogonal coordinate axes (it can also be non-orthogonal, but it is rare in practice). For example, given **x** and **y** axis, the **z** axis can be determined using the right-hand (or left-hand) rule by $\mathbf{x} \times \mathbf{y}$. According to different definitions, the coordinate system is divided into left-handed and right-handed. The third axis of the left-hand rule is opposite to the right-hand rule. Most 3D libraries use right-handed coordinates (such as OpenGL, 3DS Max, etc.), and some libraries use left-handed coordinates (such as Unity, Direct3D, etc.).

Based on basic linear algebra knowledge, we can talk about the operations between vectors/vectors, vectors/numbers, such as scalar multiplication, vector addition, subtraction, inner product, outer product, and so on. Multiplication, addition, and subtraction are fairly basic and intuitive. For example, the result of adding two vectors is to add their respective coordinates, and the same for subtraction, and so on. I won't go into details here. Inner and outer products may be somewhat unfamiliar to the reader, and their calculations are given here. For $\mathbf{a}, \mathbf{b} \in \mathbb{R}^3$, in the common definition,[3] the inner product of **a**, **b** can be written as:

$$\mathbf{a} \cdot \mathbf{b} = \mathbf{a}^T\mathbf{b} = \sum_{i=1}^{3} a_i b_i = |\mathbf{a}|\,|\mathbf{b}| \cos\langle \mathbf{a}, \mathbf{b} \rangle, \tag{2.2}$$

---

[1] Just a reminder here, the base is a set of linearly independent vectors in the space, normally being orthogonal and has unit-length.

[2] We use column vectors in this book which is same as most of the mathematics books.

[3] The inner product also has more general definitions, but this book only discusses the usual inner product.

where $\langle \mathbf{a}, \mathbf{b} \rangle$ refers to the angle between the vector $\mathbf{a}, \mathbf{b}$. The inner product can also describe the projection relationship between vectors. The outer product is like this:

$$\mathbf{a} \times \mathbf{b} = \begin{Vmatrix} \mathbf{e}_1 & \mathbf{e}_2 & \mathbf{e}_3 \\ a_1 & a_2 & a_3 \\ b_1 & b_2 & b_3 \end{Vmatrix} = \begin{bmatrix} a_2 b_3 - a_3 b_2 \\ a_3 b_1 - a_1 b_3 \\ a_1 b_2 - a_2 b_1 \end{bmatrix} = \begin{bmatrix} 0 & -a_3 & a_2 \\ a_3 & 0 & -a_1 \\ -a_2 & a_1 & 0 \end{bmatrix} \mathbf{b} \triangleq \mathbf{a}^\wedge \mathbf{b}. \quad (2.3)$$

The result of the outer product is a vector whose direction is perpendicular to the two vectors, and the length is $|\mathbf{a}|\,|\mathbf{b}| \sin\langle \mathbf{a}, \mathbf{b} \rangle$, which is also the area of the quadrilateral of the two vectors. From the outer product operations, we introduce the $^\wedge$ operator here, which means writing $\mathbf{a}$ as a *skew-symmetric matrix*.[4] You can take $^\wedge$ as a skew-symmetric symbol. It turns the outer product $\mathbf{a} \times \mathbf{b}$ into the multiplication of the matrix and the vector $\mathbf{a}^\wedge \mathbf{b}$, which is a linear operation. This symbol will be used frequently in the following sections. It is a one-to-one mapping, meaning that for any vector, it corresponds to a unique anti-symmetric matrix, and vice versa:

$$\mathbf{a}^\wedge = \begin{bmatrix} 0 & -a_3 & a_2 \\ a_3 & 0 & -a_1 \\ -a_2 & a_1 & 0 \end{bmatrix}. \quad (2.4)$$

At the same time, note that the vector operations such as addition, subtraction, inner and outer products can be calculated even when we do not have their coordinates. For example, although the inner product can be expressed by the sum of the two vectors' product when we know the coordinates, the length and angle can also be calculated even if their coordinates are unknown. Therefore, the inner product result of the two vectors is independent of the selection of the coordinate system.

2.1.2 *Euclidean Transforms Between Coordinate Systems*

We often define a variety of coordinate systems in the real scene. In robotics, you define one coordinate system for each link and joint; in 3D mapping, we also define a coordinate system for each cuboid and cylinder. If we consider a moving robot, it is common practice to set a stationary inertial coordinate system (or world coordinate system), such as the x_W, y_W, z_W defined in Fig. 2.1. Meanwhile, the camera or robot is a moving coordinate system, such as the coordinate system defined by x_C, y_C, z_C. We might ask: a vector $\mathbf{p}$ in the camera system may have coordinates $\mathbf{p}_c$; and in the world coordinate system, its coordinates maybe $\mathbf{p}_w$. Then what is the conversion between these two coordinates? It is necessary to first obtain the coordinate values of the point in the camera system and then use the transform rule to do the coordinate

[4] Skew-symmetric matrix means $\mathbf{A}$ satisfies $\mathbf{A}^T = -\mathbf{A}$.

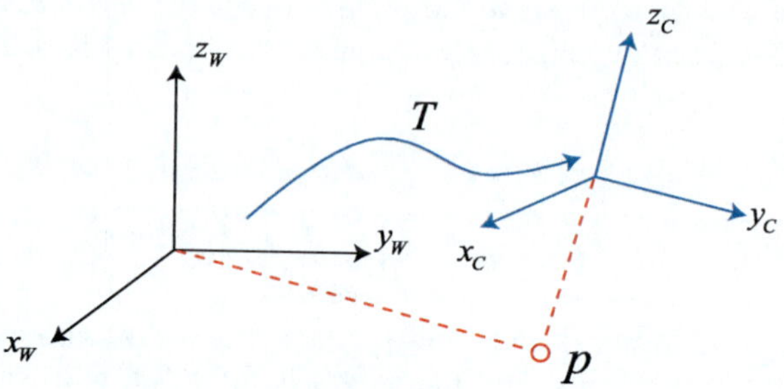

Fig. 2.1 Coordinate transform. For the same vector $\mathbf{p}$, its coordinates in the world $\mathbf{p}_W$ and the coordinates in the camera system $\mathbf{p}_C$ are different. This transformation relationship is described by the transform matrix $\mathbf{T}$

transform. We need a mathematical way to describe this transformation. As we will see later, we can describe it with a transform matrix $\mathbf{T}$.

Intuitively, the motion between two coordinate systems consists of a rotation plus a translation, which is called *rigid body motion*. Obviously, the camera movement is rigid. During the rigid body motion, the length and angle of the vector will not change. Imagine you throw your phone into the air and,[5] there may only be differences in spatial position and orientation. But the length and the angle of each face will not change. The phone will not be squashed like an eraser or be stretched during this motion. At this point, we say that the phone's motion is *Euclidean*.

The Euclidean transform consists of rotation and translation. Let's first consider the rotation. We have a unit-length orthogonal base $(\mathbf{e}_1, \mathbf{e}_2, \mathbf{e}_3)$. After a rotation it becomes $(\mathbf{e}'_1, \mathbf{e}'_2, \mathbf{e}'_3)$. Then, for the same vector $\mathbf{a}$ (the vector does not move with the rotation of the coordinate system), its coordinates in these two coordinate systems are $[a_1, a_2, a_3]^\mathrm{T}$ and $[a'_1, a'_2, a'_3]^\mathrm{T}$. Because the vector itself has not changed, according to the definition of coordinates, there are:

$$[\mathbf{e}_1, \mathbf{e}_2, \mathbf{e}_3] \begin{bmatrix} a_1 \\ a_2 \\ a_3 \end{bmatrix} = [\mathbf{e}'_1, \mathbf{e}'_2, \mathbf{e}'_3] \begin{bmatrix} a'_1 \\ a'_2 \\ a'_3 \end{bmatrix}. \tag{2.5}$$

To describe the relationship between the two coordinates, we multiply the left and right sides of the above equation by $\begin{bmatrix} \mathbf{e}_1^T \\ \mathbf{e}_2^T \\ \mathbf{e}_3^T \end{bmatrix}$, then the matrix on the left becomes an identity matrix, so:

$$\begin{bmatrix} a_1 \\ a_2 \\ a_3 \end{bmatrix} = \underbrace{\begin{bmatrix} \mathbf{e}_1^T\mathbf{e}'_1 & \mathbf{e}_1^T\mathbf{e}'_2 & \mathbf{e}_1^T\mathbf{e}'_3 \\ \mathbf{e}_2^T\mathbf{e}'_1 & \mathbf{e}_2^T\mathbf{e}'_2 & \mathbf{e}_2^T\mathbf{e}'_3 \\ \mathbf{e}_3^T\mathbf{e}'_1 & \mathbf{e}_3^T\mathbf{e}'_2 & \mathbf{e}_3^T\mathbf{e}'_3 \end{bmatrix}}_{\text{rotation matrix}} \begin{bmatrix} a'_1 \\ a'_2 \\ a'_3 \end{bmatrix} \triangleq \mathbf{R}\mathbf{a}'. \tag{2.6}$$

[5] Please don't put it into practice because you may regret doing that.

We take the intermediate matrix out and define it as a matrix $\mathbf{R}$. This matrix consists of the inner product between the two sets of bases, describing the same vector's coordinate transformation relationship before and after the rotation. As long as the rotation is the same, this matrix is the same. It can be said that the matrix $\mathbf{R}$ describes the rotation itself. So we call it the *rotation matrix*. Meanwhile, the components of the matrix are the inner product of the two coordinate system bases. Since the base vector's length is 1, it is actually the cosine of the angle between the base vectors. So this matrix is also called *direction cosine matrix*. We will just call it the rotation matrix in the following.

The rotation matrix has some special properties. In fact, it is an *orthogonal* matrix with a determinant of 1.[6, 7] Conversely, an orthogonal matrix with a determinant of 1 is also a rotation matrix. So, you can define a set of n dimensional rotation matrices as follows:

$$\mathrm{SO}(n) = \{\mathbf{R} \in \mathbb{R}^{n \times n} | \mathbf{R}\mathbf{R}^T = \mathbf{I}, \det(\mathbf{R}) = 1\}. \tag{2.7}$$

$\mathrm{SO}(n)$ refers to the *special orthogonal group*. We leave the contents of the "group" to the next lecture. This set consists of a rotation matrix of n dimensional space, in particular, SO(3) refers to the rotation of the three-dimensional space. In this way, we can talk directly about the rotation transformation between the two coordinate systems without having to start from the bases.

Since the rotation matrix is orthogonal, its inverse (i.e., transpose) describes an opposite rotation. According to the above definition, there are:

$$\mathbf{a}' = \mathbf{R}^{-1}\mathbf{a} = \mathbf{R}^T\mathbf{a}. \tag{2.8}$$

Obviously the $\mathbf{R}^T$ represents an opposite rotation.

In the Euclidean transformation, there is a translation in addition to the rotation. Consider the vector $\mathbf{a}$ in the world coordinate system, after a rotation (depicted by $\mathbf{R}$) and a translation of $\mathbf{t}$, we get $\mathbf{a}'$. Then we can put the rotation and translation together, and have:

$$\mathbf{a}' = \mathbf{R}\mathbf{a} + \mathbf{t}, \tag{2.9}$$

where $\mathbf{t}$ is called a translation vector. Compared to the rotation, the translation part simply adds the translation vector to the coordinates after the rotation, which is very simple. By the above formula, we completely describe the coordinate transformation relationship using a rotation matrix $\mathbf{R}$ and a translation vector $\mathbf{t}$. In practice, we may define the coordinate system 1 and 2, then the vector $\mathbf{a}$ under the two coordinates is $\mathbf{a}_1, \mathbf{a}_2$. The relationship between the two systems should be:

$$\mathbf{a}_1 = \mathbf{R}_{12}\mathbf{a}_2 + \mathbf{t}_{12}. \tag{2.10}$$

---

[6] Orthogonal matrix is a matrix whose inverse is its transpose. The orthogonality of the rotation matrix can be derived directly from the definition.

[7] The determinant is 1 is artificially defined. In fact, its determinant is ± 1, but the rotation with determinant -1 is called *improper rotation*, that is, one rotation plus one reflection in 3D space.

Here $\mathbf{R}_{12}$ means "rotation of the vector from system 2 to system 1". Since the vector is multiplied to the right of the rotation matrix, its subscript is *read from right to left*. This is just a customary way of writing this book. Coordinate transformations are easy to confuse, especially if multiple coordinate systems exist. Similarly, if we want to express "rotation matrix from 1 to 2", we write it as $\mathbf{R}_{21}$. The reader must be clear about the notation here, because different books may have different notations. Some notations about rotation will be denoted as the top left/subscript, and the text will be written on the right side.

About $\mathbf{t}_{12}$, readers may just take it as a translation vector without wondering about its physical meaning. In fact, it corresponds to a vector from the system 1's origin pointing to system 2's origin, and the coordinates are taken under system 1. So I suggest readers understand it as "a vector from 1 to 2". But the reverse $\mathbf{t}_{21}$, which is a vector from 2's origin to 1's origin, whose *coordinates are taken in system* 2, is not equal to $-\mathbf{t}_{12}$. It is also related to the rotation of the two systems.[8] Therefore, when beginners ask the question "What are my coordinates?", we need to clearly explain this sentence's meaning. Here "my coordinates" normally refers to the vector from the world system W pointing to the origin of the camera system C, and then take the coordinates in the world's base. Corresponding to the mathematical symbol, it should be the value of $\mathbf{t}_{WC}$. For the same reason, it is not $-\mathbf{t}_{CW}$, but actually $-\mathbf{R}_{CW}^T\mathbf{t}_{CW}$.

2.1.3 Transform Matrix and Homogeneous Coordinates

The formula (2.9) fully expresses the rotation and translation of Euclidean space, but there is still a small problem: the transformation relationship here is not a linear relationship. Suppose we made two transformations: $\mathbf{R}_1, \mathbf{t}_1$ and $\mathbf{R}_2, \mathbf{t}_2$:

$$\mathbf{b} = \mathbf{R}_1\mathbf{a} + \mathbf{t}_1, \quad \mathbf{c} = \mathbf{R}_2\mathbf{b} + \mathbf{t}_2.$$

So, the transformation from $\mathbf{a}$ to $\mathbf{c}$ is:

$$\mathbf{c} = \mathbf{R}_2\left(\mathbf{R}_1\mathbf{a} + \mathbf{t}_1\right) + \mathbf{t}_2.$$

This form is not elegant after multiple transformations. Therefore, we introduce homogeneous coordinates and transformation matrices, rewriting the form (2.9):

$$\begin{bmatrix} \mathbf{a}' \\ 1 \end{bmatrix} = \begin{bmatrix} \mathbf{R} & \mathbf{t} \\ \mathbf{0}^T & 1 \end{bmatrix} \begin{bmatrix} \mathbf{a} \\ 1 \end{bmatrix} \triangleq \mathbf{T} \begin{bmatrix} \mathbf{a} \\ 1 \end{bmatrix}. \tag{2.11}$$

This is a mathematical trick: we add 1 at the end of a 3D vector and turn it into a 4D vector called *homogeneous coordinates*. For this four-dimensional vector, we

[8] Although they are indeed inverse relations from the vector level, the coordinates of the two vectors are not the opposite. Can you find out why it looks like this?

can write the rotation and translation in one matrix, making the whole relationship a linear relationship. In this formula, the matrix $\mathbf{T}$ is called *transform matrix*.

We temporarily use $\tilde{\mathbf{a}}$ to represent the homogeneous coordinates of $\mathbf{a}$. Then, relying on homogeneous coordinates and transformation matrices, the superposition of the two transformations can have a good form:

$$\tilde{\mathbf{b}} = \mathbf{T}_1\tilde{\mathbf{a}}, \ \tilde{\mathbf{c}} = \mathbf{T}_2\tilde{\mathbf{b}} \quad \Rightarrow \tilde{\mathbf{c}} = \mathbf{T}_2\mathbf{T}_1\tilde{\mathbf{a}}. \tag{2.12}$$

But the symbols that distinguish between homogeneous and non-homogeneous coordinates are annoying, because here we only need to add 1 at the end of the vector or remove 1 to turn it into a normal vector.[9] So, without ambiguity, we will write it directly as $\mathbf{b} = \mathbf{T}\mathbf{a}$, and by default we just assume a homogeneous coordinate conversion is made if needed.[10]

The transformation matrix $\mathbf{T}$ has a special structure: the upper left corner is the rotation matrix, the right side is the translation vector, the lower-left corner is $\mathbf{0}$ vector, and the lower right corner is 1. This set of transform matrix is also known as the *special Euclidean group*:

$$\mathrm{SE}(3) = \left\{ \mathbf{T} = \begin{bmatrix} \mathbf{R} & \mathbf{t} \\ \mathbf{0}^T & 1 \end{bmatrix} \in \mathbb{R}^{4\times4} | \mathbf{R} \in \mathrm{SO}(3), \mathbf{t} \in \mathbb{R}^3 \right\}. \tag{2.13}$$

Like SO(3), the inverse of the transform matrix represents an inverse transformation:

$$\mathbf{T}^{-1} = \begin{bmatrix} \mathbf{R}^T & -\mathbf{R}^T\mathbf{t} \\ \mathbf{0}^T & 1 \end{bmatrix}. \tag{2.14}$$

Again, we use the notation of $\mathbf{T}_{12}$ to represent a transformation from 2 to 1. Moreover, to keep the symbol concise, the symbols of the homogeneous coordinates and the ordinary coordinates are not deliberately distinguished in later sections in the case of no ambiguity. For example, when we write $\mathbf{Ta}$, we use homogeneous coordinates (otherwise we can't calculate it). When we write $\mathbf{Ra}$, we use non-homogeneous coordinates. If they are written in the same equation, it is assumed that the conversion from homogeneous coordinates to normal coordinates is already done—because the conversion between homogeneous and non-homogeneous coordinates is actually very easy. In C++ programs, we can simply do this with *operator overloading* to ensure that the operations are correct.

Let's take a review now. First, we introduce the vector and its coordinate representation and introduce the operation between the vectors; then, the motion between the coordinate systems is described by the Euclidean transformation, which consists of translation and rotation. The rotation can be described by the rotation matrix SO(3),

[9] But the purpose of the homogeneous coordinates is not limited to this, we will come back to it in Chap. 6.

[10] Note that if homogeneous coordinate transformation is not performed, the matrix multiplication here does not make sense.

while the translation is directly described by an $\mathbb{R}^3$ vector. Finally, if the translation and rotation are placed in a matrix, the transformation matrix SE(3) is formed.

2.2 Practice: Use *Eigen*

The practical part of this lecture has two sections. In the first part, we will explain how to use *Eigen* to represent matrices and vectors and then extend to the calculation of rotation matrix and transformation matrix. The code for this section is in "slambook2/ch3/useEigen".

Eigen[11] is a C++ open-source linear algebra library. It provides fast linear algebra operations on matrices, as well as functions such as solving linear equations. Many upper-level software libraries also use *Eigen* for matrix operations, including *g2o*, Sophus, and others. Let's learn about *Eigen*'s programming.

Eigen may not have been installed on your PC. Please enter the following command to install it:

Listing 2.1 Terminal input:

```
sudo apt-get install libeigen3-dev
```

Most of the commonly used libraries in our book are available in the Ubuntu software center. With the apt command, we can easily install *Eigen*. Looking back at the previous lesson, we know that a library consists of header files and library files. The *Eigen* header file's default location should be */usr/include/eigen3/*. If you are not sure, you can find it by entering the following command:

Listing 2.2 Terminal input:

```
sudo locate eigen3
```

Eigen is a special library built with pure header files (this is the amazing part!). This means you can only locate its header files, not binary files like .so or .a. When you use it, you only need to import *Eigen*'s header file. You don't need to link the library file (because it doesn't have any library files). Now let's write a piece of code below to actually practice the use of *Eigen*:

Listing 2.3 slambook2/ch3/useEigen/eigenMatrix.cpp

```
#include <iostream>
using namespace std;

#include <ctime>
// Eigen core
#include <Eigen/Core>
// Algebraic operations of dense matrices (inverse, eigenvalues, etc.)
#include <Eigen/Dense>
using namespace Eigen;

```

[11] Official home page: http://eigen.tuxfamily.org/index.php?title=Main_Page.

```
#define MATRIX_SIZE 50

/****************************
* This program demonstrates the use of the basic Eigen type
****************************/

int main(int argc, char **argv) {
  // All vectors and matrices in Eigen are Eigen::Matrix, which is a
      template
  // class. Its first three parameters are: data type, row, column Declare
      a 2*3
  // float matrix
  Matrix<float, 2, 3> matrix_23;

  // At the same time, Eigen provides many built-in types via typedef, but
      the
  // bottom layer is still Eigen::Matrix. For example, Vector3d is
      essentially
  // Eigen::Matrix<double, 3, 1>, which is a three-dimensional vector.
  Vector3d v_3d;
  // This is the same
  Matrix<float, 3, 1> vd_3d;

  // Matrix3d is essentially Eigen::Matrix<double, 3, 3>
  Matrix3d matrix_33 = Matrix3d::Zero(); // initialized to zero
  // If you are not sure about the size of the matrix, you can use a matrix
      of
  // dynamic size
  Matrix<double, Dynamic, Dynamic> matrix_dynamic;
  // simpler
  MatrixXd matrix_x;
  // There are still many types of this kind. We don't list them one by one
      .

  // Here is the operation of the Eigen matrix
  // input data (initialization)
  matrix_23 << 1, 2, 3, 4, 5, 6;
  // output
  cout << "matrix 2x3 from 1 to 6: \n" << matrix_23 << endl;

  // Use () to access elements in the matrix
  cout << "print matrix 2x3: " << endl;
  for (int i = 0; i < 2; i++) {
    for (int j = 0; j < 3; j++)
    cout << matrix_23(i, j) << "\t";
    cout << endl;
  }

  // We can easily multiply a matrix with a vector (but actually still
      matrices and matrices)
  v_3d << 3, 2, 1;
  vd_3d << 4, 5, 6;

  // In Eigen you can't mix two different types of matrices, like this is
  // wrong Matrix<double, 2, 1> result_wrong_type = matrix_23 * v_3d;
  // It should be explicitly converted
  Matrix<double, 2, 1> result = matrix_23.cast<double>() * v_3d;
  cout << "[1,2,3;4,5,6]*[3,2,1]=" << result.transpose() << endl;

  Matrix<float, 2, 1> result2 = matrix_23 * vd_3d;
  cout << "[1,2,3;4,5,6]*[4,5,6]: " << result2.transpose() << endl;

  // Also you can't misjudge the dimensions of the matrix
  // Try canceling the comments below to see what Eigen will report.
  // Eigen::Matrix<double, 2, 3> result_wrong_dimension =
  // matrix_23.cast<double>() * v_3d;

```

```
  // some matrix operations
  // The basic operations are not demonstrated, just use +-*/ operators.
  Matrix_33 = Matrix3d::Random(); // Random Number Matrix
  cout << "random matrix: \n" << matrix_33 << endl;
  cout << "transpose: \n" << matrix_33.transpose() << endl;
  cout << "sum: " << matrix_33.sum() << endl;
  cout << "trace: " << matrix_33.trace() << endl;
  cout << "times 10: \n" << 10 * matrix_33 << endl;
  cout << "inverse: \n" << matrix_33.inverse() << endl;
  cout << "det: " << matrix_33.determinant() << endl;

  // Eigenvalues
  // Real symmetric matrix can guarantee successful diagonalization
  SelfAdjointEigenSolver<Matrix3d> eigen_solver(matrix_33.transpose() *
    matrix_33);
  cout << "Eigen values = \n" << eigen_solver.eigenvalues() << endl;
  cout << "Eigen vectors = \n" << eigen_solver.eigenvectors() << endl;

  // Solving equations
  // We solve the equation of matrix_NN * x = v_Nd
  // The size of N is defined in the previous macro, which is generated by a
  // random number Direct inversion is the most direct, but the amount of
  // inverse operations is large.

  Matrix<double, MATRIX_SIZE, MATRIX_SIZE> matrix_NN =
    MatrixXd::Random(MATRIX_SIZE, MATRIX_SIZE);
  matrix_NN =
    matrix_NN * matrix_NN.transpose(); // Guarantee semi-positive definite
  Matrix<double, MATRIX_SIZE, 1> v_Nd = MatrixXd::Random(MATRIX_SIZE, 1);

  Clock_t time_stt = clock(); // timing
  // Direct inversion
  Matrix<double, MATRIX_SIZE, 1> x = matrix_NN.inverse() * v_Nd;
  cout << "time of normal inverse is "
    << 1000 * (clock() - time_stt) / (double)CLOCKS_PER_SEC << "ms" << endl;
  cout << "x = " << x.transpose() << endl;

  // Usually solved by matrix decomposition, such as QR decomposition, the speed
  // will be much faster
  time_stt = clock();
  x = matrix_NN.colPivHouseholderQr().solve(v_Nd);
  cout << "time of Qr decomposition is "
    << 1000 * (clock() - time_stt) / (double)CLOCKS_PER_SEC << "ms" << endl;
  cout << "x = " << x.transpose() << endl;

  // For positive definite matrices, you can also use cholesky decomposition to
  // solve equations.
  time_stt = clock();
  x = matrix_NN.ldlt().solve(v_Nd);
  cout << "time of ldlt decomposition is "
    << 1000 * (clock() - time_stt) / (double)CLOCKS_PER_SEC << "ms" << endl;
  cout << "x = " << x.transpose() << endl;

  return 0;
}
```

This example demonstrates the basic operations and operations of the *Eigen* matrix. To compile it, you need to specify the header file directory of *Eigen* in the *CMakeLists.txt*:

Listing 2.4 slambook2/ch3/useEigen/CMakeLists.txt

```
# Add header file
include_directories( "/usr/include/eigen3" )
```

Because the *Eigen* library only has header files, we don't need to link the program to the library with the target_link_libraries statement. However, for most other libraries, you still need to use the link command. The approach here is not necessarily the best because others may have *Eigen* installed in different locations, then you must manually modify the header file directory here. In the rest of the book, we will use the find_ package command to search the library, but we keep it simple in this lecture. After compiling this program, run it, and you can see the output of each matrix.

Listing 2.5 Terminal output:

```
% build/eigenMatrix
matrix 2x3 from 1 to 6:
1 2 3
4 5 6
print matrix 2x3:
1 2 3
4 5 6
[1,2,3;4,5,6]*[3,2,1]=10 28
[1,2,3;4,5,6]*[4,5,6]: 32 77
random matrix:
0.680375 0.59688 -0.329554
-0.211234 0.823295 0.536459
0.566198 -0.604897 -0.444451
transpose:
0.680375 -0.211234 0.566198
0.59688 0.823295 -0.604897
-0.329554 0.536459 -0.444451
sum: 1.61307
trace: 1.05922
times 10:
6.80375 5.9688 -3.29554
-2.11234 8.23295 5.36459
5.66198 -6.04897 -4.44451
inverse:
-0.198521 2.22739 2.8357
1.00605 -0.555135 -1.41603
-1.62213 3.59308 3.28973
it: 0.208598
```

Since the detailed comments are given in the code, we will not elaborate on each line of the statements. This book will only describe several important places (the latter part will also keep this style).

1. Please enter the above code by yourself if you are a beginner in C++ (not including comments). At least compile and run the above program for once.
2. KDevelop may not prompt C++ member functions, which is caused by its incompleteness. Please type the above contents just like what they are. Do not care if KDevelop prompts any errors. Clion's completion may be better.
3. Eigen's matrix is very similar to MATLAB, and almost all data is treated as a matrix. However, to achieve better efficiency, you need to specify the size and type of the *Eigen* matrix. For matrices that we know the size at compile-time, they

are processed faster than dynamically changing matrices. Therefore, data such as rotation matrices and transformation matrices can be determined at compile times by their size and data type.

4. The matrix implementation inside *Eigen* is more complicated. I won't introduce it here. We hope you can use *Eigen*'s matrix just like the built-in data types such as float and double.
5. The *Eigen* matrix does not support automatic type promotion, which is quite different from C++'s built-in data types. In a C++ program, we can add and multiply a float variable and double variable, and the compiler will automatically *cast* the data type to the most appropriate one. In *Eigen*, for performance reasons, you must explicitly convert the matrix type. And if you forget to do this, *Eigen* will (not very friendly) prompt you with a very long "YOU MIXED DIFFERENT NUMERIC TYPES ..." compilation error. You can try to find out which part of the error message this message appears in. If the error message is too long, it is best to save it to a file and find it.
6. At the same time, the calculation process also needs to ensure the correctness of the matrix dimension. Otherwise, there will be a "YOU MIXED MATRICES OF DIFFERENT SIZES" error. Please don't complain about this kind of error prompts. For C++ template meta-programming, it is very fortunate to have readable error information. Later, if you find some compilation error about *Eigen*, you can directly look for the uppercase part and figure out the problem.
7. Our example only covers the very basic matrix operations. You can learn more about *Eigen* by reading the *Eigen* official website tutorial: http://eigen.tuxfamily.org/dox-devel/modules.html. Only the simplest part is demonstrated here. It is not equal to the fact that you can understand *Eigen*.

In the last piece of code, the efficiency of inversion and QR decomposition is compared. You can look at the time difference on your own machine. Is there a significant difference between the two methods?

2.3 Rotation Vectors and the Euler Angles

2.3.1 Rotation Vectors

Now let's return to the theoretical part. With a rotation matrix to describe the rotation, is it enough to use a 4×4 transformation matrix to represent a 6-degree-of-freedom 3D rigid body motion? Obviously, the matrix representation has at least the following disadvantages:

1. SO(3) has a rotation matrix of 9 quantities, but a 3D rotation only has 3 degrees of freedom. Therefore the matrix expression is redundant. Similarly, the transformation matrix expresses a 6 degree-of-freedom transformation with 16 quantities. So, is there a more compact representation?

2. The rotation matrix itself has constraints: it must be an orthogonal matrix with a determinant of 1. The same is true for the transformation matrix. These constraints make the solution more difficult when you want to estimate or optimize a rotation matrix/transform matrix.

Therefore, we hope that there is a way to describe rotation and translation more compactly. For example, is it feasible to express the rotation with a three-dimensional vector and express transformation with a six-dimensional vector? Obviously, a rotation can be described by a rotation axis and a rotation angle. Thus, we can use a vector whose direction is parallel with the axis of rotation, and the length is equal to the angle of rotation, which is called the *rotation vector* (or angle-axis/axis-angle). Only a three-dimensional vector is needed here to describe the rotation. Similarly, we may also use a rotation vector and a translation vector to express a transformation for a transformation matrix. The variable dimension at this time is exactly six dimensions.

Consider a rotation represented by $\mathbf{R}$. If described by a rotation vector, assuming that the rotation axis is a unit-length vector $\mathbf{n}$ and the angle is θ, then the vector $\theta\mathbf{n}$ can also describe this rotation. So, we have to ask, what is the connection between the two expressions? In fact, it is not difficult to derive their conversion relationship. The conversion from the rotation vector to the rotation matrix is shown by the *Rodrigues' formula*. Since the derivation process is a little complicated, it is not described here. Only the result of the conversion is given[12]:

$$\mathbf{R} = \cos\theta\mathbf{I} + (1 - \cos\theta)\,\mathbf{n}\mathbf{n}^T + \sin\theta\mathbf{n}^\wedge. \tag{2.15}$$

The symbol $^\wedge$ is a vector to skew-symmetric conversion, see the formula (2.3). Conversely, we can also calculate the conversion from a rotation matrix to a rotation vector. For the corner θ, taking the *trace* of both sides,[13] we have:

$$\begin{aligned}
\operatorname{tr}(\mathbf{R}) &= \cos\theta \operatorname{tr}(\mathbf{I}) + (1 - \cos\theta)\operatorname{tr}\left(\mathbf{n}\mathbf{n}^T\right) + \sin\theta \operatorname{tr}(\mathbf{n}^\wedge)\\
&= 3\cos\theta + (1 - \cos\theta)\\
&= 1 + 2\cos\theta.
\end{aligned} \tag{2.16}$$

Therefore:

$$\theta = \arccos\left(\frac{\operatorname{tr}(\mathbf{R}) - 1}{2}\right). \tag{2.17}$$

Regarding the axis $\mathbf{n}$, since the rotation axis does not change after the rotation, we have:

$$\mathbf{R}\mathbf{n} = \mathbf{n}. \tag{2.18}$$

Therefore, the axis $\mathbf{n}$ is the eigenvector corresponding to the matrix $\mathbf{R}$'s eigenvalue 1. Solving this equation and normalizing it gives the axis of rotation. By the way,

---

[12] For interested readers, please refer to https://en.wikipedia.org/wiki/Rodrigues%27_rotation_formula, in fact, the next chapter will give a brief proof from the Lie algebra view.

[13] The *trace* is to find the sum of the diagonal elements of the matrix.

the two conversion formulas here will still appear in the next lecture, and you will find that they are exactly the correspondence between Lie group and Lie algebra on SO(3).

2.3.2 Euler Angles

Let's talk about the Euler angle.

Whether it is a rotation matrix or a rotation vector, although they can describe the rotation, it is hard to imagine what this rotation is like only with those numbers. When they change, we don't know which direction the object is turning. The Euler angle provides a very intuitive way to describe the rotation—it uses three primal axes to decompose a rotation into three rotations around different axes. Humans can easily understand the process of rotating around a single axis. However, due to the variety of decomposition methods, there are many alternative and confusing definitions of Euler angles. For example, we can first rotate around the X axis, then the Y axis, and finally around the Z axis, and in this way, we get a rotation like XYZ order. Similarly, you can define rotation orders such as ZYZ and ZYX. You also need to distinguish whether it is rotated around the *fixed axis* or around the *axis after rotation*, which will also give a different definition.

This uncertainty in the axis orders brings many practical difficulties. Fortunately, in specific research areas, Euler angles usually have a uniform definition. You may have heard the words "pitch angle" and "yaw angle" of an aircraft. One of the most commonly used Euler angles is the yaw-pitch-roll angles. Since it is equivalent to the rotation of the ZYX axis, we take the ZYX Euler angle as an example. Suppose the front of a rigid body (toward our direction) is the X axis, the right side is the Y axis, and the top is the Z axis, as shown by Fig. 2.2. Then, the ZYX angle is equivalent to decompose any rotation into the following three axes:

1. Rotate around the Z axis of the object to get the yaw angle $\theta_{\text{yaw}} = y$;
2. Rotate around the Y axis of *after rotation* to get the pitch angle $\theta_{\text{pitch}} = p$;
3. Rotate around the X axis of *after rotation* to get the roll angle $\theta_{\text{roll}} = r$.

In this way, we can use a three-dimensional vector such as $[y, p, r]^T$ to describe any rotation. This vector is very intuitive. We can imagine the rotation process from this vector. The other Euler angles are also decomposed into three axes to obtain a three-dimensional vector, but the axes and order may be different. The ypr angle introduced here is a widely used one, and only a few Euler angles have such a famous name as ypr. Different Euler angles are referred to in the order of the axes of rotation. For example, the rotation order of the ypr angle is ZYX. Similarly, there are Euler angles like XYZ, ZYZ—but they don't have a specific name. It is worth mentioning that most areas have their own coordinate directions and habits when using Euler angles, not necessarily the same as we said here.

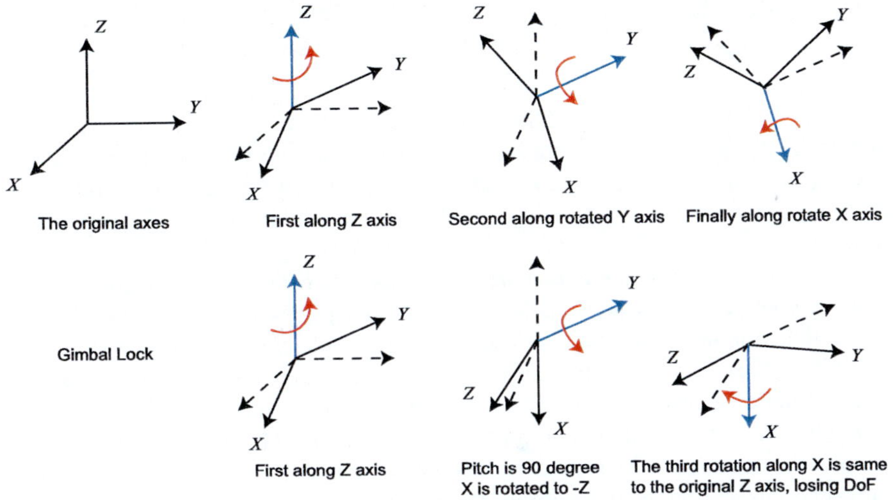

Fig. 2.2 Euler angles. The top is defined for the ZYX order (ypr order). The bottoms shows when pitch $= 90°$, the third rotation is using the same axis as the first one, causing the system to lose a degree of freedom. If you don't understand the gimbal lock, please take a look at the related videos and it will be more convenient to understand

A major drawback of Euler Angle is that it encounters the famous *Gimbal lock*[14]): in the ypr's case, when the pitch angle is $\pm 90°$, the first rotation and the third rotation will use the same axis, causing the system to lose a degree of freedom (from 3 rotations to 2 rotations). This is called the singularity problem and also exists in other forms of Euler angles. In theory, it can be proved that as long as you want to use three real numbers to express the three-dimensional rotation, you will inevitably encounter the singularity problem.[15] Due to this fact, Euler angles are not suitable for interpolation or iterations, and are often only used in human-computer interaction. We rarely use Euler angles to express poses directly in the SLAM program, nor do we use Euler angles to describe rotation in filtering or optimization (because it has singularity). However, if you want to verify that your algorithm is correct or not, converting to Euler angles can help you quickly determine if the results are correct or not. In some cases where the main body is mainly 2D motion (such as sweepers, self-driving vehicles), we can also decompose the rotation into three Euler angles and then take one of them (such as the yaw angle) as the orientation output.

---

[14] See https://en.wikipedia.org/wiki/Gimbal_lock.

[15] The rotation vector also has a singularity, which occurs when the angle θ exceeds 2π. Obviously, rotating 2π is the same with no rotation.

2.4 Quaternions

The rotation matrix describes 3 degrees of freedom with 9 quantities, of course, with redundancy; the Euler angles and the rotation vectors are compact but suffer from the singularity. In fact, we *cannot* find a three-dimensional vector description without singularity [18]. This is somewhat similar to using two coordinates to represent the Earth's surface (such as longitude and latitude), which also has the singularity (longitude is meaningless when the latitude is $\pm 90°$).

Recall the complex number that we have studied before. We use the complex set $\mathbb{C}$ to represent the vector on the 2D complex plane, and the complex multiplication with a unit complex number can represent the rotation on the 2D plane: for example, multiplying the complex i is equivalent to rotating a complex vector counterclockwise by 90°. Similarly, when expressing a three-dimensional space rotation, there is also an algebra similar to a complex number: the *quaternions*. Quaternions are extended complex numbers found by Hamilton. It is both compact and not singular. If we must find some shortcomings, the quaternion is not intuitive enough, and its operation is a bit more complicated.

Comparing quaternions to complex numbers can help you understand quaternions faster. For example, when we want to rotate the vector of a complex plane by θ, we can multiply this complex vector by $\mathrm{e}^{i\theta}$, which is represented by polar coordinates. It can also be written in the usual form like the famous Euler equation:

$$\mathrm{e}^{i\theta} = \cos\theta + i\sin\theta. \tag{2.19}$$

This is a unit-length complex number. Therefore, in two dimensions, the rotation can be described by a unit complex number. Similarly, we will see that 3D rotation can be described by a unit quaternion.

A quaternion $\mathbf{q}$ has a real part and three imaginary parts. We write the real part in the front (and there are also some books where the real part is in the last), like this:

$$\mathbf{q} = q_0 + q_1 i + q_2 j + q_3 k, \tag{2.20}$$

where i, j, k are three imaginary parts of the quaternion. These three imaginary parts satisfy the following relationship:

$$\begin{cases} i^2 = j^2 = k^2 = -1 \\ ij = k, ji = -k \\ jk = i, kj = -i \\ ki = j, ik = -j \end{cases}. \tag{2.21}$$

If we look at i, j, k as three axes, they look the same as complex numbers when multiplying with themselves. And look the same as the outer product when multiplying with the others.

We can also use a scalar and a vector to express quaternions:

$$\mathbf{q} = [s, \mathbf{v}]^T, \quad s = q_0 \in \mathbb{R}, \quad \mathbf{v} = [q_1, q_2, q_3]^T \in \mathbb{R}^3,$$

Here, s is the real part of the quaternion, and $\mathbf{v}$ is its imaginary part. If the imaginary part of a quaternion is $\mathbf{0}$, it is called *real quaternion*. Conversely, if its real part is 0, it is called *imaginary quaternion*.

We can use a unit quaternion to represent any rotation in 3D space, but this expression is subtly different from the complex numbers. In the complex, multiplying by i means rotating 90°. Does this mean that in the quaternion, multiplied by i is rotating around the i axis by 90° ? So, does $ij = k$ means, first rotating around the i by 90°, then around j by 90°, is equivalent to rotating around k by 90° ? Readers can use a cell phone to simulate that, then you will find that this is not the right case. The correct situation should be that multiplying i corresponds to rotating 180°, in order to guarantee the nature of $ij = k$. And $i^2 = -1$ means that after rotating 360° around the i axis, we get an opposite thing. This object has to be rotated by 720 ° to be equal to its original appearance.

This seems a bit mysterious. The complete explanation needs too many extra things. Let's calm down and come back to the quaternions. At least, we know that a unit quaternion can express the rotation of a three-dimensional space. So what are the properties of the quaternions? And how can they operate with each other?

2.4.1 *Quaternion Operations*

Quaternions are very similar to complex numbers, and a series of quaternion operations can be performed. We can easily plus, minus, multiplies to quaternions just like doing with two complex numbers. Assume there are two quaternions $\mathbf{q}_a, \mathbf{q}_b$, whose vectors are represented as $[s_a, \mathbf{v}_a]^{\mathrm{T}}, [s_b, \mathbf{v}_b]^{\mathrm{T}}$, or the original quaternion is expressed as:

$$\mathbf{q}_a = s_a + x_a i + y_a j + z_a k, \quad \mathbf{q}_b = s_b + x_b i + y_b j + z_b k.$$

Then, their operations can be expressed as follows.

1. **Addition and subtraction**. The addition and subtraction of the quaternion $\mathbf{q}_a, \mathbf{q}_b$ is:

$$\mathbf{q}_a \pm \mathbf{q}_b = [s_a \pm s_b, \mathbf{v}_a \pm \mathbf{v}_b]^T. \tag{2.22}$$

2. **Multiplication**. Quaternion multiplication is the multiplication of each item of $\mathbf{q}_a$ with each item of $\mathbf{q}_b$. The imaginary part is done according to formula (2.21):

$$\begin{aligned}\mathbf{q}_a\mathbf{q}_b = &\ s_a s_b - x_a x_b - y_a y_b - z_a z_b \\ &+ (s_a x_b + x_a s_b + y_a z_b - z_a y_b)\, i \\ &+ (s_a y_b - x_a z_b + y_a s_b + z_a x_b)\, j \\ &+ (s_a z_b + x_a y_b - y_a x_b + z_a s_b)\, k.\end{aligned} \tag{2.23}$$

The scalar form is a little complicated, but the vector form is more concise:

$$\mathbf{q}_a\mathbf{q}_b = \left[s_a s_b - \mathbf{v}_a^T \mathbf{v}_b, s_a \mathbf{v}_b + s_b \mathbf{v}_a + \mathbf{v}_a \times \mathbf{v}_b\right]^T. \tag{2.24}$$

Under this multiplication definition, the product of two real quaternions is still real, which is also consistent with the real number multiplication. However, note that due to the existence of the last outer product, quaternion multiplication is usually not commutative unless $\mathbf{v}_a$ and $\mathbf{v}_b$ at $\mathbb{R}^3$ are parallel, which means the outer product term is zero.

3. **Length**. The length of a quaternion is defined as:

$$\|\mathbf{q}_a\| = \sqrt{s_a^2 + x_a^2 + y_a^2 + z_a^2}. \tag{2.25}$$

It can be verified that the length of the product is the product of the lengths. This makes the unit quaternion keep unit-length when multiplied by another unit quaternion:

$$\|\mathbf{q}_a\mathbf{q}_b\| = \|\mathbf{q}_a\|\|\mathbf{q}_b\|. \tag{2.26}$$

4. **Conjugate**. The conjugate of a quaternion is to take the imaginary part as the opposite:

$$\mathbf{q}_a^* = s_a - x_a i - y_a j - z_a k = [s_a, -\mathbf{v}_a]^T. \tag{2.27}$$

We get a real quaternion if the quaternion is multiplied by its conjugate. The real part is the square of its length:

$$\mathbf{q}^*\mathbf{q} = \mathbf{q}\mathbf{q}^* = [s^2 + \mathbf{v}^T\mathbf{v}, \mathbf{0}]^T. \tag{2.28}$$

5. **Inverse**. The inverse of a quaternion is:

$$\mathbf{q}^{-1} = \mathbf{q}^*/\|\mathbf{q}\|^2. \tag{2.29}$$

According to this definition, the product of the quaternion and its inverse is the real quaternion $\mathbf{1}$:

$$\mathbf{q}\mathbf{q}^{-1} = \mathbf{q}^{-1}\mathbf{q} = \mathbf{1}. \tag{2.30}$$

If $\mathbf{q}$ is a unit quaternion, its inverse and conjugate are the same. So the inverse of the product has properties similar to matrices:

$$(\mathbf{q}_a\mathbf{q}_b)^{-1} = \mathbf{q}_b^{-1}\mathbf{q}_a^{-1}. \tag{2.31}$$

6. **Scalar Multiplication**. Similar to vectors, quaternions can be multiplied by numbers:

$$k\mathbf{q} = [ks, k\mathbf{v}]^T. \tag{2.32}$$

2.4.2 Use Quaternion to Represent a Rotation

We can use a quaternion to express the rotation of a point. Suppose a spatial 3D point $\mathbf{p} = [x, y, z]^T \in \mathbb{R}^3$, and a rotation is specified by a unit quaternion $\mathbf{q}$. The 3D point $\mathbf{p}$ is rotated to become $\mathbf{p}'$. If we use matrix, then there is $\mathbf{p}' = \mathbf{Rp}$. And if we use quaternion to describe rotation, how do we operate a 3D vector with a quaternion?

First, we extend the 3D point to an imaginary quaternion:

$$\mathbf{p} = [0, x, y, z]^T = [0, \mathbf{v}]^T.$$

We just put the three coordinates into the imaginary part and leave the real part to be zero. Then, the rotated point $\mathbf{p}'$ can be expressed as such a product:

$$\mathbf{p}' = \mathbf{q}\mathbf{p}\mathbf{q}^{-1}. \tag{2.33}$$

The multiplication here is the quaternion multiplication, and the result is also a quaternion. Finally, we take the imaginary part of $\mathbf{p}'$ and get the coordinates of the point after the rotation. It can be easily verified (we leave as an exercise here) that the real part of the calculation is 0, so it is a pure imaginary quaternion.

2.4.3 Conversion of Quaternions to Other Rotation Representations

An arbitrary unit quaternion describes a rotation, which can also be described by a rotation matrix or a rotation vector. Now let's examine the conversion relationship between quaternions and rotation vectors/matrices. Before that, we have to say that quaternion multiplication can also be written as a matrix multiplication. Let $\mathbf{q} = [s, \mathbf{v}]^T$, then define the following symbols $^+$ and $^\oplus$ as [19]:

$$\mathbf{q}^+ = \begin{bmatrix} s & -\mathbf{v}^T \\ \mathbf{v} & s\mathbf{I} + \mathbf{v}^\wedge \end{bmatrix}, \quad \mathbf{q}^\oplus = \begin{bmatrix} s & -\mathbf{v}^T \\ \mathbf{v} & s\mathbf{I} - \mathbf{v}^\wedge \end{bmatrix}, \tag{2.34}$$

These two symbols map the quaternion to a 4×4 matrix. Then the quaternion multiplication can be written in the form of a matrix:

$$\mathbf{q}_1^+\mathbf{q}_2 = \begin{bmatrix} s_1 & -\mathbf{v}_1^T \\ \mathbf{v}_1 & s_1\mathbf{I} + \mathbf{v}_1^\wedge \end{bmatrix} \begin{bmatrix} s_2 \\ \mathbf{v}_2 \end{bmatrix} = \begin{bmatrix} -\mathbf{v}_1^T\mathbf{v}_2 + s_1 s_2 \\ s_1\mathbf{v}_2 + s_2\mathbf{v}_1 + \mathbf{v}_1^\wedge\mathbf{v}_2 \end{bmatrix} = \mathbf{q}_1\mathbf{q}_2 \tag{2.35}$$

Note that the left side is matrix multiplication and the right side is quaternion multiplication. Similar for $^\oplus$, we get:

$$\mathbf{q}_1\mathbf{q}_2 = \mathbf{q}_1^+\mathbf{q}_2 = \mathbf{q}_2^\oplus\mathbf{q}_1. \tag{2.36}$$

Then, consider the problem of using a quaternion to rotate a spatial point. According to the previous section, we have:

$$\begin{aligned} \mathbf{p}' &= \mathbf{q}\mathbf{p}\mathbf{q}^{-1} = \mathbf{q}^{+}\mathbf{p}^{+}\mathbf{q}^{-1} \\ &= \mathbf{q}^{+}\mathbf{q}^{-1^{\oplus}}\mathbf{p}. \end{aligned} \tag{2.37}$$

Substituting the matrix corresponding to two symbols, we get:

$$\mathbf{q}^{+}\left(\mathbf{q}^{-1}\right)^{\oplus} = \begin{bmatrix} s & -\mathbf{v}^T \\ \mathbf{v} & s\mathbf{I} + \mathbf{v}^{\wedge} \end{bmatrix} \begin{bmatrix} s & \mathbf{v}^T \\ -\mathbf{v} & s\mathbf{I} + \mathbf{v}^{\wedge} \end{bmatrix} = \begin{bmatrix} 1 & \mathbf{0} \\ \mathbf{0}^T & \mathbf{v}\mathbf{v}^T + s^2\mathbf{I} + 2s\mathbf{v}^{\wedge} + (\mathbf{v}^{\wedge})^2 \end{bmatrix}. \tag{2.38}$$

Since $\mathbf{p}'$ and $\mathbf{p}$ are both imaginary quaternions, so in fact that the bottom right corner of the matrix gives the transformation formula *from quaternion to rotation matrix* :

$$\mathbf{R} = \mathbf{v}\mathbf{v}^T + s^2\mathbf{I} + 2s\mathbf{v}^{\wedge} + (\mathbf{v}^{\wedge})^2. \tag{2.39}$$

In order to obtain the conversion formula of the quaternion to the rotation vector, we take the trace on both sides of the above formula:

$$\begin{aligned} \operatorname{tr}(\mathbf{R}) &= \operatorname{tr}(\mathbf{v}\mathbf{v}^T) + 3s^2 + 2s \cdot 0 + \operatorname{tr}((\mathbf{v}^{\wedge})^2) \\ &= v_1^2 + v_2^2 + v_3^2 + 3s^2 - 2(v_1^2 + v_2^2 + v_3^2) \\ &= (1 - s^2) + 3s^2 - 2(1 - s^2) \\ &= 4s^2 - 1. \end{aligned} \tag{2.40}$$

Also obtained by the formula (2.17):

$$\begin{aligned} \theta &= \arccos\left(\frac{\operatorname{tr}(\mathbf{R}) - 1}{2}\right) \\ &= \arccos(2s^2 - 1), \end{aligned} \tag{2.41}$$

which means:

$$\cos\theta = 2s^2 - 1 = 2\cos^2\frac{\theta}{2} - 1, \tag{2.42}$$

and so we have:

$$\theta = 2\arccos s. \tag{2.43}$$

For the rotation axis, if we replace $\mathbf{p}$ with the imaginary part of $\mathbf{q}$ in the formula (2.38), it is easy to know the imaginary part of $\mathbf{q}$ is not moving when it is rotated, that is, it constitutes exactly the rotation axis. So we get the rotation axis just by normalizing $\mathbf{q}$'s imaginary part. In summary, the conversion formula from quaternion to rotation vector can be written as follows:

$$\begin{cases} \theta = 2 \arccos q_0 \\ \left[n_x, n_y, n_z\right]^T = [q_1, q_2, q_3]^T / \sin \frac{\theta}{2} \end{cases}. \tag{2.44}$$

And for converting from other representations to quaternions, we only need to reversely follow the above steps. In actual programming, the library usually prepares for the conversion between various forms for us. Whether it's a quaternion, a rotation matrix, or an angle-axis, it can always be used to describe the same rotation. We should choose the most convenient form in practice without having to stick to a particular form. In the subsequent practices and exercises, we will demonstrate the transition between various expressions to deepen the reader's impression.

2.5 Affine and Projective Transformation

In addition to the Euclidean transformation, there are several other transformations in the 3D space, where the Euclidean is the simplest. Some of them are related to camera geometry. We will introduce them in the following chapters, so here we only list their basic properties. The Euclidean transformation keeps the vector's length and angle, which is equivalent to moving or rotating a rigid body without changing its appearance. The other transformations will change their shape, and they all have similar matrix representations.

1. **Similarity transformation.**
   The similarity transformation has one more degree of freedom than the Euclidean transformation, which allows the object to be uniformly scaled, and its matrix is expressed as:
   $$\mathbf{T}_S = \begin{bmatrix} s\mathbf{R} & \mathbf{t} \\ \mathbf{0}^T & 1 \end{bmatrix}. \tag{2.45}$$
   Note that the rotation part has an extra scaling factor s, which means that we can evenly scale the three coordinates of $x,\ y,\ and\ z$ of a vector after it is rotated. Due to the scaling, a similarity transformation no longer keeps the volume of the transformed boy unchanged. You can imagine a cube with a side length of 1 transforming into a side with a length of 10 (but still being a cube). The set of three-dimensional similarity transform is also called *similarity transform group*, which is denoted as Sim(3).
2. **Affine transformation.**
   The matrix form of the affine transformation is as follows:
   $$\mathbf{T}_A = \begin{bmatrix} \mathbf{A} & \mathbf{t} \\ \mathbf{0}^T & 1 \end{bmatrix}. \tag{2.46}$$
   Unlike the Euclidean transformation, the affine transformation requires only $\mathbf{A}$ to be an invertible matrix, not necessarily an orthogonal matrix. An affine trans-

Table 2.1 Comparison of common transformation properties

Transform name	Matrix form	Degrees of freedom	Invariance
Euclidean	$\begin{bmatrix} \mathbf{R} & \mathbf{t} \\ \mathbf{0}^T & 1 \end{bmatrix}$	6	Length, angle, volume
Similarity	$\begin{bmatrix} s\mathbf{R} & \mathbf{t} \\ \mathbf{0}^T & 1 \end{bmatrix}$	7	Volume ratio
Affine	$\begin{bmatrix} \mathbf{A} & \mathbf{t} \\ \mathbf{0}^T & 1 \end{bmatrix}$	12	Parallelism, volume ratio
Perspective	$\begin{bmatrix} \mathbf{A} & \mathbf{t} \\ \mathbf{a}^T & v \end{bmatrix}$	15	Plane intersection and tangency

formation is also called an orthogonal projection. After the affine transformation, the cube is no longer square, but the faces are still parallelograms.

3. **Perspective transformation.**
Perspective transformation is the most general transformation. Its matrix form is:

$$\mathbf{T}_P = \begin{bmatrix} \mathbf{A} & \mathbf{t} \\ \mathbf{a}^T & v \end{bmatrix}. \tag{2.47}$$

Its upper left corner is the invertible matrix $\mathbf{A}$. The upper right corner is the translation $\mathbf{t}$, and the lower-left corner is the scale $\mathbf{a}^T$. Since the homogeneous coordinates are used, when $v \neq 0$, we can divide the entire matrix by v to get a matrix with a bottom right corner of 1; otherwise, we get a matrix with a lower right corner of 0. Therefore, the 2D perspective transformation has a total of 8 degrees of freedom, and 3D has a total of 15 degrees of freedom. Perspective transformation is the most general transformation that has been said so far. The transformation from the real world to a camera photo can be seen as a perspective transformation. The reader can imagine what a square tile would look like in a photo: first, it is no longer square. Second, since the close part is larger than the far-away part, it is not even a parallelogram but an irregular quadrilateral.

Table 2.1 summarizes the properties of several transformations currently covered. Note that in the "invariance", there is an inclusion relationship from top to bottom. For example, in addition to maintaining volume, the Euclidean transformation also keeps the parallelism, intersection, and others.

We will introduce later that the transformation from the real world to the camera photo is a perspective transformation. If the focal length of the camera is infinity, then this transformation is affine. However, before we go into the camera model's details, let's do some experiments to have a rough impression of these transformations.

2.6 Practice: *Eigen* Geometry Module

2.6.1 Data Structure of the Eigen Geometry Module

Now, let's actually practice the various rotation expressions mentioned earlier. We will use quaternions, Euler angles, and rotation matrices in *Eigen* to demonstrate how they are transformed. We will also provide a visualization program to help the reader understand the relationship between these transformations.

Listing 2.6 slambook2/ch3/useGeometry/useGeometry.cpp

```
#include <iostream>
#include <cmath>
using namespace std;

#include <Eigen/Core>
#include <Eigen/Geometry>

using namespace Eigen;
// This program demonstrates how to use the Eigen geometry module

int main(int argc, char **argv) {
  // The Eigen/Geometry module provides a variety of rotation and translation representations
  // 3D rotation matrix directly using Matrix3d or Matrix3f
  Matrix3d rotation_matrix = Matrix3d::Identity();
  // The rotation vector uses AngleAxis, the underlying layer is not directly Matrix, but the operation can be treated as a matrix (because the operator is overloaded)
  AngleAxisd rotation_vector(M_PI / 4, Vector3d(0, 0, 1)); // Rotate 45 degrees along the Z axis
  cout.precision(3);
  cout << "rotation matrix = \n " << rotation_vector.matrix() << endl; // convert to matrix with matrix()
  // can also be assigned directly
  rotation_matrix = rotation_vector.toRotationMatrix();
  // coordinate transformation with AngleAxis
  Vector3d v(1, 0, 0);
  Vector3d v_rotated = rotation_vector * v;
  cout << "(1,0,0) after rotation (by angle axis) = " << v_rotated.transpose() << endl;
  // Or use a rotation matrix
  v_rotated = rotation_matrix * v;
  cout << "(1,0,0) after rotation (by matrix) = " << v_rotated.transpose() << endl;

  // Euler angle: You can convert the rotation matrix directly into Euler angles
  Vector3d euler_angles = rotation_matrix.eulerAngles(2, 1, 0); // ZYX order, ie roll pitch yaw order
  cout << "yaw pitch roll = " << euler_angles.transpose() << endl;

  // Euclidean transformation matrix using Eigen::Isometry
  Isometry3d T = Isometry3d::Identity(); // Although called 3d, it is essentially a 4*4 matrix
  T.rotate(rotation_vector); // Rotate according to rotation_vector
  T.pretranslate(Vector3d(1, 3, 4)); // Set the translation vector to (1,3,4)
  cout << "Transform matrix = \n" << T.matrix() << endl;

  // Use the transformation matrix for coordinate transformation
  Vector3d v_transformed = T * v; // Equivalent to R*v+t
```

```
    cout << "v tranformed = " << v_transformed.transpose() << endl;

    // For affine and projective transformations, use Eigen::Affine3d and Eigen::Projective3d.

    // Quaternion
    // You can assign AngleAxis directly to quaternions, and vice versa
    Quaterniond q = Quaterniond(rotation_vector);
    cout << "quaternion from rotation vector = " << q.coeffs().transpose() << endl;
    // Note that the order of coeffs is (x, y, z, w), w is the real part, the first three are the imaginary part
    // can also assign a rotation matrix to it
    q = Quaterniond(rotation_matrix);
    cout << "quaternion from rotation matrix = " << q.coeffs().transpose() << endl;
    // Rotate a vector with a quaternion and use overloaded multiplication
    V_rotated = q * v; // Note that the math is qvq^{-1}
    cout << "(1,0,0) after rotation = " << v_rotated.transpose() << endl;
    // expressed by regular vector multiplication, it should be calculated as follows
    cout << "should be equal to " << (q * Quaterniond(0, 1, 0, 0) * q.inverse()).coeffs().transpose() << endl;

    return 0;
}
```

The various forms of expression in *Eigen* are summarized below. Note that each type has both single and double precision types and, as before, cannot be automatically converted by the compiler. Taking the double-precision as an example, you can simply change the last "d" to "f" to use a single-precision data structure.

- Rotation matrix (3×3): Eigen::Matrix3d.
- Rotation vector (3×1): Eigen::AngleAxisd.
- Euler angle (3×1): Eigen::Vector3d.
- Quaternion (4×1): Eigen::Quaterniond.
- Euclidean transformation matrix (4×4): Eigen::Isometry3d.
- Affine transform (4×4): Eigen::Affine3d.
- Perspective transformation (4×4): Eigen::Projective3d.

This program can be compiled by referring to the corresponding "CMakeLists.txt" in the code. This program demonstrates how to use the rotation matrix, rotation vectors (AngleAxis), Euler angles, and quaternions in *Eigen*. We use these rotations to rotate a vector $\mathbf{v}$ and find that the result is the same. At the same time, it also demonstrates how to convert these expressions in the program. Readers who want to learn more about *Eigen*'s geometry modules can refer to http://eigen.tuxfamily.org/dox/group__TutorialGeometry.html.

Note that the code has some subtle differences from the mathematical representation. For example, by operator overloading in C++, quaternions and three-dimensional vectors can directly be multiplied, but mathematically, the vector needs to be converted into an imaginary quaternion, as we talked about in the last section, and then quaternion multiplication is used for calculation. The same applies to the transformation matrix multiplying with a three-dimensional vector. In general, the usage in the program is more flexible than the mathematical formula.

2.6.2 Coordinate Transformation Example

Let's take a small example to demonstrate the coordinate transformation.

Example 1. The robot No.1 and the robot No.2 are located in the world coordinate system. We use the world coordinate system as W, robot coordinate system as R_1 and R_2. The pose of the robot 1 is $\mathbf{q}_1 = [0.35, 0.2, 0.3, 0.1]^T$, $\mathbf{t}_1 = [0.3, 0.1, 0.1]^T$. The pose of the robot 2 is $\mathbf{q}_2 = [-0.5, 0.4, -0.1, 0.2]^T$, $\mathbf{t}_2 = [-0.1, 0.5, 0.3]^T$. Here $\mathbf{q}$ and $\mathbf{t}$ express $\mathbf{T}_{R_k,W}$, $k = 1, 2$, which is the world to the robot transform matrix. Now, assume that robot 1 sees a point in its own coordinate system with coordinates of $\mathbf{p}_{R_1} = [0.5, 0, 0.2]^T$. We want to find the coordinates of the vector in the robot 2's coordinate system.

This is a very simple but representative example. In real scenarios you often need to convert coordinates between different parts of the same robot or between different robots. Below we write a program to demonstrate this calculation.

Listing 2.7 slambook2/ch3/examples/coordinateTransform.cpp

```
#include<iostream>
#include<vector>
#include<algorithm>
#include<Eigen/Core>
#include<Eigen/Geometry>

using namespace std;
using namespace Eigen;

int main(int argc, char** argv) {
  Quaterniond q1(0.35, 0.2, 0.3, 0.1), q2(-0.5, 0.4, -0.1, 0.2);
  q1.normalize();
  q2.normalize();
  Vector3d t1(0.3, 0.1, 0.1), t2(-0.1, 0.5, 0.3);
  Vector3d p1(0.5, 0, 0.2);

  Isometry3d T1w(q1), T2w(q2);
  T1w.pretranslate (t1);
  T2w.pretranslate (t2);

  Vector3d p2 = T2w * T1w.inverse() * p1;
  cout << endl << p2.transpose() << endl;
  return 0;
}
```

The answer to the program is $[-0.0309731, 0.73499, 0.296108]^T$, and the calculation process is very simple, just by calculating

$$\mathbf{p}_{R_2} = \mathbf{T}_{R_2,W}\mathbf{T}_{W,R_1}\mathbf{p}_{R_1}.$$

Note that the quaternion needs to be normalized before use.

2.7 Visualization Demo

2.7.1 *Plotting Trajectory*

If you are new to rotation and translation concepts, you may find that their form looks a little complicated. There are so many representation methods, and we need to convert to a preferred one if necessary. Fortunately, although the rotation and transformation matrix values may not be intuitive enough, we can easily draw them in a 3D window.

In this section, we demonstrate two visual examples. First, let's say that we recorded the trajectory of a robot somehow, and now we want to draw it in a figure. Suppose the trajectory file is stored in a text file called "trajectory.txt", and each line is stored in the following format:

$$\text{time}, t_x, t_y, t_z, q_x, q_y, q_z, q_w,$$

where time refers the recording time of this pose, $\mathbf{t}$ is translation, $\mathbf{q}$ is the quaternion, all recorded in the world coordinate system to the robot coordinate system. Below we read these tracks from the file and display them in a window. In principle, if we just talk about "robot pose", then we can use any one of $\mathbf{T}_{WR}$ or $\mathbf{T}_{RW}$ because they are just the inverse of each other. It means that knowing one of them makes it easy to get the other. If we want to store *robot's trajectory*, then saving $\mathbf{T}_{WR}$ or $\mathbf{T}_{RW}$ doesn't make much difference.

When drawing the trajectory, we should draw the "trajectory" as a sequence of points, which is similar to the "trajectory" we imagined. Strictly speaking, these are actually the coordinates of the robot's origin in the world coordinate system. Consider the origin of the robot coordinate system, i.e., $\mathbf{O}_R$, then the $\mathbf{O}_W$ at this time is the coordinates of the origin in the world coordinate system:

$$\mathbf{O}_W = \mathbf{T}_{WR}\mathbf{O}_R = \mathbf{t}_{WR}. \tag{2.48}$$

This is exactly the translation part of $\mathbf{T}_{WR}$. So, you can see the robot's position directly from $\mathbf{T}_{WR}$. Therefore, in most of the public datasets, the trajectory file stores $\mathbf{T}_{WR}$ instead of $\mathbf{T}_{RW}$.

Finally, we need a library that supports 3D drawing. Many libraries support 3D drawing, such as the famous Matlab, python matplotlib, OpenGL, etc. In Linux, a widely used library in SLAM is the OpenGL-based *Pangolin* library,[16] which provides simple OpenGL drawing operations in a window. In the second edition of the book, we used git's submodule feature to manage the third-party libraries that this book relies on. Readers can go directly to the "3rdparty" folder to install the required libraries, and git guarantees that you are using the same version with us.

[16] See https://github.com/stevenlovegrove/Pangolin.

Listing 2.8 slambook2/ch3/examples/plotTrajectory.cpp

```
#include <pangolin/pangolin.h>
#include <Eigen/Core>
#include <unistd.h>

using namespace std;
using namespace Eigen;

// path to trajectory file
string trajectory_file = "./examples/trajectory.txt";

void DrawTrajectory(vector<Isometry3d, Eigen::aligned_allocator<Isometry3d
    >>);

int main(int argc, char **argv) {
  vector<Isometry3d, Eigen::aligned_allocator<Isometry3d>> poses;
  ifstream fin(trajectory_file);
  if (!fin) {
    cout << "cannot find trajectory file at " << trajectory_file << endl;
    return 1;
  }

  while (!fin.eof()) {
    double time, tx, ty, tz, qx, qy, qz, qw;
    fin >> time >> tx >> ty >> tz >> qx >> qy >> qz >> qw;
    Isometry3d Twr(Quaterniond(qw, qx, qy, qz));
    Twr.pretranslate(Vector3d(tx, ty, tz));
    poses.push_back(Twr);
  }
  cout << "read total " << poses.size() << " pose entries" << endl;

  // draw trajectory in pangolin
  DrawTrajectory(poses);
  return 0;
}

void DrawTrajectory(vector<Isometry3d, Eigen::aligned_allocator<Isometry3d
    >> poses) {
  // create pangolin window and plot the trajectory
  pangolin::CreateWindowAndBind("Trajectory Viewer", 1024, 768);
  glEnable(GL_DEPTH_TEST);
  glEnable(GL_BLEND);
  glBlendFunc(GL_SRC_ALPHA, GL_ONE_MINUS_SRC_ALPHA);

  pangolin::OpenGlRenderState s_cam(
  pangolin::ProjectionMatrix(1024, 768, 500, 500, 512, 389, 0.1, 1000),
  pangolin::ModelViewLookAt(0, -0.1, -1.8, 0, 0, 0, 0.0, -1.0, 0.0)
  );

  pangolin::View &d_cam = pangolin::CreateDisplay()
  .SetBounds(0.0, 1.0, 0.0, 1.0, -1024.0f / 768.0f)
  .SetHandler(new pangolin::Handler3D(s_cam));

  while (pangolin::ShouldQuit() == false) {
    glClear(GL_COLOR_BUFFER_BIT | GL_DEPTH_BUFFER_BIT);
    d_cam.Activate(s_cam);
    glClearColor(1.0f, 1.0f, 1.0f, 1.0f);
    glLineWidth(2);
    for (size_t i = 0; i < poses.size(); i++) {
      // draw three axes of each pose
      Vector3d Ow = poses[i].translation();
      Vector3d Xw = poses[i] * (0.1 * Vector3d(1, 0, 0));
      Vector3d Yw = poses[i] * (0.1 * Vector3d(0, 1, 0));
      Vector3d Zw = poses[i] * (0.1 * Vector3d(0, 0, 1));
      glBegin(GL_LINES);
      glColor3f(1.0, 0.0, 0.0);
```

```
      glVertex3d (Ow [0], Ow [1], Ow [2]);
      glVertex3d (Xw [0], Xw [1], Xw [2]);
      glColor3f(0.0, 1.0, 0.0);
      glVertex3d (Ow [0], Ow [1], Ow [2]);
      glVertex3d (Yw [0], Yw [1], Yw [2]);
      glColor3f(0.0, 0.0, 1.0);
      glVertex3d (Ow [0], Ow [1], Ow [2]);
      glVertex3d (Zw [0], Zw [1], Zw [2]);
      glEnd();
    }
    // draw a connection
    for (size_t i = 0; i < poses.size() - 1; i++) {
      glColor3f(0.0, 0.0, 0.0);
      glBegin(GL_LINES);
      auto p1 = poses[i], p2 = poses[i + 1];
      glVertex3d(p1.translation()[0], p1.translation()[1], p1.translation()
          [2]);
      glVertex3d(p2.translation()[0], p2.translation()[1], p2.translation()
          [2]);
      glEnd();
    }
    pangolin::FinishFrame();
    usleep(5000); // sleep 5 ms
  }
}
```

This program demonstrates how to draw a 3D pose in *Pangolin*. We draw the three axes of each pose in red, green, and blue (actually, we calculate each axis's world coordinates) and then connect the poses with black lines. The result is shown in Fig. 2.3.

2.7.2 Displaying Camera Pose

In addition to displaying the trajectory, we can also display the camera's pose in the 3D window. In slambook2/ch3/visualizeGeometry, we visualize various expressions of camera poses (see Fig. 2.4). When the reader uses the mouse to move the camera, the box on the left side will display the rotation matrix, translation, Euler angle,

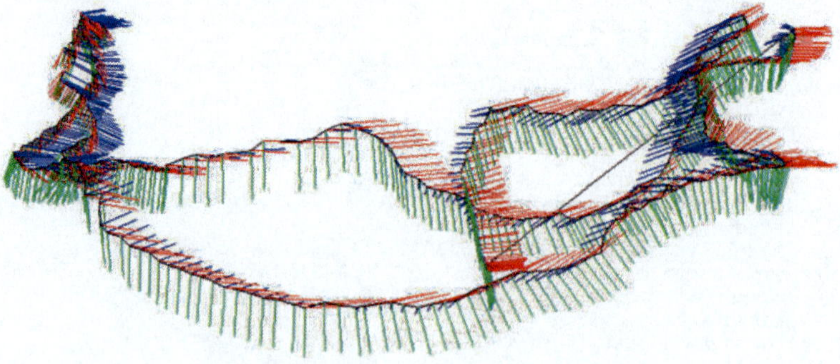

Fig. 2.3 Results of pose visualization

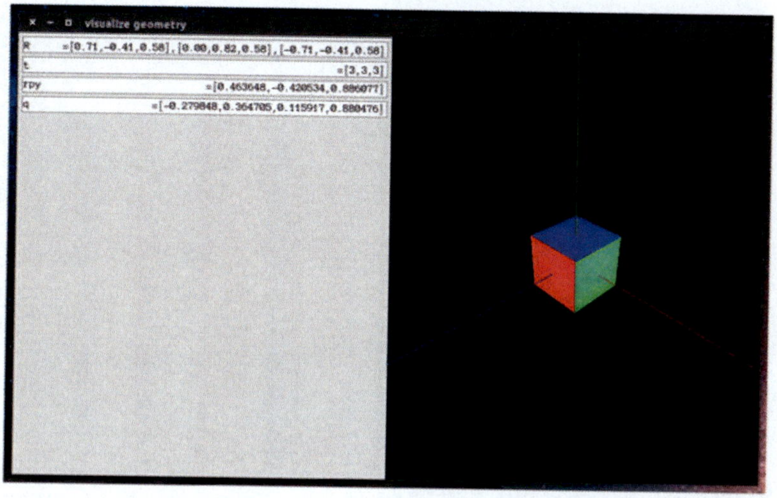

Fig. 2.4 Visualization program for rotation matrix, Euler angle, quaternion

and quaternion of the camera pose in real-time. You can see how the data changes. According to our experience, it is hard to infer the exact rotation from quaternions or matrices. However, although the rotation matrix or transformation matrix is not intuitive, it is not difficult to visually display them. This program uses the *Pangolin* library as a 3D display library. Please refer to "Readme.txt" to compile the program.

Exercises

1. Verify that the rotation matrix is an orthogonal matrix.
2. Prove the Rodrigues formula.
3. Verify that after the quaternion rotates a point, the result is a imaginary quaternion (the real part is zero), so it still corresponds to a three-dimensional space point, see (2.33).
4. Draw a table that summarizes the conversion relationship of the rotation matrix, rotation angle, Euler angle and quaternion.
5. Suppose there is a large *Eigen* matrix, we want to know the value in the top left 3×3 blocks, and then assign it to $\mathbf{I}_{3\times 3}$. Please implement it in C++.
6. When does a general linear equation $\mathbf{Ax} = \mathbf{b}$ has a unique solution of $\mathbf{x}$? How to solve it numerically? Can you implement it in *Eigen*?

Chapter 3
Lie Group and Lie Algebra

Goal of Study

1 Learn the concept of Lie group, Lie algebra, and their applications of SO(3), SE(3) and the corresponding Lie algebras.
2 Learn the meaning and usage of the BCH (Baker-Campbell-Hausdorff) formula.
3 Learn the perturbation model on Lie algebra.
4 Use Sophus to perform operations on Lie algebras.

In the last lecture, we introduced the description of rigid body motion in the three-dimensional world, including the rotation matrix, rotation vector, Euler angle, quaternion, and so on. We focused on the representation of rotation, but in SLAM, we have to estimate and optimize them in addition to the representation. Because the pose is unknown in SLAM, we need to solve the problem of which camera pose best matches the current observation. A typical way is to build it into an optimization problem, solving the optimal $\mathbf{R}, \mathbf{t}$ and minimizing the error.

As mentioned before, the rotation matrix itself is a constrained (orthogonal, and the determinant is 1) matrix. When used as optimization variables, it introduces additional constraints on matrices that make optimization difficult. Through the transformation relationship between Lie group and Lie algebra, we can turn the pose estimation into an unconstrained optimization problem and simplify the solution. Considering that the reader may not have the basic knowledge of Lie Group and Lie algebra, we will start with the most basic knowledge.

3.1 Basics of Lie Group and Lie Algebra

In the last lecture, we introduced the definition of the rotation matrix and the transformation matrix. At the time, we said that the three-dimensional rotation matrix

X. Gao and T. Zhang, *Introduction to Visual SLAM*,
https://doi.org/10.1007/978-981-16-4939-4_3

constitutes the *special orthogonal group* SO(3), and the transformation matrix constitutes the *special Euclidean group* SE(3). They are written like this:

$$\mathrm{SO}(3) = \{\mathbf{R} \in \mathbb{R}^{3\times3} | \mathbf{R}\mathbf{R}^T = \mathbf{I}, \det(\mathbf{R}) = 1\}. \tag{3.1}$$

$$\mathrm{SE}(3) = \left\{ \mathbf{T} = \begin{bmatrix} \mathbf{R} & \mathbf{t} \\ \mathbf{0}^T & 1 \end{bmatrix} \in \mathbb{R}^{4\times4} | \mathbf{R} \in \mathrm{SO}(3), \mathbf{t} \in \mathbb{R}^3 \right\}. \tag{3.2}$$

However, at that time, we did not explain the meaning of the group in detail. Readers should note that both the rotation matrix and the transformation matrix are not closed to addition. In other words, for any two rotation matrices $\mathbf{R}_1, \mathbf{R}_2$, according to the definition, their addition is no longer a rotation matrix:

$$\mathbf{R}_1 + \mathbf{R}_2 \notin \mathrm{SO}(3), \quad \mathbf{T}_1 + \mathbf{T}_2 \notin \mathrm{SE}(3). \tag{3.3}$$

You can also say that the two matrices do not have a well-defined addition operator, or the matrix addition is not closed in these two sets. The multiplication is the only one closed operation in these sets:

$$\mathbf{R}_1\mathbf{R}_2 \in \mathrm{SO}(3), \quad \mathbf{T}_1\mathbf{T}_2 \in \mathrm{SE}(3). \tag{3.4}$$

We know that matrix multiplication corresponds to the composition of two rotations or transformations. For a set that only has one "well-defined" operation, we call it a *group*.

3.1.1 Group

For the following contents, we need to talk a little bit about abstract algebra. I think this is a necessary condition for discussing Lie Group and Lie Algebra, but in fact, except for the students of mathematics and physics, most of the students will not have this knowledge in undergraduate classes. So let's look at some basic concepts first.

A group is an algebraic structure of one set plus one operator. We denote the set as A and the operation as $\cdot$, then the group can be denoted as $G = (A, \cdot)$. We say G is a *group* if the operation satisfies the following conditions:

1 **Closure**: $\forall a_1, a_2 \in A, \ a_1 \cdot a_2 \in A$.
2 **Combination**: $\forall a_1, a_2, a_3 \in A, \ (a_1 \cdot a_2) \cdot a_3 = a_1 \cdot (a_2 \cdot a_3)$.
3 **Unit element**: $\exists a_0 \in A, \ \text{s.t.} \ \forall a \in A, \ a_0 \cdot a = a \cdot a_0 = a$.
4 **Inverse element**: $\forall a \in A, \ \exists a^{-1} \in A, \ st \ a \cdot a^{-1} = a_0$.

It is easy to verify that the rotation matrix set with the normal matrix multiplication form a group, the same for the transformation matrix with matrix multiplication. They can be called rotation matrix and transformation matrix groups. Other common groups include the addition of integers $(\mathbb{Z}, +)$, the rational numbers with multiplication after removing 0 $(\mathbb{Q}\backslash 0, \cdot)$, etc. Common groups in the matrix are:

- General Linear group $\mathrm{GL}(n)$. The invertible matrix of $n \times n$ with matrix multiplication.
- Special Orthogonal Group $\mathrm{SO}(n)$. Or the rotation matrix group, where SO(2) and SO(3) is the most common.
- Special Euclidean group $\mathrm{SE}(n)$. Or the n dimensional transformation described earlier, such as SE(2) and SE(3).

The group structure guarantees that the group's operations have very good properties. The group theory is the theory that studies the various structures and properties of the groups. Readers interested in group theory can refer to any of the modern algebra books. *Lie Group* refers to a group with continuous (smooth) properties. Discrete groups like the integer group $\mathbb{Z}$ have no continuous properties, so they are not Lie groups. And obviously, $\mathrm{SO}(n)$ and $\mathrm{SE}(n)$ are continuous in real space because we can intuitively imagine that a rigid body moving continuously in the space, so they are all Lie Groups. Since SO(3) and SE(3) are especially important for camera pose estimation, we mainly discuss these two Lie groups. However, strictly discussing the concepts of "continuous" and "smooth" requires knowledge of analysis and topology. We don't want to write this book into a mathematics book, so only some important conclusions directly related to SLAM are introduced. If the reader is interested in the theoretical nature of Lie Groups, please refer to books like [20].

We usually have two ways to introduce the Lie Groups or Lie Algebras. The first is to directly introduce Lie group and Lie algebra and then present to the reader that each Lie group corresponds to a Lie algebra. But, in this case, the reader may think that Lie algebra seems to be a symbol that jumps out with no reason and does not know its physical meaning. So, I will take a little time to draw the Lie algebra from the rotation matrix, similar to the way of [21] and [22]. Let's start with the simpler SO(3), leading to the Lie algebra $\mathfrak{so}(3)$ above SO(3).

3.1.2 Introduction of the Lie Algebra

Consider an arbitrary rotation matrix $\mathbf{R}$, we know that it satisfies:

$$\mathbf{R}\mathbf{R}^T = \mathbf{I}. \tag{3.5}$$

Now, we say that $\mathbf{R}$ is the rotation of a camera that changes continuously over time, which is a function of time: $\mathbf{R}(t)$. Since it is still a rotation matrix, we have

$$\mathbf{R}(t)\mathbf{R}(t)^T = \mathbf{I}.$$

Deriving time on both sides of the equation yields (we use $\dot{\mathbf{R}}$ to represent the derivative of $\mathbf{R}$ on time t, just like many other control books):

$$\dot{\mathbf{R}}(t)\mathbf{R}(t)^T + \mathbf{R}(t)\dot{\mathbf{R}}(t)^T = 0.$$

Move the second term to right and commute the matrices by using the transposed relation:

$$\dot{\mathbf{R}}(t)\mathbf{R}(t)^T = -\left(\dot{\mathbf{R}}(t)\mathbf{R}(t)^T\right)^T. \tag{3.6}$$

It can be seen that $\dot{\mathbf{R}}(t)\mathbf{R}(t)^T$ is a *skew-symmetric* matrix. Recall that we introduced the $\wedge$ symbol in the cross product formula (2.3), which turns a vector into a skew-symmetric matrix. Similarly, for any skew-symmetric matrix, we can also find a unique vector corresponding to it. Let this operation be represented by the symbol $\vee$:

$$\mathbf{a}^{\wedge} = \mathbf{A} = \begin{bmatrix} 0 & -a_3 & a_2 \\ a_3 & 0 & -a_1 \\ -a_2 & a_1 & 0 \end{bmatrix}, \quad \mathbf{A}^{\vee} = \mathbf{a}. \tag{3.7}$$

So, since $\dot{\mathbf{R}}(t)\mathbf{R}(t)^T$ is a skew-symmetric matrix, we can find a three-dimensional vector $\boldsymbol{\phi}(t) \in \mathbb{R}^3$ corresponds to it:

$$\dot{\mathbf{R}}(t)\mathbf{R}(t)^T = \boldsymbol{\phi}(t)^{\wedge}.$$

Right multiply with $\mathbf{R}(t)$ on both sides. Since $\mathbf{R}$ is an orthogonal matrix, we have:

$$\dot{\mathbf{R}}(t) = \boldsymbol{\phi}(t)^{\wedge}\mathbf{R}(t) = \begin{bmatrix} 0 & -\phi_3 & \phi_2 \\ \phi_3 & 0 & -\phi_1 \\ -\phi_2 & \phi_1 & 0 \end{bmatrix} \mathbf{R}(t). \tag{3.8}$$

It can be seen that we can take the time derivative of a ration matrix just by multiplying a $\boldsymbol{\phi}^{\wedge}(t)$ matrix on the left. Consider at time $t_0 = 0$ that the rotation matrix is $\mathbf{R}(0) = \mathbf{I}$. According to the derivative definition, we use the first-order Taylor expansion around $t = t_0$ to write $\mathbf{R}(t)$ as:

$$\begin{aligned} \mathbf{R}(t) &\approx \mathbf{R}(t_0) + \dot{\mathbf{R}}(t_0)(t - t_0) \\ &= \mathbf{I} + \boldsymbol{\phi}(t_0)^{\wedge}(t). \end{aligned} \tag{3.9}$$

We see that $\boldsymbol{\phi}$ reflects the derivative of $\mathbf{R}$, so it is called the *tangent space* near the origin of SO(3).

The above formula is a differential equation for $\mathbf{R}$, and with the initial value $\mathbf{R}(0) = \mathbf{I}$, we have solution like:

$$\mathbf{R}(t) = \exp\left(\boldsymbol{\phi}_0^{\wedge} t\right). \tag{3.10}$$

where we note $\boldsymbol{\phi}(t_0) = \boldsymbol{\phi}_0$.

The reader can verify that the above equation holds for both the differential equation and the initial value. This means that around $t = 0$, the rotation matrix can be calculated from $\exp(\boldsymbol{\phi}_0^{\wedge} t)$.[1] We see that the rotation matrix $\mathbf{R}$ is associated with another skew-symmetric matrix $\boldsymbol{\phi}_0^{\wedge} t$ through an exponential relationship. But what is the exponential of a matrix? Here we have two questions that need to be clarified:

1 Given $\mathbf{R}$ at a certain moment, we can find a $\boldsymbol{\phi}$ that describes the local derivative relationship of $\mathbf{R}$. How are they correlated with each other? We will say that $\boldsymbol{\phi}$ corresponds to the Lie algebra $\mathfrak{so}(3)$ on SO(3);
2 Second, when a vector $\boldsymbol{\phi}$ is given, how is $\exp(\boldsymbol{\phi}^{\wedge})$ calculated? Conversely, given $\mathbf{R}$, is there an opposite operation to calculate $\boldsymbol{\phi}$? In fact, this is the exponential/logarithmic mapping between Lie group and Lie algebra.

Let's solve these two problems below.

3.1.3 The Definition of Lie Algebra

Now let's give the strict definition of Lie Algebra. Each Lie group has a Lie algebra corresponding to it. Lie algebra describes the local structure of the Lie group around its origin point, or in other words, is the tangent space. The general definition of Lie algebra is listed as follows:

A Lie algebra consists of a set $\mathbb{V}$, a scalar field $\mathbb{F}$, and a binary operation $[,]$. If they satisfy the following properties, then $(\mathbb{V}, \mathbb{F}, [,])$ is a Lie algebra, denoted as $\mathfrak{g}$.

1 **Closure**: $\forall \mathbf{X}, \mathbf{Y} \in \mathbb{V}; [\mathbf{X}, \mathbf{Y}] \in \mathbb{V}$.
2 **Bilinear composition**: $\forall \mathbf{X}, \mathbf{Y}, \mathbf{Z} \in \mathbb{V}; a, b \in \mathbb{F}$, we have:

$$[a\mathbf{X} + b\mathbf{Y}, \mathbf{Z}] = a[\mathbf{X}, \mathbf{Z}] + b[\mathbf{Y}, \mathbf{Z}], \quad [\mathbf{Z}, a\mathbf{X} + b\mathbf{Y}] = a[\mathbf{Z}, \mathbf{X}] + b[\mathbf{Z}, \mathbf{Y}].$$

3 **Reflexive**[2]: $\forall \mathbf{X} \in \mathbb{V}; [\mathbf{X}, \mathbf{X}] = \mathbf{0}$.
4 **Jacobi identity**: $\forall \mathbf{X}, \mathbf{Y}, \mathbf{Z} \in \mathbb{V}; [\mathbf{X}, [\mathbf{Y}, \mathbf{Z}]] + [\mathbf{Z}, [\mathbf{X}, \mathbf{Y}]] + [\mathbf{Y}, [\mathbf{Z}, \mathbf{X}]] = \mathbf{0}$.

The binary operations $[,]$ are called *Lie brackets*. At first glance, we require a lot of properties about the Lie bracket. Compared to the simpler binary operations in the group, the Lie bracket expresses the difference between the two elements. It does not require a combination law but requires the element and itself to be zero after the brackets. For example, the cross product $\times$ defined on the 3D vector $\mathbb{R}^3$ is a kind of Lie bracket, so $\mathfrak{g} = (\mathbb{R}^3, \mathbb{R}, \times)$ constitutes a Lie algebra. Readers can try to substitute the cross product into the four properties to verify the above conclusion.

---

[1] At this point we have not explained what this exp means and how it works. We will talk about its definition and calculation process right after this section.

[2] Reflexive means that an element operates with itself results in zero.

3.1.4 *Lie Algebra* $\mathfrak{so}(3)$

The previously mentioned $\boldsymbol{\phi}$ is actually a kind of Lie algebra. The Lie algebra corresponding to SO(3) is a vector defined on $\mathbb{R}^3$, which we will denote as $\boldsymbol{\phi}$. According to the previous derivation, each $\boldsymbol{\phi}$ can generate a skew-symmetric matrix:

$$\boldsymbol{\Phi} = \boldsymbol{\phi}^\wedge = \begin{bmatrix} 0 & -\phi_3 & \phi_2 \\ \phi_3 & 0 & -\phi_1 \\ -\phi_2 & \phi_1 & 0 \end{bmatrix} \in \mathbb{R}^{3\times3}. \tag{3.11}$$

Under this definition, the two vectors $\boldsymbol{\phi}_1, \boldsymbol{\phi}_2$'s Lie bracket is:

$$[\boldsymbol{\phi}_1, \boldsymbol{\phi}_2] = (\mathbf{fl}_1\mathbf{fl}_2 - \mathbf{fl}_2\mathbf{fl}_1)^\vee. \tag{3.12}$$

Readers can verify that the Lie bracket under this definition satisfy the above properties. Since the vector $\boldsymbol{\phi}$ is one-to-one with the skew-symmetric matrix, we say the elements of $\mathfrak{so}(3)$ are three-dimensional vectors or three-dimensional skew-symmetric matrices, without any ambiguity:

$$\mathfrak{so}(3) = \left\{\boldsymbol{\phi} \in \mathbb{R}^3 \text{ or } \boldsymbol{\Phi} = \boldsymbol{\phi}^\wedge \in \mathbb{R}^{3\times3}\right\}. \tag{3.13}$$

Some books also use the symbol $\widehat{\boldsymbol{\phi}}$ to represent $\boldsymbol{\phi}^\wedge$, but the meaning is the same. At this point, we have made it clear about the contents of $\mathfrak{so}(3)$. They are just a set of 3D vectors that can express the derivative of the rotation matrix. Its relationship to SO(3) is given by the exponential map:

$$\mathbf{R} = \exp(\boldsymbol{\phi}^\wedge). \tag{3.14}$$

The exponential map will be introduced later. Since we have introduced $\mathfrak{so}(3)$, we will first look at the corresponding Lie algebra on SE(3).

3.1.5 *Lie Algebra* $\mathfrak{se}(3)$

For SE(3), it also has a corresponding Lie algebra $\mathfrak{se}(3)$. To save space, we won't start by taking time derivatives. Similar to $\mathfrak{so}(3)$, $\mathfrak{se}(3)$ is located in the $\mathbb{R}^6$ space:

$$\mathfrak{se}(3) = \left\{\boldsymbol{\xi} = \begin{bmatrix} \boldsymbol{\rho} \\ \boldsymbol{\phi} \end{bmatrix} \in \mathbb{R}^6, \boldsymbol{\rho} \in \mathbb{R}^3, \boldsymbol{\phi} \in \mathfrak{so}(3), \boldsymbol{\xi}^\wedge = \begin{bmatrix} \boldsymbol{\phi}^\wedge & \boldsymbol{\rho} \\ \mathbf{0}^T & 0 \end{bmatrix} \in \mathbb{R}^{4\times4}\right\}. \tag{3.15}$$

We write each $\mathfrak{se}(3)$ element as $\boldsymbol{\xi}$, which is a six-dimensional vector. The first three dimensions are "translation part" (but keep in mind that the meaning is *different* from the translation in the matrix), which is denoted as $\boldsymbol{\rho}$; the second part is a rotation part

$\boldsymbol{\phi}$, which is essentially a $\mathfrak{so}(3)$ element.[3] At the same time, we extended the meaning of the $^\wedge$ symbol. In $\mathfrak{se}(3)$, a six-dimensional vector is converted to a four-dimensional matrix also using the $^\wedge$ symbol, but no longer a skew-symmetric one:

$$\boldsymbol{\xi}^\wedge = \begin{bmatrix} \boldsymbol{\phi}^\wedge & \boldsymbol{\rho} \\ \mathbf{0}^T & 0 \end{bmatrix} \in \mathbb{R}^{4\times 4}. \tag{3.16}$$

We still use the $^\wedge$ and $^\vee$ symbols to refer to the relationship from "vector to matrix" and "matrix to vector" to maintain the consistency with $\mathfrak{so}(3)$. They are still one-to-one correspondence. The readers can simply take $\mathfrak{se}(3)$ as a "vector consisting of a translation plus a $\mathfrak{so}(3)$ element" (although $\boldsymbol{\rho}$ is not the direct translation).

Finally, the Lie algebra $\mathfrak{se}(3)$ also has a Lie bracket similar to $\mathfrak{so}(3)$:

$$[\boldsymbol{\xi}_1, \boldsymbol{\xi}_2] = \left(\boldsymbol{\xi}_1^\wedge \boldsymbol{\xi}_2^\wedge - \boldsymbol{\xi}_2^\wedge \boldsymbol{\xi}_1^\wedge\right)^\vee. \tag{3.17}$$

The reader can verify that it satisfies the definition of Lie algebra (we'll leave it as an exercise). So far we have seen two important Lie algebras $\mathfrak{so}(3)$ and $\mathfrak{se}(3)$.

3.2 Exponential and Logarithmic Mapping

3.2.1 Exponential Map of SO(3)

Now consider the second question: How to calculate $\exp(\boldsymbol{\phi}^\wedge)$? In other words, it is an exponential map of a matrix. Again, we will first discuss the exponential mapping of $\mathfrak{so}(3)$ and then the case of $\mathfrak{se}(3)$.

The exponential of an arbitrary matrix can be written as a Taylor expansion if it has converged, whose result is still a matrix:

$$\exp(\mathbf{A}) = \sum_{n=0}^{\infty} \frac{1}{n!}\mathbf{A}^n. \tag{3.18}$$

Similarly, for any element in $\boldsymbol{\phi} \in \mathfrak{so}(3)$, we can also define its exponential map in this way:

$$\exp(\boldsymbol{\phi}^\wedge) = \sum_{n=0}^{\infty} \frac{1}{n!}(\boldsymbol{\phi}^\wedge)^n. \tag{3.19}$$

But this definition cannot be calculated directly because we don't want to calculate the infinite power of a matrix. Below we derive a convenient way to calculate the exponential mapping. Since $\boldsymbol{\phi}$ is a three-dimensional vector, we can define its length

[3] Please note that in some books the authors may put the rotation in the front and the translation in the back, which has no significant difference.

and direction, denoted as θ and $\mathbf{n}$, respectively. So we have $\boldsymbol{\phi} = \theta \mathbf{n}$, where $\mathbf{n}$ is a unit-length direction vector, i.e., $\|\mathbf{n}\| = 1$. First, for such a unit-length vector $\mathbf{n}$, there are two properties:

$$\mathbf{n}^\wedge \mathbf{n}^\wedge = \begin{bmatrix} -n_2^2 - n_3^2 & n_1 n_2 & n_1 n_3 \\ n_1 n_2 & -n_1^2 - n_3^2 & n_2 n_3 \\ n_1 n_3 & n_2 n_3 & -n_1^2 - n_2^2 \end{bmatrix} = \mathbf{n}\mathbf{n}^T - \mathbf{I}, \tag{3.20}$$

as well as

$$\mathbf{n}^\wedge \mathbf{n}^\wedge \mathbf{n}^\wedge = \mathbf{n}^\wedge (\mathbf{n}\mathbf{n}^T - \mathbf{I}) = \underbrace{\mathbf{n}^\wedge \mathbf{n}}_{\text{zero}} \mathbf{n}^T - \mathbf{n}^\wedge = -\mathbf{n}^\wedge. \tag{3.21}$$

These two formulas provide a way to handle the high-order $\mathbf{n}^\wedge$ items. Now we can write the exponential map as:

$$\begin{aligned}
\exp\left(\boldsymbol{\phi}^\wedge\right) &= \exp\left(\theta \mathbf{n}^\wedge\right) = \sum_{n=0}^{\infty} \frac{1}{n!} (\theta \mathbf{n}^\wedge)^n \\
&= \mathbf{I} + \theta \mathbf{n}^\wedge + \frac{1}{2!} \theta^2 \mathbf{n}^\wedge \mathbf{n}^\wedge + \frac{1}{3!} \theta^3 \mathbf{n}^\wedge \mathbf{n}^\wedge \mathbf{n}^\wedge + \frac{1}{4!} \theta^4 (\mathbf{n}^\wedge)^4 + \ldots \\
&= \mathbf{n}\mathbf{n}^T - \mathbf{n}^\wedge \mathbf{n}^\wedge + \theta \mathbf{n}^\wedge + \frac{1}{2!} \theta^2 \mathbf{n}^\wedge \mathbf{n}^\wedge - \frac{1}{3!} \theta^3 \mathbf{n}^\wedge - \frac{1}{4!} \theta^4 (\mathbf{n}^\wedge)^2 + \ldots \\
&= \mathbf{n}\mathbf{n}^T + \underbrace{\left(\theta - \frac{1}{3!}\theta^3 + \frac{1}{5!}\theta^5 - \ldots\right)}_{\sin\theta} \mathbf{n}^\wedge - \underbrace{\left(1 - \frac{1}{2!}\theta^2 + \frac{1}{4!}\theta^4 - \ldots\right)}_{\cos\theta} \mathbf{n}^\wedge \mathbf{n}^\wedge \\
&= \mathbf{n}^\wedge \mathbf{n}^\wedge + \mathbf{I} + \sin\theta \mathbf{n}^\wedge - \cos\theta \mathbf{n}^\wedge \mathbf{n}^\wedge \\
&= (1 - \cos\theta) \mathbf{n}^\wedge \mathbf{n}^\wedge + \mathbf{I} + \sin\theta \mathbf{n}^\wedge \\
&= \cos\theta \mathbf{I} + (1 - \cos\theta) \mathbf{n}\mathbf{n}^T + \sin\theta \mathbf{n}^\wedge.
\end{aligned}$$

Finally, we get a very familiar equation:

$$\exp(\theta \mathbf{n}^\wedge) = \cos\theta \mathbf{I} + (1 - \cos\theta) \mathbf{n}\mathbf{n}^T + \sin\theta \mathbf{n}^\wedge. \tag{3.22}$$

Recall the previous lesson. This equation is exactly the same as Rodrigues' formula, i.e., Eq. (2.15). This shows that $\mathfrak{so}(3)$ is actually the *rotation vector*, and the exponential map is just Rodrigues' formula. Through them, we map any vector in $\mathfrak{so}(3)$ to a rotation matrix in SO(3). Conversely, if we define a logarithmic map, we can also map the elements in SO(3) to $\mathfrak{so}(3)$:

$$\boldsymbol{\phi} = \ln\left(\mathbf{R}\right)^\vee = \left(\sum_{n=0}^{\infty} \frac{(-1)^n}{n+1} (\mathbf{R} - \mathbf{I})^{n+1}\right)^\vee. \tag{3.23}$$

Just like the exponential mapping, we have to use the Taylor series to expand the logarithmic mapping. In Lecture 3, we have already introduced how to calculate the corresponding Lie algebra according to the rotation matrix, that is, using the formula (2.17), and use the properties of the trace to solve the rotation angle and the rotation axis separately, which is more convenient.

Now, we've introduced the calculation method of exponential mapping. Readers may ask, what is the property of the exponential mapping? Can I find a unique $\boldsymbol{\phi}$ for any $\mathbf{R}$? Unfortunately, the exponential map is just a surjective map, not injective. This means that for each element in SO(3) we can find a $\mathfrak{so}(3)$ element corresponding to it; however, there may be multiple $\mathfrak{so}(3)$ elements corresponding to the same SO(3) element. At least for the rotation angle θ, we know that rotating multiple 360° will give the same rotation - it has periodicity. However, if we fix the rotation angle between $\pm\pi$, then the Lie group and the Lie algebra elements are one-to-one correspondence.

The conclusion of SO(3) and $\mathfrak{so}(3)$ seems to be in our expectation. It is very similar to the rotation vector we talked about earlier, and the exponential mapping is the Rodrigues formula. The derivative of the rotation matrix can be specified by the rotation vector, which guides how to perform calculus operations in the rotation matrix.

3.2.2 *Exponential Map of* SE(3)

The exponential map on $\mathfrak{se}(3)$ is described below. To save space, we no longer deduct the exponential mapping in detail like $\mathfrak{so}(3)$. The exponential mapping on $\mathfrak{se}(3)$ is as follows:

$$\exp\left(\boldsymbol{\xi}^{\wedge}\right) = \begin{bmatrix} \sum\limits_{n=0}^{\infty} \frac{1}{n!}\left(\boldsymbol{\phi}^{\wedge}\right)^n & \sum\limits_{n=0}^{\infty} \frac{1}{(n+1)!}\left(\boldsymbol{\phi}^{\wedge}\right)^n \boldsymbol{\rho} \\ \mathbf{0}^T & 1 \end{bmatrix} \tag{3.24}$$

$$\triangleq \begin{bmatrix} \mathbf{R} & \mathbf{J}\boldsymbol{\rho} \\ \mathbf{0}^T & 1 \end{bmatrix} = \mathbf{T}. \tag{3.25}$$

With a little patience, you can derive the Taylor expansion from the practice of $\mathfrak{so}(3)$. Let $\boldsymbol{\phi} = \theta\mathbf{a}$, where $\mathbf{a}$ is the unit vector, then:

$$\begin{aligned}\sum_{n=0}^{\infty} \frac{1}{(n+1)!}(\boldsymbol{\phi}^\wedge)^n &= \mathbf{I} + \frac{1}{2!}\theta \mathbf{a}^\wedge + \frac{1}{3!}\theta^2 (\mathbf{a}^\wedge)^2 + \frac{1}{4!}\theta^3 (\mathbf{a}^\wedge)^3 + \frac{1}{5!}\theta^4 (\mathbf{a}^\wedge)^4 \cdots \\ &= \frac{1}{\theta}\left(\frac{1}{2!}\theta^2 - \frac{1}{4!}\theta^4 + \cdots\right)(\mathbf{a}^\wedge) + \frac{1}{\theta}\left(\frac{1}{3!}\theta^3 - \frac{1}{5}\theta^5 + \cdots\right)(\mathbf{a}^\wedge)^2 + \mathbf{I} \\ &= \frac{1}{\theta}(1 - \cos\theta)(\mathbf{a}^\wedge) + \frac{\theta - \sin\theta}{\theta}\left(\mathbf{a}\mathbf{a}^T - \mathbf{I}\right) + \mathbf{I} \\ &= \frac{\sin\theta}{\theta}\mathbf{I} + \left(1 - \frac{\sin\theta}{\theta}\right)\mathbf{a}\mathbf{a}^T + \frac{1 - \cos\theta}{\theta}\mathbf{a}^\wedge \triangleq \mathbf{J}.\end{aligned} \tag{3.26}$$

From the results, we can see the $\mathbf{R}$ in the upper left corner of the exponential map of $\boldsymbol{\xi}$ is just the well-known SO(3), which means the rotation part of $\boldsymbol{\xi}$ is just the rotation part in $\mathfrak{so}(3)$. The $\mathbf{J}$ in the upper right corner is given by the above derivation:

$$\mathbf{J} = \frac{\sin\theta}{\theta}\mathbf{I} + \left(1 - \frac{\sin\theta}{\theta}\right)\mathbf{a}\mathbf{a}^T + \frac{1 - \cos\theta}{\theta}\mathbf{a}^\wedge. \tag{3.27}$$

This formula is somewhat similar to the Rodrigues formula but not exactly the same. We see that after passing the exponential map, the translation part is multiplied by a linear jacobian matrix $\mathbf{J}$. Please pay attention to the $\mathbf{J}$ here, as it will be used later.

Similarly, although we can also derive the logarithmic mapping analytically, there is a more trouble-free way to find the corresponding vector on $\mathfrak{so}(3)$ according to the transformation matrix $\mathbf{T}$: from the upper left corner $\mathbf{R}$ we can calculate the rotation vector, while $\mathbf{t}$ the upper right corner satisfies:

$$\mathbf{t} = \mathbf{J}\boldsymbol{\rho}. \tag{3.28}$$

Since $\mathbf{J}$ can be obtained from $\boldsymbol{\phi}$, $\boldsymbol{\rho}$ can also be solved by this linear equation. Now, we have clarified the definition of Lie group and Lie algebra and their mutual conversion relationship, as summarized in Fig. 3.1. If the reader doesn't understand everything, please go back to a few pages to read through the derivation again.

3.3 Lie Algebra Derivation and Perturbation Model

3.3.1 BCH Formula and Its Approximation

A major motivation for using Lie algebra is to do optimization. The derivative is a very necessary part of the optimization process (we will talk about it in detail in lecture 5). Let's consider the problem below. Although we have already understood the relationship between Lie group and Lie algebra on SO(3) and SE(3), but what happens in $\mathfrak{so}(3)$ when two matrices are multiplied in SO(3)? Conversely, when we

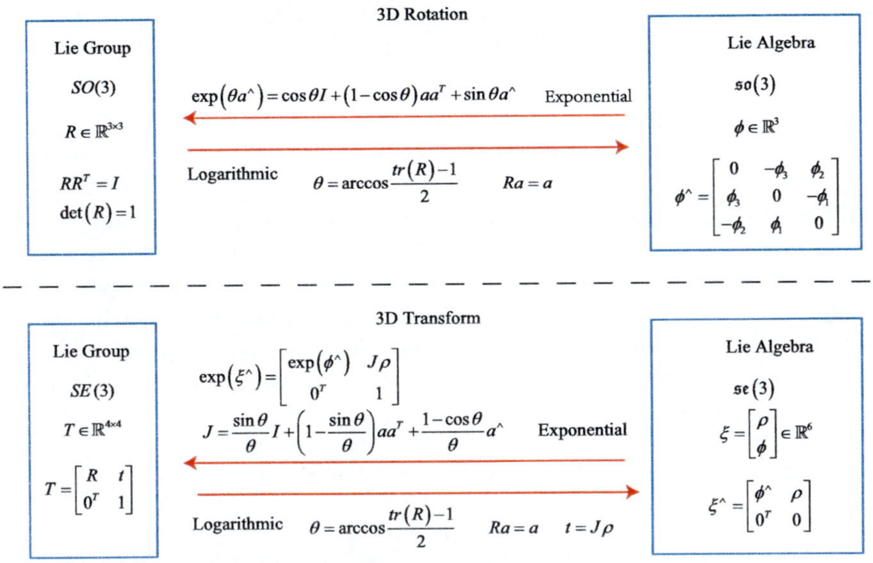

Fig. 3.1 The correspondence between SO(3), SE(3), $\mathfrak{so}(3)$, $\mathfrak{se}(3)$

add two vectors in $\mathfrak{so}(3)$, does SO(3) correspond to the product of the two matrices? If we write it out, it should be:

$$\exp\left(\boldsymbol{\phi}_1^\wedge\right)\exp\left(\boldsymbol{\phi}_2^\wedge\right) = \exp\left(\left(\boldsymbol{\phi}_1 + \boldsymbol{\phi}_2\right)^\wedge\right)?$$

If $\boldsymbol{\phi}_1, \boldsymbol{\phi}_2$ are scalars, then this is obviously true; but here we calculate the exponential map of *matrices* instead of scalars. In other words, we are studying whether the following formula holds:

$$\ln\left(\exp\left(\mathbf{A}\right)\exp\left(\mathbf{B}\right)\right) = \mathbf{A} + \mathbf{B}\ ?$$

for matrices. Unfortunately, this formula is not true in the matrix. The complete form of the product is given by the Baker-Campbell-Hausdorff formula (BCH formula).[4] Due to the complexity of its complete form, we only give the first few items of its expansion:

$$\ln\left(\exp\left(\mathbf{A}\right)\exp\left(\mathbf{B}\right)\right) = \mathbf{A} + \mathbf{B} + \frac{1}{2}\left[\mathbf{A}, \mathbf{B}\right] + \frac{1}{12}\left[\mathbf{A}, \left[\mathbf{A}, \mathbf{B}\right]\right] - \frac{1}{12}\left[\mathbf{B}, \left[\mathbf{A}, \mathbf{B}\right]\right] + \cdots \tag{3.29}$$

where [] is the Lie brackets. The BCH formula tells us that how to deal with the product of two matrices: they produce some extra Lie brackets compared with the scalar form. In particular, consider the case of SO(3) and $\ln\left(\exp\left(\boldsymbol{\phi}_1^\wedge\right)\exp\left(\boldsymbol{\phi}_2^\wedge\right)\right)^\vee$,

[4] See https://en.wikipedia.org/wiki/Baker-Campbell-Hausdorff_formula.

when $\boldsymbol{\phi}_1$ or $\boldsymbol{\phi}_2$ is small. Small items with more than quadratic can be ignored when taking derivatives. At this time, BCH has a linear approximation[5]:

$$\ln\left(\exp\left(\boldsymbol{\phi}_1^{\wedge}\right)\exp\left(\boldsymbol{\phi}_2^{\wedge}\right)\right)^{\vee} \approx \begin{cases} \mathbf{J}_l(\boldsymbol{\phi}_2)^{-1}\boldsymbol{\phi}_1 + \boldsymbol{\phi}_2 & \text{when } \boldsymbol{\phi}_1 \text{ is a small amount,} \\ \mathbf{J}_r(\boldsymbol{\phi}_1)^{-1}\boldsymbol{\phi}_2 + \boldsymbol{\phi}_1 & \text{when } \boldsymbol{\phi}_2 \text{ is a small amount.} \end{cases} \tag{3.30}$$

Take the first approximation as an example. This formula tells us that left multiplying a tiny rotation matrix $\mathbf{R}_1$ on a rotation matrix $\mathbf{R}_2$ (whose Lie algebra is $\boldsymbol{\phi}_1$ and $\boldsymbol{\phi}_2$, respectively), in $\mathfrak{so}(3)$ it can be approximated by adding a $\mathbf{J}_l\left(\boldsymbol{\phi}_2\right)^{-1}\boldsymbol{\phi}_1$ to the original Lie algebra $\boldsymbol{\phi}_2$. Similarly, the second approximation describes the case where $\mathbf{R}_1$ is right multiplied by a small rotation. Therefore, under the BCH approximation, the Lie algebra is divided into a left-multiplying approximation and a right-multiplying approximation. We must pay attention to whether the left model or the right model is used in daily usage. This book takes the left multiplication as an example. The jacobian in our left model $\mathbf{J}_l$ is exactly the content of the form (3.27) :

$$\mathbf{J}_l = \mathbf{J} = \frac{\sin\theta}{\theta}\mathbf{I} + \left(1 - \frac{\sin\theta}{\theta}\right)\mathbf{a}\mathbf{a}^T + \frac{1-\cos\theta}{\theta}\mathbf{a}^{\wedge}. \tag{3.31}$$

Its inverse is:

$$\mathbf{J}_l^{-1} = \frac{\theta}{2}\cot\frac{\theta}{2}\mathbf{I} + \left(1 - \frac{\theta}{2}\cot\frac{\theta}{2}\right)\mathbf{a}\mathbf{a}^T - \frac{\theta}{2}\mathbf{a}^{\wedge}, \tag{3.32}$$

when θ is not zero (in that case we take both $\mathbf{J}_l$ and its inverse as identity). To get the right jacobian we only need to take a negative sign for the argument:

$$\mathbf{J}_r(\boldsymbol{\phi}) = \mathbf{J}_l(-\boldsymbol{\phi}). \tag{3.33}$$

In this way, we've made it clear about the relationship between Lie group multiplication and Lie algebra addition.

For the convenience, we restate the meaning of the BCH approximation. Suppose we have a rotation $\mathbf{R}$, the corresponding Lie algebra is $\boldsymbol{\phi}$. We give it a small perturbation to the left, denoted as $\Delta\mathbf{R}$, and so that the corresponding Lie algebra is $\Delta\boldsymbol{\phi}$. Then, on Lie group, the result is $\Delta\mathbf{R}\cdot\mathbf{R}$, and on the Lie algebra, according to the BCH approximation, it is $\mathbf{J}_l^{-1}(\boldsymbol{\phi})\Delta\boldsymbol{\phi} + \boldsymbol{\phi}$. Putting them together, we can simply write:

$$\exp\left(\Delta\boldsymbol{\phi}^{\wedge}\right)\exp\left(\boldsymbol{\phi}^{\wedge}\right) = \exp\left(\left(\boldsymbol{\phi} + \mathbf{J}_l^{-1}\left(\boldsymbol{\phi}\right)\Delta\boldsymbol{\phi}\right)^{\wedge}\right). \tag{3.34}$$

Conversely, if we do addition on Lie algebra by adding $\boldsymbol{\phi}$ with $\Delta\boldsymbol{\phi}$, we can approximate the multiplication on the Lie group as:

$$\exp\left(\left(\boldsymbol{\phi} + \Delta\boldsymbol{\phi}\right)^{\wedge}\right) = \exp\left(\left(\mathbf{J}_l\Delta\boldsymbol{\phi}\right)^{\wedge}\right)\exp\left(\boldsymbol{\phi}^{\wedge}\right) = \exp\left(\boldsymbol{\phi}^{\wedge}\right)\exp\left(\left(\mathbf{J}_r\Delta\boldsymbol{\phi}\right)^{\wedge}\right). \tag{3.35}$$

[5] We are not going to do the detailed derivation of BCH approximation, see [5] if you are interested.

This provides a theoretical basis for calculus on Lie algebra. Similarly, for SE(3), there is a similar BCH approximation:

$$\exp\left(\Delta\xi^{\wedge}\right)\exp\left(\xi^{\wedge}\right) \approx \exp\left(\left(\mathcal{J}_l^{-1}\Delta\xi + \xi\right)^{\wedge}\right), \tag{3.36}$$

$$\exp\left(\xi^{\wedge}\right)\exp\left(\Delta\xi^{\wedge}\right) \approx \exp\left(\left(\mathcal{J}_r^{-1}\Delta\xi + \xi\right)^{\wedge}\right). \tag{3.37}$$

Here the $\mathcal{J}_l$ and $\mathcal{J}_r$ are more complicated 6×6 matrices. Readers can find its detailed contents in [5]. Since we did not use these two jacobians matrices in the calculation (we will see in the next subsection), the exact form is omitted here.

3.3.2 Derivative on SO(3)

Now let's talk about how to compute the derivation if our target function is related to a rotation or a transform, which has a powerful practical meaning since we usually have these functions to optimize in solving the SLAM problem. Assume we want to estimate a pose described by SO(3) or SE(3) elements. Our robot observes a point with the world coordinate $\mathbf{p}$ and generates an observation data $\mathbf{z}$, which can be written as:

$$\mathbf{z} = \mathbf{T}\mathbf{p} + \mathbf{w}, \tag{3.38}$$

where $\mathbf{w}$ is the noise (and is unknown). Because of the noise, the real observed data is not absolutely the same as the one we computed from the observation model, so we can calculate the error of predicted observation with the real one:

$$\mathbf{e} = \mathbf{z} - \mathbf{T}\mathbf{p}. \tag{3.39}$$

Suppose we have N points in total, then we find a best $\mathbf{T}$ to make the error minimized:

$$\min_{\mathbf{T}} J(\mathbf{T}) = \sum_{i=1}^{N} \left\| \mathbf{z}_i - \mathbf{T}\mathbf{p}_i \right\|_2^2. \tag{3.40}$$

Because of the noise, the real observed data is not absolutely the same as the one we computed from the observation model, so we can calculate the error of predicted observation with the real one: To solve such an optimized problem (which is a least square problem), we need to calculate the derivative of J by $\mathbf{T}$. We leave the least square problem to the next section. Here we just want to clarify that we normally have some functions that have rotations or transforms as their variables. We have to adjust those rotations or transforms to find a better/best estimation. But, as we mentioned before, since SO(3) and SE(3) do not have a well-defined addition (they are just groups), so the derivatives cannot be defined in their common form. If we treat the **R** or **T** as common matrices, we have to introduce the constraints into our optimization.

However, from the perspective of Lie algebra, since it consists of vectors, it has a good addition operation. Therefore, there are two ways to solve the problem of derivation using Lie algebra:

1 Assume we add a infinitesimal amount on Lie algebra, then compute the change of the object function.
2 Assume we multiply an infinitesimal perturbation on the Lie group left multiplication or right multiplication, use Lie algebra to describe the perturbation, and then compute the derivative on this perturbation. This is called as left perturbation or right perturbation model.

The first method corresponds to the normal derivation model of the Lie algebra, and the second corresponds to the perturbation model. Let's discuss the similarities and differences between these two approaches.

3.3.3 Derivative Model

First, consider the situation on SO(3). Suppose we rotate a space point **p** and get **Rp**. To calculate the derivative of the point coordinates by the rotation, we informally write it as[6]:

$$\frac{\partial (\mathbf{Rp})}{\partial \mathbf{R}}.$$

Since SO(3) has no addition operator, it cannot be calculated by the common derivative definition. Let the Lie algebra corresponding to **R** be $\boldsymbol{\phi}$, and we will calculate instead of the common derivative[7]:

$$\frac{\partial \left(\exp\left(\boldsymbol{\phi}^{\wedge}\right)\mathbf{p}\right)}{\partial \boldsymbol{\phi}}.$$

According to the definition of the derivative, we have:

[6] Please note that the derivative cannot be defined by matrix differentiation. Here we just write it for convenience.

[7] Strictly speaking, in matrix differentiation, we can only compute the derivative of a row vector to a column vector, whose result is a matrix. However, in this book, we write the derivative of the column vector to the column vector for convenience. The reader can think that the numerator is transposed first, and after the computation, the final result is also transposed (see Appendix B for details). This makes the formula look simple; otherwise, we have to add a transpose to each equation. In this sense, we can use equations like $\mathrm{d}(\mathbf{Ax})/\mathrm{d}\mathbf{x} = \mathbf{A}$.

$$
\begin{aligned}
\frac{\partial\left(\exp\left(\boldsymbol{\phi}^{\wedge}\right)\mathbf{p}\right)}{\partial\boldsymbol{\phi}} &= \lim_{\delta\boldsymbol{\phi}\to\mathbf{0}} \frac{\exp\left((\boldsymbol{\phi}+\delta\boldsymbol{\phi})^{\wedge}\right)\mathbf{p}-\exp\left(\boldsymbol{\phi}^{\wedge}\right)\mathbf{p}}{\delta\boldsymbol{\phi}} \\
&= \lim_{\delta\boldsymbol{\phi}\to\mathbf{0}} \frac{\exp\left((\mathbf{J}_l\delta\boldsymbol{\phi})^{\wedge}\right)\exp\left(\boldsymbol{\phi}^{\wedge}\right)\mathbf{p}-\exp\left(\boldsymbol{\phi}^{\wedge}\right)\mathbf{p}}{\delta\boldsymbol{\phi}} \\
&= \lim_{\delta\boldsymbol{\phi}\to\mathbf{0}} \frac{\left(\mathbf{I}+(\mathbf{J}_l\delta\boldsymbol{\phi})^{\wedge}\right)\exp\left(\boldsymbol{\phi}^{\wedge}\right)\mathbf{p}-\exp\left(\boldsymbol{\phi}^{\wedge}\right)\mathbf{p}}{\delta\boldsymbol{\phi}} \\
&= \lim_{\delta\boldsymbol{\phi}\to\mathbf{0}} \frac{(\mathbf{J}_l\delta\boldsymbol{\phi})^{\wedge}\exp\left(\boldsymbol{\phi}^{\wedge}\right)\mathbf{p}}{\delta\boldsymbol{\phi}} \\
&= \lim_{\delta\boldsymbol{\phi}\to\mathbf{0}} \frac{-\left(\exp\left(\boldsymbol{\phi}^{\wedge}\right)\mathbf{p}\right)^{\wedge}\mathbf{J}_l\delta\boldsymbol{\phi}}{\delta\boldsymbol{\phi}} = -(\mathbf{R}\mathbf{p})^{\wedge}\mathbf{J}_l.
\end{aligned}
$$

The second line is BCH approximation. The third line is Taylor's approximation after throwing the high-order terms (but we still write the equal sign here because the limit is taken). The fourth to the fifth line treat the skew-symmetric symbol as a cross-product so that $\mathbf{a}\times\mathbf{b}=-\mathbf{b}\times\mathbf{a}$. Thus, we compute the derivative of the rotated point relative to the addition in Lie algebra:

$$
\frac{\partial(\mathbf{R}\mathbf{p})}{\partial\boldsymbol{\phi}} = (-\mathbf{R}\mathbf{p})^{\wedge}\mathbf{J}_l. \tag{3.41}
$$

However, since there is still a very complicated form of $\mathbf{J}_l$, we don't want to calculate it. The perturbation model described below provides a more straightforward way to compute derivatives.

3.3.4 Perturbation Model

Another way to do this is to perturb $\mathbf{R}$ by $\Delta\mathbf{R}$ and see the change of the result relative to the disturbance. This disturbance can be multiplied on the left or on the right. The final result will be slightly different. Let's take the left perturbation as an example. Let the left perturbation $\Delta\mathbf{R}$ correspond to the Lie algebra as $\boldsymbol{\varphi}$. Then, for $\boldsymbol{\varphi}$, that is:

$$
\frac{\partial(\mathbf{R}\mathbf{p})}{\partial\boldsymbol{\varphi}} = \lim_{\boldsymbol{\varphi}\to\mathbf{0}} \frac{\exp\left(\boldsymbol{\varphi}^{\wedge}\right)\exp\left(\boldsymbol{\phi}^{\wedge}\right)\mathbf{p}-\exp\left(\boldsymbol{\phi}^{\wedge}\right)\mathbf{p}}{\boldsymbol{\varphi}}. \tag{3.42}
$$

The derivation of this formula is simpler than the above:

$$\begin{aligned}\frac{\partial\left(\mathbf{Rp}\right)}{\partial\boldsymbol{\varphi}} &= \lim_{\boldsymbol{\varphi}\to\mathbf{0}} \frac{\exp\left(\boldsymbol{\varphi}^\wedge\right)\exp\left(\boldsymbol{\phi}^\wedge\right)\mathbf{p}-\exp\left(\boldsymbol{\phi}^\wedge\right)\mathbf{p}}{\boldsymbol{\varphi}}\\ &= \lim_{\boldsymbol{\varphi}\to\mathbf{0}} \frac{\left(\mathbf{I}+\boldsymbol{\varphi}^\wedge\right)\exp\left(\boldsymbol{\phi}^\wedge\right)\mathbf{p}-\exp\left(\boldsymbol{\phi}^\wedge\right)\mathbf{p}}{\boldsymbol{\varphi}}\\ &= \lim_{\boldsymbol{\varphi}\to\mathbf{0}} \frac{\boldsymbol{\varphi}^\wedge\mathbf{Rp}}{\boldsymbol{\varphi}} = \lim_{\boldsymbol{\varphi}\to\mathbf{0}} \frac{-(\mathbf{Rp})^\wedge\boldsymbol{\varphi}}{\boldsymbol{\varphi}} = -(\mathbf{Rp})^\wedge.\end{aligned}$$

It can be seen that the calculation of a Jacobian $\mathbf{J}_l$ is omitted compared to Lie algebra's direct derivation. This makes the perturbation model more practical. Please keep in mind the derivative here since we will use it in the pose estimation sections.

3.3.5 *Derivative on* SE(3)

Finally, we give the perturbation model on SE(3), and skip the derivative model. Suppose a point $\mathbf{p}$ is transformed by $\mathbf{T}$ (corresponding to Lie algebra $\boldsymbol{\xi}$), and the result is $\mathbf{Tp}$.[8] Now, give $\mathbf{T}$ a left perturbation $\Delta\mathbf{T} = \exp\left(\delta\boldsymbol{\xi}^\wedge\right)$, whose Lie algebra is $\delta\boldsymbol{\xi} = [\delta\boldsymbol{\rho}, \delta\boldsymbol{\phi}]^T$, then:

$$\begin{aligned}\frac{\partial\left(\mathbf{Tp}\right)}{\partial\delta\boldsymbol{\xi}} &= \lim_{\delta\boldsymbol{\xi}\to\mathbf{0}} \frac{\exp\left(\delta\boldsymbol{\xi}^\wedge\right)\exp\left(\boldsymbol{\xi}^\wedge\right)\mathbf{p}-\exp\left(\boldsymbol{\xi}^\wedge\right)\mathbf{p}}{\delta\boldsymbol{\xi}}\\ &= \lim_{\delta\boldsymbol{\xi}\to\mathbf{0}} \frac{\left(\mathbf{I}+\delta\boldsymbol{\xi}^\wedge\right)\exp\left(\boldsymbol{\xi}^\wedge\right)\mathbf{p}-\exp\left(\boldsymbol{\xi}^\wedge\right)\mathbf{p}}{\delta\boldsymbol{\xi}}\\ &= \lim_{\delta\boldsymbol{\xi}\to\mathbf{0}} \frac{\delta\boldsymbol{\xi}^\wedge\exp\left(\boldsymbol{\xi}^\wedge\right)\mathbf{p}}{\delta\boldsymbol{\xi}}\\ &= \lim_{\delta\boldsymbol{\xi}\to\mathbf{0}} \frac{\begin{bmatrix}\delta\boldsymbol{\phi}^\wedge & \delta\boldsymbol{\rho}\\ \mathbf{0}^T & 0\end{bmatrix}\begin{bmatrix}\mathbf{Rp}+\mathbf{t}\\ 1\end{bmatrix}}{\delta\boldsymbol{\xi}}\\ &= \lim_{\delta\boldsymbol{\xi}\to\mathbf{0}} \frac{\begin{bmatrix}\delta\boldsymbol{\phi}^\wedge\left(\mathbf{Rp}+\mathbf{t}\right)+\delta\boldsymbol{\rho}\\ \mathbf{0}^T\end{bmatrix}}{[\delta\boldsymbol{\rho}, \delta\boldsymbol{\phi}]^T} = \begin{bmatrix}\mathbf{I} & -(\mathbf{Rp}+\mathbf{t})^\wedge\\ \mathbf{0}^T & \mathbf{0}^T\end{bmatrix} \triangleq (\mathbf{Tp})^\odot.\end{aligned}$$

We define the final result as an operator $\odot$,[9] which transforms a spatial point of homogeneous coordinates into a matrix of 4×6. This equation requires a little explanation about matrix differentiation. Assuming that $\mathbf{a}, \mathbf{b}, \mathbf{x}, \mathbf{y}$ are column vectors, then in our book, there are following rules:

---

[8] Please note that to make multiplication make sense, $\mathbf{p}$ must use homogeneous coordinates.

[9] I will read it as "Duang", like a stone falling into a well.

$$\frac{\mathrm{d}\begin{bmatrix}\mathbf{a}\\\mathbf{b}\end{bmatrix}}{\mathrm{d}\begin{bmatrix}\mathbf{x}\\\mathbf{y}\end{bmatrix}} = \left(\frac{\mathrm{d}[\mathbf{a},\mathbf{b}]^T}{\mathrm{d}\begin{bmatrix}\mathbf{x}\\\mathbf{y}\end{bmatrix}}\right)^T = \begin{bmatrix}\frac{\mathrm{da}}{\mathrm{dx}} & \frac{\mathrm{db}}{\mathrm{dx}}\\\frac{\mathrm{da}}{\mathrm{dy}} & \frac{\mathrm{db}}{\mathrm{dy}}\end{bmatrix}^T = \begin{bmatrix}\frac{\mathrm{da}}{\mathrm{dx}} & \frac{\mathrm{da}}{\mathrm{dy}}\\\frac{\mathrm{db}}{\mathrm{dx}} & \frac{\mathrm{db}}{\mathrm{dy}}\end{bmatrix} \tag{3.43}$$

Substituting this into the last line, you can get the final result. So far, we have introduced the differential operation on Lie group Lie algebra. In the following chapters, we will apply this knowledge to solve practical problems.

3.4 Practice: Sophus

3.4.1 Basic Usage of Sophus

We have introduced the basic knowledge of Lie algebra, and now it is time to consolidate what we have learned through practical exercises. Let's discuss how to manipulate Lie algebra in a program. In lecture 3, we saw that *Eigen* provided geometry modules but did not support Lie algebra. A better Lie algebra library is the Sophus library maintained by Strasdat (https://github.com/strasdat/Sophus).[10] The Sophus library supports SO(3) and SE(3), which are mainly discussed in this chapter. In addition, it also contains two-dimensional motion SO(2), SE(2) and the similar transformation of Sim(3). It is developed directly on top of *Eigen*, and we don't need to install additional dependencies. Readers can get Sophus directly from GitHub, or the Sophus source code is also available in our book's code directory "slambook2/3rdparty". For historical reasons, earlier versions of Sophus only provided double-precision Lie group/Lie algebra classes. Subsequent versions have been rewritten as template classes so that different precision of Lie group/Lie algebra can be used in the Sophus from the template class. But, at the same time, it increases the difficulty of use. In the second edition of this book, we use the Sophus library with templates. The Sophus provided in the 3rdparty of this book is the template version, which should have been copied to your computer during the downloading process. Sophus itself is also a CMake project. Presumably, you already know how to compile the CMake project, so I won't go into details here. The Sophus library only needs to be compiled, no need to install it.

Let's demonstrate the SO(3) and SE(3) operations in the Sophus library:

Listing 3.1 slambook/ch4/useSophus.cpp

```
#include <iostream>
#include <cmath>
#include <Eigen/Core>
#include <Eigen/Geometry>
#include "sophus/se3.hpp"

using namespace std;
```

[10] Sophus Lie first proposed the Lie algebra. The library is named after him.

```
using namespace Eigen;

/// This program demonstrates the basic usage of Sophus
int main(int argc, char **argv) {
  // Rotation matrix with 90 degrees along Z axis
  Matrix3d R = AngleAxisd(M_PI / 2, Vector3d(0, 0, 1)).toRotationMatrix();
  // or quaternion
  Quaterniond q(R);
  Sophus::SO3d SO3_R(R); // Sophus::SO3d can be constructed from rotation matrix
  Sophus::SO3d SO3_q(q); // or quaternion
  // they are equivalent of course
  cout << "SO(3) from matrix:\n" << SO3_R.matrix() << endl;
  cout << "SO(3) from quaternion:\n" << SO3_q.matrix() << endl;
  cout << "they are equal" << endl;

  // Use logarithmic map to get the Lie algebra
  Vector3d so3 = SO3_R.log();
  cout << "so3 = " << so3.transpose() << endl;
  // hat is from vector to skew-symmetric matrix
  cout << "so3 hat=\n" << Sophus::SO3d::hat(so3) << endl;
  // inversely from matrix to vector
  cout << "so3 hat vee= " << Sophus::SO3d::vee(Sophus::SO3d::hat(so3)).transpose() << endl;

  // update by perturbation model
  Vector3d update_so3(1e-4, 0, 0); // this is a small update
  Sophus::SO3d SO3_updated = Sophus::SO3d::exp(update_so3) * SO3_R;
  cout << "SO3 updated = \n" << SO3_updated.matrix() << endl;

  cout << "*******************************" << endl;
  // Similar for SE(3)
  Vector3d t(1, 0, 0); // translation 1 along X
  Sophus::SE3d SE3_Rt(R, t); // construction SE3 from R,t
  Sophus::SE3d SE3_qt(q, t); // or q,t
  cout << "SE3 from R,t= \n" << SE3_Rt.matrix() << endl;
  cout << "SE3 from q,t= \n" << SE3_qt.matrix() << endl;
  // Lie Algebra is 6d vector, we give a typedef
  typedef Eigen::Matrix<double, 6, 1> Vector6d;
  Vector6d se3 = SE3_Rt.log();
  cout << "se3 = " << se3.transpose() << endl;
  // The output shows Sophus puts the translation at first in se(3), then rotation.
  // Save as SO(3) wehave hat and vee
  cout << "se3 hat = \n" << Sophus::SE3d::hat(se3) << endl;
  cout << "se3 hat vee = " << Sophus::SE3d::vee(Sophus::SE3d::hat(se3)).transpose() << endl;

  // Finally the update
  Vector6d update_se3;
  update_se3.setZero();
  update_se3(0, 0) = 1e-4d;
  Sophus::SE3d SE3_updated = Sophus::SE3d::exp(update_se3) * SE3_Rt;
  cout << "SE3 updated = " << endl << SE3_updated.matrix() << endl;

  return 0;
}
```

The demo is divided into two parts. The first half introduces the operation on SO(3), and the second half is SE(3). We demonstrate how to construct SO(3), SE(3) objects as well as the exponential/logarithm mapping. And then, we update the lie group elements when we know the updated amount. If the reader has a good understanding of this lecture's content, then this program should not be difficult for you. To compile it, add the following lines to "CMakeLists.txt":

Listing 3.2 slambook2/ch4/useSophus/CMakeLists.txt

```
# we use find_package to make CMake find sophus
find_package( Sophus REQUIRED )
include_directories( ${Sophus_INCLUDE_DIRS} ) # sohpus is header only

add_executable( useSophus useSophus.cpp )
```

The *find_package* is a command provided by CMake to find the header and library files of a library. If CMake can find it, it will provide the variables for the directory where the header and library files are located. In the example of Sophus, it is Sophus_INCLUDE_DIRS. The template-based Sophus library, like *Eigen*, contains only header files and no source files. Based on them, we can introduce the Sophus library into our own CMake project. Readers are asked to see the output of this program on their own, consistent with our previous derivation.

3.4.2 Example: Evaluating the Trajectory

In practical engineering, we often need to evaluate the difference between the estimated trajectory of an algorithm and the real trajectory to evaluate the algorithm's accuracy. The real (or ground-truth) trajectory is often obtained by some higher precision systems, and the estimated one is calculated by the algorithm to be evaluated. In the last lecture, we demonstrated how to display a trajectory stored in a file. In this section, we will consider how to calculate the error of two trajectories. Consider an estimated trajectory $\mathbf{T}_{\text{esti},i}$ and the real trajectory $\mathbf{T}_{\text{gt},i}$, where $i = 1, \ldots, N$; then we can define some error indicators to describe the difference between them.

There are many kinds of error indicators. The common used one is *absolute trajectory error*, which is like:

$$\mathrm{ATE}_{\text{all}} = \sqrt{\frac{1}{N}\sum_{i=1}^{N} \| \log(\mathbf{T}_{\text{gt},i}^{-1}\mathbf{T}_{\text{esti},i})^{\vee} \|_2^2}, \tag{3.44}$$

This is actually the root-mean-squared error (RMSE) for each pose in Lie algebra. This error can describe both the rotation and translation errors. At the same time, some literature only consider the translation error [23], so we can define the *average translational error*:

$$\mathrm{ATE}_{\text{trans}} = \sqrt{\frac{1}{N}\sum_{i=1}^{N} \|\mathrm{trans}(\mathbf{T}_{\text{gt},i}^{-1}\mathbf{T}_{\text{esti},i})\|_2^2}, \tag{3.45}$$

where the function "trans" represents the translation of the internal variables of the parentheses. From the perspective of the entire trajectory, if we have a rotation error, it will also affect the subsequent translation. So both indicators are applicable in practice.

In addition to this, relative error indicators can also be defined. For example, consider the movement from i to the time of $i + \Delta t$, then the relative pose error (RPE) can be defined as:

$$\mathrm{RPE}_{\mathrm{all}} = \sqrt{\frac{1}{N-\Delta t}\sum_{i=1}^{N-\Delta t} \| \log\left(\left(\mathbf{T}_{\mathrm{gt},i}^{-1} \mathbf{T}_{\mathrm{gt},i+\Delta t} \right)^{-1} \left(\mathbf{T}_{\mathrm{esti},i}^{-1} \mathbf{T}_{\mathrm{esti},i+\Delta t} \right) \right)^{\vee} \|_2^2}, \tag{3.46}$$

Similarly, you can only take the translation part:

$$\mathrm{RPE}_{\mathrm{trans}} = \sqrt{\frac{1}{N-\Delta t}\sum_{i=1}^{N-\Delta t} \| \mathrm{trans}\left(\left(\mathbf{T}_{\mathrm{gt},i}^{-1} \mathbf{T}_{\mathrm{gt},i+\Delta t} \right)^{-1} \left(\mathbf{T}_{\mathrm{esti},i}^{-1} \mathbf{T}_{\mathrm{esti},i+\Delta t} \right) \right) \|_2^2}. \tag{3.47}$$

This part of the calculation is easy to implement with the Sophus library. Below we demonstrate the calculation of the absolute trajectory error. In this example, we have two trajectories: "groundtruth.txt" and "estimated.txt". The following code will read the two trajectories, calculate the error, and display it in a 3D window. For the sake of brevity, the code for the trajectory plotting has been omitted, as we have done similar work in the previous section.

Listing 3.3 slambook/ch4/example/trajectoryError.cpp (part)

```
#include <iostream>
#include <fstream>
#include <unistd.h>
#include <pangolin/pangolin.h>
#include <sophus/se3.hpp>

using namespace Sophus;
using namespace std;

string groundtruth_file = "./example/groundtruth.txt";
string estimated_file = "./example/estimated.txt";

typedef vector<Sophus::SE3d, Eigen::aligned_allocator<Sophus::SE3d>>
    TrajectoryType;

void DrawTrajectory(const TrajectoryType &gt, const TrajectoryType &esti);

TrajectoryType ReadTrajectory(const string &path);

int main(int argc, char **argv) {
  TrajectoryType groundtruth = ReadTrajectory(groundtruth_file);
  TrajectoryType estimated = ReadTrajectory(estimated_file);
  assert(!groundtruth.empty() && !estimated.empty());
  assert(groundtruth.size() == estimated.size());

  // compute rmse
  double rmse = 0;
  for (size_t i = 0; i < estimated.size(); i++) {
    Sophus::SE3d p1 = estimated[i], p2 = groundtruth[i];
    double error = (p2.inverse() * p1).log().norm();
    rmse += error * error;
  }
  rmse = rmse / double(estimated.size());
  rmse = sqrt(rmse);
```

```
  cout << "RMSE = " << rmse << endl;

  DrawTrajectory(groundtruth, estimated);
  return 0;
}

TrajectoryType ReadTrajectory(const string &path) {
  ifstream fin(path);
  TrajectoryType trajectory;
  if (!fin) {
    cerr << "trajectory " << path << " not found." << endl;
    return trajectory;
  }

  while (!fin.eof()) {
    double time, tx, ty, tz, qx, qy, qz, qw;
    fin >> time >> tx >> ty >> tz >> qx >> qy >> qz >> qw;
    Sophus::SE3d p1(Eigen::Quaterniond(qx, qy, qz, qw), Eigen::Vector3d(tx,
         ty, tz));
    trajectory.push_back(p1);
  }
  return trajectory;
}
```

The result of this program is 2.207, and the image is shown as Fig. 3.2. You can also try to remove the rotating part and only calculates the error of the translation part. In fact, in this example, we have helped the reader to do some pre-processing tasks, including time alignment of the trajectory and external parameter estimation. These contents have not been mentioned yet, and we will talk about them in the future.

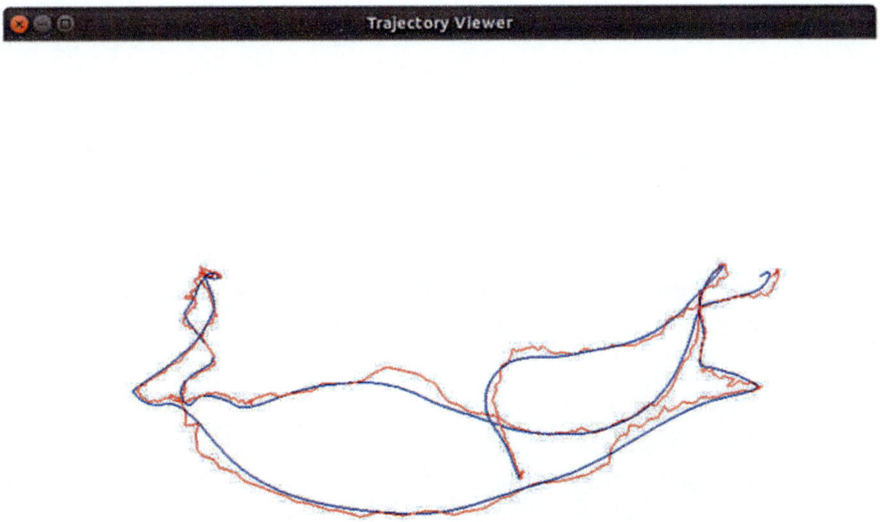

Fig. 3.2 Calculates the error between the estimated trajectory and the real trajectory

3.5 Similar Transform Group and Its Lie Algebra

Finally, we would like to mention the similar transform group Sim(3) used in monocular vision, and the corresponding Lie algebra $\mathfrak{sim}(3)$. If you are only interested in stereo or RGB-D SLAM, you can skip this section.

We have already introduced the concept of scale ambiguity. If SE(3) is used in the monocular SLAM to represent the pose, then the scale in the entire SLAM process will change due to scale uncertainty and scale drift, which is what SE(3) not reflects. Therefore, in the case of monocular SLAM, we generally express the scale factor explicitly. In mathematical terms, for the point $\mathbf{p}$ in space, a *similar transformation* is passed in the camera coordinate system instead of the Euclidean transformation:

$$\mathbf{p}' = \begin{bmatrix} s\mathbf{R} & \mathbf{t} \\ \mathbf{0}^T & 1 \end{bmatrix} \mathbf{p} = s\mathbf{R}\mathbf{p} + \mathbf{t}. \tag{3.48}$$

In the similarity transformation, we express the scale as s. It also acts on top of the three coordinates of $\mathbf{p}$ and scales $\mathbf{p}$ once. Similar to SO(3), SE(3), the similarity transform also forms a group on matrix multiplication, called the similarity transform group Sim(3):

$$\mathrm{Sim}(3) = \left\{ \mathbf{S} = \begin{bmatrix} s\mathbf{R} & \mathbf{t} \\ \mathbf{0}^T & 1 \end{bmatrix} \in \mathbb{R}^{4\times4} \right\}. \tag{3.49}$$

Similarly, Sim(3) also has corresponding Lie algebra, exponential mapping, logarithmic mapping, and so on. The Lie algebra $\mathfrak{sim}(3)$ element is a 7-dimensional vector $\boldsymbol{\zeta}$. Its first 6 dimensions are the same as $\mathfrak{se}(3)$, followed by a σ to denote the scale.

$$\mathfrak{sim}\,(3) = \left\{ \boldsymbol{\zeta} | \boldsymbol{\zeta} = \begin{bmatrix} \boldsymbol{\rho} \\ \boldsymbol{\phi} \\ \sigma \end{bmatrix} \in \mathbb{R}^7, \boldsymbol{\zeta}^\wedge = \begin{bmatrix} \sigma\mathbf{I} + \boldsymbol{\phi}^\wedge & \boldsymbol{\rho} \\ \mathbf{0}^T & 0 \end{bmatrix} \in \mathbb{R}^{4\times4} \right\}. \tag{3.50}$$

It has an additional σ compared with $\mathfrak{se}(3)$. The Sim(3) and $\mathfrak{sim}(3)$ are still associated with exponential maps and logarithm maps. The exponential mapping is:

$$\exp\left(\boldsymbol{\zeta}^\wedge\right) = \begin{bmatrix} \mathrm{e}^\sigma \exp\left(\boldsymbol{\phi}^\wedge\right) & \mathbf{J}_s\boldsymbol{\rho} \\ \mathbf{0}^T & 1 \end{bmatrix}, \tag{3.51}$$

where $\mathbf{J}_s$ is:

$$\begin{aligned}\mathbf{J}_s =& \frac{\mathrm{e}^\sigma - 1}{\sigma}\mathbf{I} + \frac{\sigma \mathrm{e}^\sigma \sin\theta + (1 - \mathrm{e}^\sigma \cos\theta)\,\theta}{\sigma^2 + \theta^2}\mathbf{a}^\wedge \\ &+ \left(\frac{\mathrm{e}^\sigma - 1}{\sigma} - \frac{(\mathrm{e}^\sigma \cos\theta - 1)\,\sigma + (\mathrm{e}^\sigma \sin\theta)\theta}{\sigma^2 + \theta^2}\right)\mathbf{a}^\wedge\mathbf{a}^\wedge.\end{aligned}$$

Through exponential mapping, we can find the relationship between Lie algebra and Lie group. For the Lie algebra $\boldsymbol{\zeta}$, its correspondence with the Lie group is:

$$s = \mathrm{e}^{\sigma}, \ \mathbf{R} = \exp(\boldsymbol{\phi}^{\wedge}), \ \mathbf{t} = \mathbf{J}_s \boldsymbol{\rho}. \tag{3.52}$$

The rotation is consistent with SO(3). In the translation part, we need to multiply a Jacobian $\mathcal{J}$ in $\mathfrak{se}(3)$, and the similarly transformed Jacobi is more complicated. For the scale factor, you can see that s in the Lie group is the exponential function of σ in the Lie algebra.

The BCH approximation of Sim(3) is similar to SE(3). We can discuss a derivative of $\mathbf{S}$ after a similar transformation of $\mathbf{Sp}$ relative to $\mathbf{S}$. Similarly, there are two ways of the differential model and perturbation model, and the perturbation model is the simpler one. We omit the derivation process and directly give the results of the perturbation model. Let $\mathbf{Sp}$ a small perturbation $\exp(\boldsymbol{\zeta}^{\wedge})$ on the left and ask for $\mathbf{Sp}$ The derivative of the disturbance. Since $\mathbf{Sp}$ is a 4-dimensional homogeneous coordinate, $\boldsymbol{\zeta}$ is a 7-dimensional vector, which should have 4×7 Jacobian. For convenience, remember the first three-dimensional composition vector $\mathbf{q}$ of $\mathbf{Sp}$, then:

$$\frac{\partial \mathbf{Sp}}{\partial \boldsymbol{\zeta}} = \begin{bmatrix} \mathbf{I} & -\mathbf{q}^{\wedge} & \mathbf{q} \\ \mathbf{0}^T & \mathbf{0}^T & 0 \end{bmatrix}. \tag{3.53}$$

We will end here about the contents of Sim(3). For more detailed information on Sim(3), please refer to the literature [24].

3.6 Summary

This lecture introduces Lie group SO(3) and SE(3), and their corresponding Lie algebras $\mathfrak{so}(3)$ and $\mathfrak{se}(3)$. We introduce the expression and transformation of poses on them, and then through the linear approximation of BCH, we can perturb and predict the pose. This lays the theoretical foundation for optimizing the pose afterward because we need to adjust the estimate of a certain pose frequently to reduce the corresponding error. Only after we have figured out how to adjust and update the pose can we continue to the next step.

The content of this lecture may be more theoretical. After all, it is not as good as computer vision. Compared to the mathematics textbooks that explain Lie group Lie algebra, since we only care about practical content, the process is very streamlined, and the speed is relatively fast. The reader must understand the content of this lecture, which is the basis for solving many subsequent problems, especially the pose estimation part.

It should be mentioned that in addition to the Lie algebra, the rotation can also be expressed by quaternion, Euler angle, etc., but the subsequent processing is troublesome. In practice, you can also use SO(3) plus a translation instead of SE(3) to avoid some Jacobian calculations.

Exercises

1 Verify SO(3), SE(3), and Sim(3) are groups on matrix multiplication.
2 Verify that $(\mathbb{R}^3, \mathbb{R}, \times)$ constitutes a Lie algebra.
3 Verify that $\mathfrak{so}(3)$ and $\mathfrak{se}(3)$ satisfy the requirements of Lie algebra.
4 Show that:

$$\mathbf{R}\mathbf{p}^\wedge\mathbf{R}^T = (\mathbf{R}\mathbf{p})^\wedge.$$

5 Show that:

$$\mathbf{R}\exp(\mathbf{p}^\wedge)\mathbf{R}^T = \exp((\mathbf{R}\mathbf{p})^\wedge).$$

This is called the *adjoint* property on SO(3). Similarly, there is an adjoint property on SE(3):

$$\mathbf{T}\exp(\boldsymbol{\xi}^\wedge)\mathbf{T}^{-1} = \exp\left((\mathrm{Ad}(\mathbf{T})\boldsymbol{\xi})^\wedge\right), \tag{3.54}$$

where

$$\mathrm{Ad}(\mathbf{T}) = \begin{bmatrix} \mathbf{R} & \mathbf{t}^\wedge\mathbf{R} \\ \mathbf{0} & \mathbf{R} \end{bmatrix}. \tag{3.55}$$

6 Follow the derivation of the left perturbation and derives the derivatives of SO(3) and SE(3) under the right perturbation.
7 Search how CMake's *find_package* works. What optional parameters does it have? What are the prerequisites for CMake to find a library?

Chapter 4
Cameras and Images

Check for updates

Goal of Study

1. Learn the models of the pinhole camera, intrinsics, extrinsics, and distortion.
2. Learn how to project a spatial point into image planes.
3. Learn the basic image process in OpenCV.

In the previous two lectures, we introduced how to express and optimize the robot's 6 DoF pose and partially explained the meaning of the variables and the equations of motion and observation in SLAM. This chapter will discuss "How robots observe the outside world", which belongs to the observation equation. In the camera-based visual SLAM, the observation mainly refers to the process of image projection.

We have seen a lot of photos in real life. A photo consists of millions of pixels in the computer, recording information about color or brightness. We will see a bundle of light reflected or emitted by an object in the three-dimensional world pass through the camera's optical center and is projected onto the camera's imaging plane. After the camera's light sensor receives the light, it produces a measurement, and we get the pixels, which form the photo we see. Can this process be described by mathematical equations? This lecture will first discuss the camera model, explain how the projection relationship is described, and the internal parameters in this projection process. At the same time, we will also give a brief introduction to the stereo and RGB-D cameras. Then, we introduce the basic operations of 2D images in OpenCV. Finally, an experiment of point cloud stitching is demonstrated to show the meaning of intrinsic and extrinsic parameters.

4.1 Pinhole Camera Models

The process of projecting a 3D point (in meters) to a 2D image plane (in pixels) can be described by a geometric model. Actually, several models describe this, the simplest of which is called the pinhole model. We will start with this pinhole projection. At the

X. Gao and T. Zhang, *Introduction to Visual SLAM*,
https://doi.org/10.1007/978-981-16-4939-4_4

same time, due to the presence of the lens on the camera lens, *distortion* is generated during the projection. Therefore, we will use the pinhole model plus a distortion model to describe the entire projection process.

4.1.1 Pinhole Camera Geometry

Most of us have seen the candle projection experiment in the physics class of high school: a lit candle is placed in front of a dark box, and the light of the candle is projected through a small hole in the dark box on the rear plane. Then an inverted candle image is formed on this plane. In this process, the small hole can project a candle in a three-dimensional world onto a two-dimensional imaging plane. For the same reason, we can use this simple model to explain the camera's imaging process, as shown in Fig. 4.1.

Let's take a look at the simple geometry in this model. Let $O - x - y - z$ be the camera coordinate system. Commonly we put the z axis to the front of the camera, x to the right, and y to the down (so in this figure, we should stand on the left side to see the right side). O is the camera's optical center, which is also the "hole" in the pinhole model. The 3D point P, after being projected through the hole O, falls on the physical imaging plane $O' - x' - y'$ and produces the image point P'. Let the coordinates of P be $[X, Y, Z]^T$, P' is $[X', Y', Z']^T$, and set the physical distance from the imaging plane to camera plane is f (focal length). Then, according to the similarity of the triangles, there are:

$$\frac{Z}{f} = -\frac{X}{X'} = -\frac{Y}{Y'}. \tag{4.1}$$

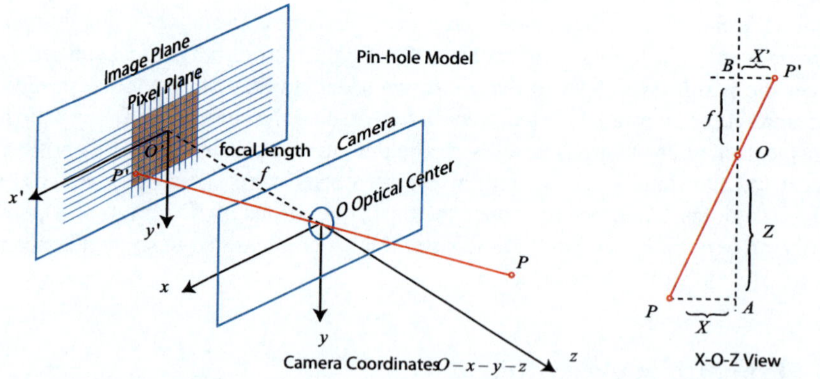

Fig. 4.1 Pinhole camera model

The negative sign indicates that the image is inverted. However, the image obtained by modern cameras is not inverted (otherwise, the usage of the camera would be very inconvenient). To make the model more realistic, we can equivalently place the imaging plane symmetrically in front of the camera, as shown by Fig. 4.2. This can remove the negative sign in the formula to make it more compact:

$$\frac{Z}{f} = \frac{X}{X'} = \frac{Y}{Y'}. \tag{4.2}$$

Put X', Y' to the left side:

$$X' = f\frac{X}{Z}, \quad Y' = f\frac{Y}{Z}. \tag{4.3}$$

Readers may ask why we can arbitrarily move the imaging plane to the front? In fact, this is just a mathematical approach to handle the camera projection, and most of the images captured by the camera are not upside-down. The camera's software will flip the picture for you, so what we actually get is the symmetric plane's image. Although the pin-hole image should be inverted from the physical principle since we have pre-processed the picture, it is not wrong to take the symmetric one. Therefore, without causing ambiguity, we often emit the minus symbol in the pin-hole model.

The formula (4.3) describes the spatial relationship between the point P and its image, where the units of all points are meters. For example, a focal length maybe 0.2 m, and X' be 0.14 m. However, in the camera, we end up with pixels, where we need to sample and quantize the pixels on the imaging plane. To describe how the sensor converts the perceived light into image pixels, we set a pixel plane $o-u-v$ fixed on the physical imaging plane. Finally, we get *pixel coordinates* of P' in the pixel plane: $[u, v]^T$.

The usual definition of the pixel coordinate system[1] is: the origin o' is in the upper left corner of the image, the u axis is parallel to the x axis, and the v axis is parallel to the y axis. Between the pixel coordinate system and the imaging plane, there is an apparent *zoom* and a *translation of the origin*. We set the pixel coordinates to scale α times on the u axis and β times on v. At the same time, the origin is translated by $[c_x, c_y]^T$. Then, the relationship between the coordinates of P' and the

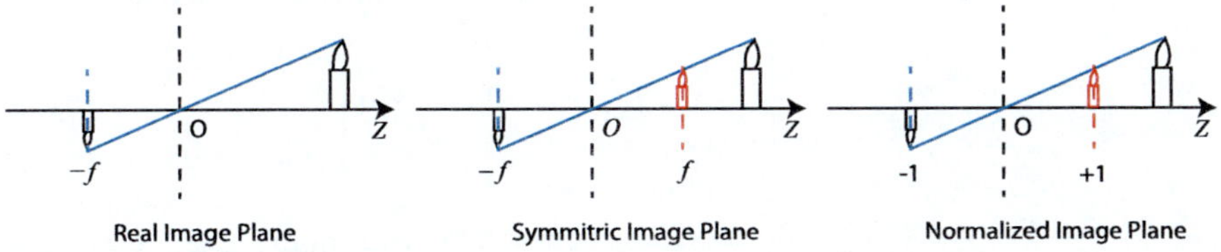

Fig. 4.2 The real, symmetric and normalized image plane

[1] Or image coordinate system, see Sect. 4.2 of this lecture.

pixel coordinate $[u, v]^T$ is:

$$\begin{cases} u = \alpha X' + c_x \\ v = \beta Y' + c_y \end{cases}. \tag{4.4}$$

Put it into (4.3) and set αf as f_x, βf as f_y:

$$\begin{cases} u = f_x \frac{X}{Z} + c_x \\ v = f_y \frac{Y}{Z} + c_y \end{cases}, \tag{4.5}$$

where f is the focal length in meters, α, *and* β is in pixels/meter, so f_x, f_y and c_x, c_y are in pixels. It would be more compact to write this form as a matrix, but we need to use homogeneous coordinates on the left and non-homogeneous coordinates on the right:

$$\begin{pmatrix} u \\ v \\ 1 \end{pmatrix} = \frac{1}{Z} \begin{pmatrix} f_x & 0 & c_x \\ 0 & f_y & c_y \\ 0 & 0 & 1 \end{pmatrix} \begin{pmatrix} X \\ Y \\ Z \end{pmatrix} \triangleq \frac{1}{Z}\mathbf{KP}. \tag{4.6}$$

Let's put Z to the left side as in most books:

$$Z \begin{pmatrix} u \\ v \\ 1 \end{pmatrix} = \begin{pmatrix} f_x & 0 & c_x \\ 0 & f_y & c_y \\ 0 & 0 & 1 \end{pmatrix} \begin{pmatrix} X \\ Y \\ Z \end{pmatrix} \triangleq \mathbf{KP}. \tag{4.7}$$

In this equation, we refer to the matrix composed of the middle quantities as the camera's inner parameter matrix (o intrinsics) $\mathbf{K}$. It is generally assumed that the camera's internal parameters are fixed after manufacturing and will not change during usage. Some camera manufacturers will tell you the intrinsics, and sometimes you need to estimate the internal parameters by yourself, which is called calibration. Because of the maturity of the calibration algorithm (such as the famous Zhang Zhengyou's calibration [25]), it will not be introduced here.[2]

There are internal parameters, and naturally, there must be something like "external parameters". In the Eq. (4.6), we use the coordinates of P in the camera coordinate system, but in fact, the coordinates of P should be its world coordinates because the camera is moving (we use the symbol $\mathbf{P}_w$). It should be converted to the camera coordinate system based on the current pose of the camera. The camera's pose is described by its rotation matrix $\mathbf{R}$ and the translation vector $\mathbf{t}$. Then there are:

$$Z\mathbf{P}_{uv} = Z \begin{bmatrix} u \\ v \\ 1 \end{bmatrix} = \mathbf{K}\left(\mathbf{R}\mathbf{P}_w + \mathbf{t}\right) = \mathbf{K}\mathbf{T}\mathbf{P}_w. \tag{4.8}$$

Note that the latter formula implies a conversion from homogeneous to non-homogeneous coordinates (can you see it?) We use homogeneous coordinates in $\mathbf{TP}$,

[2] I'm sure professor Zhang has a copy of this book now.

then convert to non-homogeneous coordinates, and then multiply it by $\mathbf{K}$. It describes the projection relationship of world coordinates to pixel coordinates of P. Among them, the camera's pose $\mathbf{R}, \mathbf{t}$ is also called the camera's *extrinsics*.[3] Compared with the intrinsic, the extrinsics may change with the camera installation and is also the target to be estimated in the SLAM if we only have a camera.

The projection process can also be viewed from another perspective. The formula (4.8) shows that we can convert a world coordinate point to the camera coordinate system first and then remove the last dimension. The depth of the point from the imaging plane of the camera is then removed, which is equivalent to the normalization on the last dimension. In this way, we get the projection of the point P on the camera *normalized plane*:

$$(\mathbf{R}\mathbf{P}_w + \mathbf{t}) = \underbrace{[X, Y, Z]^T}_{\text{Camera Coordinates}} \rightarrow \underbrace{[X/Z, Y/Z, 1]^T}_{\text{Normalized Coordinates}}. \tag{4.9}$$

The *normalized coordinates* can be seen as a point in the $z = 1$ plane in front of the camera.[4] This $z = 1$ plane is also called *normalized plane*. We normalize the coordinates and then multiply them with the intrinsic matrix, yielding the pixel coordinates. We can also consider the pixel coordinates $[u, v]^T$ as the result of quantitative measurements on points on the normalized plane. If the camera coordinates are multiplied by any non-zero constant simultaneously, the normalized coordinates are the same, which means that the depth is lost during the projection process. So, in monocular vision, the pixel's depth value cannot be obtained by a single image.

4.1.2 Distortion

In order to get a larger FoV (Field-of-View), we usually add a lens in front of the camera. The addition of the lens has an influence on the propagation of light during imaging: (1) the shape of the lens may affect the propagation way of light, (2) during the mechanical assembly, the lens and the imaging plane are not entirely parallel, which also changes the projected position.

There are some mathematical models to describe the distortion caused by the shape of the lens. In the pinhole model, a straight line keeps straight when projected onto the pixel plane. However, in real photos, the camera lens tends to make a straight line in the real environment become a curve.[5] The closer to the edge of the image, the more obvious this phenomenon is. Since the lenses actually produced are often center-

[3] In robots or autonomous vehicles, extrinsics is often defined as the transform between the camera coordinate system and the robot body coordinate system, describing "where the camera is installed".

[4] Note that in the calculation, it is necessary to check whether Z is positive because the negative Z can also get the point on the normalized plane by this method. However, the camera does not capture the scene behind the imaging plane.

[5] Yes, it is no longer straight but becomes curved. If it makes an inside curve, it is called barrel-like distortion; otherwise, it is cushion-like distortion if the curve looks outward.

symmetrical, this makes the irregular distortion generally radially symmetrical. They fall into two main categories: *barrel-like distortion* and *cushion-like distortion*, as shown by Fig. 4.3.

In barrel distortion, the radius of pixels decreases as the optical axis's distance increases, while the cushion distortion is just the opposite. In both distortions, the line that intersects the center of the image and the optical axis remains the same.

In addition to the shape of the lens, which introduces radial distortion, *tangential distortion* is introduced during assembly of the camera because the lens and the imaging surface cannot be strictly parallel, as shown by Fig. 4.4.

To better understand radial and tangential distortion, we describe them in a more rigorous mathematical form. Consider any point on the *normalized plane*, $\mathbf{p}$, whose coordinates are $[x, y]^T$, or $[r, \theta]^T$ in the form of polar coordinates, where r represents the distance between the point $\mathbf{p}$ and the origin of the coordinate system, and θ represents the angle to the horizontal axis. Radial distortion can be seen as a change in the coordinate point along the length, that is, its radius from the origin. Tangential distortion can be seen as a change in the coordinate point along the tangential direction, which means that the horizontal angle has changed. It is generally assumed that these distortions are polynomial, namely:

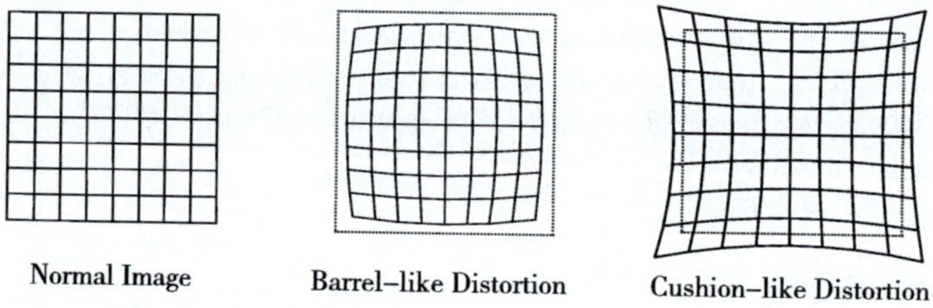

Fig. 4.3 The radical distortion

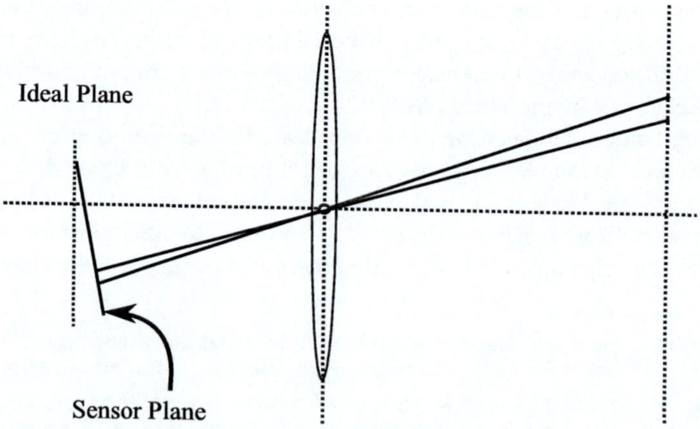

Fig. 4.4 Tangential distortion

$$
\begin{aligned}
x_{\text{distorted}} &= x(1 + k_1 r^2 + k_2 r^4 + k_3 r^6) \\
y_{\text{distorted}} &= y(1 + k_1 r^2 + k_2 r^4 + k_3 r^6)
\end{aligned}, \tag{4.10}
$$

where $[x_{\text{distorted}}, y_{\text{distorted}}]^T$ is the *normalized coordinates* of the point after distortion. On the other hand, for *tangential distortion*, we can use the other two parameters p_1, p_2 to describe it:

$$
\begin{aligned}
x_{\text{distorted}} &= x + 2p_1 xy + p_2(r^2 + 2x^2) \\
y_{\text{distorted}} &= y + p_1(r^2 + 2y^2) + 2p_2 xy
\end{aligned}. \tag{4.11}
$$

Putting (4.10) and (4.11) together we get a joint model with 5 distortion coefficients. The complete form is:

$$
\begin{cases}
x_{\text{distorted}} = x(1 + k_1 r^2 + k_2 r^4 + k_3 r^6) + 2p_1 xy + p_2(r^2 + 2x^2) \\
y_{\text{distorted}} = y(1 + k_1 r^2 + k_2 r^4 + k_3 r^6) + p_1(r^2 + 2y^2) + 2p_2 xy
\end{cases}. \tag{4.12}
$$

In the above process of correcting distortion, we used 5 distortion coefficients. In practical applications, you can flexibly choose to number of parameters, for example, only selecting k_1, p_1, p_2, or use k_1, k_2, p_1, p_2, etc.

This section models the camera's imaging process using a pinhole model and described the radial and tangential distortions caused by the lens. Researchers have proposed many other models in the existing imaging system, such as the affine model and perspective model, and there are many different types of distortion. In most visual SLAM systems, pinhole and rad-tan distortion models are sufficient, so we will not describe the other ones.

It is worth mentioning that there are two ways of undistortion (or correction). We can choose to undistort the entire image first, get the corrected image, and then discuss the spatial position of the points on the image. Alternatively, we can also extract some feature points in the distorted image and find its real position through the distortion equation. Both are feasible, but the former seems to be more common in visual SLAM. Therefore, when an image is undistorted, we can directly establish a projection relationship with the pinhole model without considering distortion. Therefore, in the discussion that follows, we can directly assume that the image has been undistorted.

Finally, let's summarize the imaging process of a monocular camera:

1. First, there is a point P in the world coordinate system, and its world coordinates are $\mathbf{P}_w$.
2. Since the camera is moving, its motion is described by $\mathbf{R}, \mathbf{t}$ or transform matrix $\mathbf{T} \in \mathrm{SE}(3)$. The camera coordinates for P are $\tilde{\mathbf{P}}_c = \mathbf{R}\mathbf{P}_w + \mathbf{t}$.
3. The $\tilde{\mathbf{P}}_c$ components are X, Y, Z, and they are projected onto the normalized plane $Z = 1$ to get the normalized coordinates: $\mathbf{P}_c = [X/Z, Y/Z, 1]^T$. Note that Z may be less than 1, indicating that the point is behind the normalization plane and it should not be projected on the camera plane.

4. If the image is distorted, the coordinates of $\mathbf{P}_c$ after distortion are calculated according to the distortion parameters.
5. Finally, the distorted coordinates of P pass through the intrinsics and we find its pixel coordinates: $\mathbf{P}_{uv} = \mathbf{K}\mathbf{P}_c$.

In summary, we have talked about four coordinates: the world coordinates, the camera coordinates, the normalized coordinates, and the pixel coordinates. Readers should clarify their relationship. They reflect the entire imaging process and will be used in the future.

4.1.3 Stereo Cameras

The pinhole camera model describes the imaging model of a single camera. However, we cannot determine the specific location of a spatial point only by a single pixel. This is because all points on the line from the camera's optical center to the normalized plane can be projected onto that pixel. Only when the depth of P is determined (such as through a binocular or RGB-D camera) can we know exactly its spatial location, as shown in Fig. 4.5.

There are many ways to measure the pixel distance (or depth). For example, the human eye can judge the object's distance according to the difference (or parallax) of the scene seen by the left and right eyes. The binocular camera principle is also the same. By simultaneously acquiring the left and right cameras' images and calculating the parallax/disparity between the images, each pixel's depth is estimated. In the following paragraph, we briefly describe the stereo camera's imaging principle (as shown in Fig. 4.6).

A binocular camera is generally composed of a left-eye camera and a right-eye camera. Of course, it can also be made up and down, but the mainstream binoculars we've seen are all left and right. Both the left and right cameras are regarded as simple pinhole cameras. They are synchronized and placed horizontally, meaning

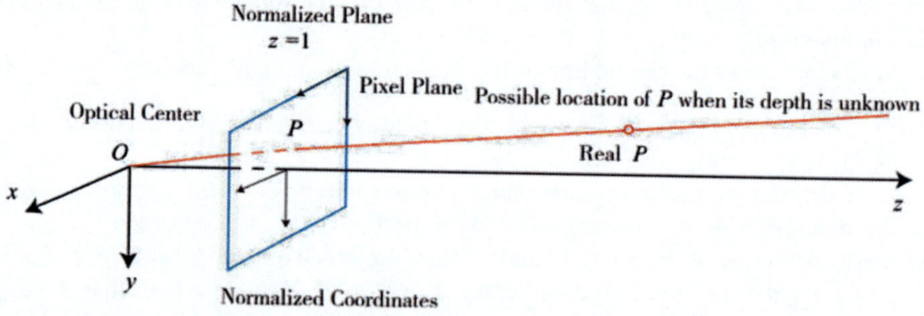

Fig. 4.5 The possible location of a single pixel

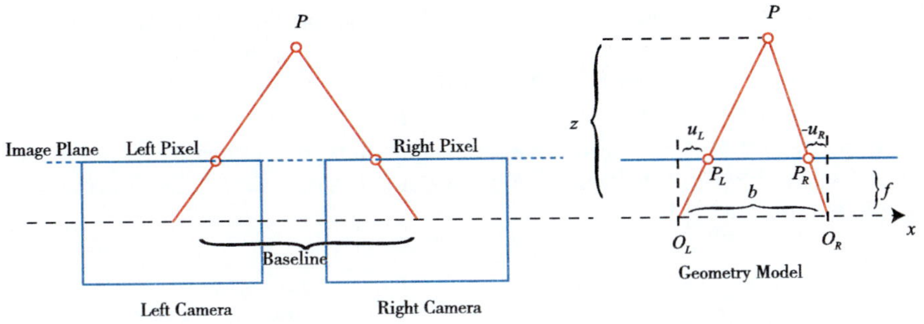

Fig. 4.6 Geometry model of stereo cameras from upside down view. The O_L, O_R are left and right optical centers. f is the focal length, u_L and u_R are pixel coordinates of the same point along the x axis. Note that u_R should be a negative value in this figure, so the physical distance should be $-u_R$

that both cameras' centers are on the same x axis. The distance between the two centers is called *baseline* (denoted as b), which is an important parameter.

Now consider a 3D point P, projected into the left-eye and the right-eye, written as P_L, P_R. Due to the presence of the camera baseline, these two imaging positions are different. Ideally, since the left and right cameras are only shifted on the x axis, the image of P also differs only on the x axis (corresponding to the u axis of the image). Take the left pixel coordinate as u_L and the right coordinate as u_R. The geometric relationship is shown on the right of Fig. 4.6. According to the similarity relationship between $\Delta P P_L P_R$ and $\Delta P O_L O_R$, there are:

$$\frac{z - f}{z} = \frac{b - u_L + u_R}{b}. \tag{4.13}$$

Rearrange it, and we have:

$$z = \frac{fb}{d}, \quad d \triangleq u_L - u_R, \tag{4.14}$$

where d is defined as the difference between the left and right figures' horizontal coordinates and is called disparity or parallax. Based on the parallax, we can estimate the distance between a pixel and the camera. Parallax is inversely proportional to distance: the larger the parallax is, the closer the distance is.[6] Simultaneously, since the parallax is at least one pixel, there is a theoretical maximum value for the binocular depth, which is determined by fb. To see the far-away things, we need a larger stereo camera; conversely, small binocular devices can only measure very close distances. By analogy, when the human eye looks at a very distant object (such as a very distant airplane), it is usually impossible to determine its distance accurately.

Although the depth's formula is simple, the real calculation of d itself is more complicated. We need to know precisely where a pixel of the left-eye image appears

[6] Readers can simulate it with your own eyes.

in the right-eye image (that is, the corresponding relationship). This also belongs to the kind of task that is "easy for humans but difficult for computers". When we want to calculate each pixel's depth in an image, the calculation amount and accuracy will become a problem, and the parallax can be calculated only in the place where the image texture is rich. Due to the calculation amount, binocular depth estimation still needs GPU or FPGA to make the distance calculation run in real-time. This will be mentioned in Chap. 11.

4.1.4 RGB-D Cameras

Compared to the binocular camera's way of calculating depth, the RGB-D camera's approach is more "active": it can actively measure each pixel's depth. The current RGB-D cameras can be divided into two categories according to their principle (see Fig. 4.7):

1. The first kind of RGB-D sensor uses structured infrared light to measure pixel distance. Many of the old RGB-D sensors are belong to this kind, for example, the Kinect 1st generation, Project Tango 1st generation, Intel RealSense, etc.
2. The second kind measures pixel distance using the *time-of-flight (ToF)*. Examples are Kinect 2 and some existing ToF sensors in cellphones.

Regardless of the type, the RGB-D camera needs to emit a light beam (usually infrared light) to the target object. In the structured light principle, the camera calculates the distance between the object and itself based on the returned structured light pattern. In the ToF principle, the camera emits a light pulse to the target and then determines the distance according to the beam's time of flight. The ToF principle is very similar to the laser sensor, except that the laser obtains the distance by scanning point by point (or line by line). The ToF camera can obtain the entire image's pixel depth, which is also the RGB-D camera's main advantage. So, if you take apart an RGB-D camera, you will usually find that there will be at least one transmitter and one receiver in addition to the ordinary camera.

After measuring the depth, the RGB-D camera usually completes the pairing between the depth and color map pixels according to each camera's position at the time of production. It outputs a pixel-to-pixel corresponding color image and depth image. We can read the color information and distance information at the same image position, calculate the 3D camera coordinates of the pixels, and generate a point cloud. RGB-D data can be processed either at the image level or the point cloud level. The second experiment of this lecture will demonstrate the point cloud construction of RGB-D cameras.

The RGB-D camera can measure the distance of each pixel in real-time. However, due to this measurement of transmitting and receiving, its range of use is limited. RGB-D cameras that use infrared light for depth measurement are susceptible to interference from infrared light emitted by daylight or other sensors, so they cannot be

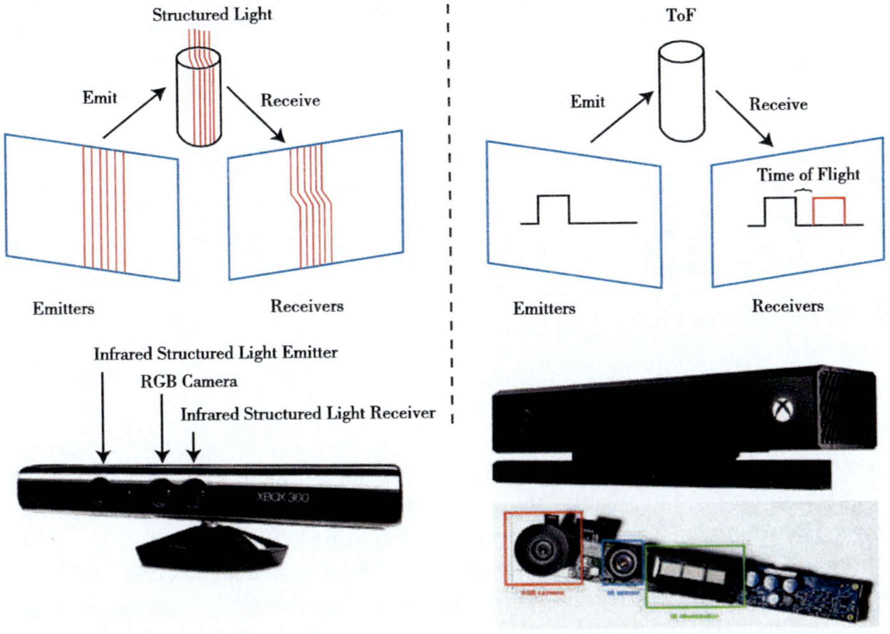

Fig. 4.7 RGB-D cameras

used outdoors. Without modulation, multiple RGB-D cameras can interfere with each other. These points' positions cannot be measured for transparent objects because they cannot receive reflected light. Also, RGB-D cameras have some disadvantages in terms of cost and power consumption.

4.2 Images

Cameras and lens convert the information in the three-dimensional world into a photo composed of pixels, which is then stored in the computer as a data source for subsequent processing. In mathematics, images can be described by a matrix; in computers, they occupy a continuous disk or memory space, which can be represented by a two-dimensional array. In this way, the program does not have to distinguish whether they are dealing with a numerical matrix or a meaningful image.

In this section, we will introduce some basic operations of computer image processing. In particular, we are going to introduce the basic steps of processing images with OpenCV and lay the foundation for subsequent chapters. Let's start with the simplest case, the grayscale image. Each pixel position (x, y) corresponds to a grayscale value of I in a grayscale image. Therefore, an image with the width of w and the height of h can be mathematically written as a function:

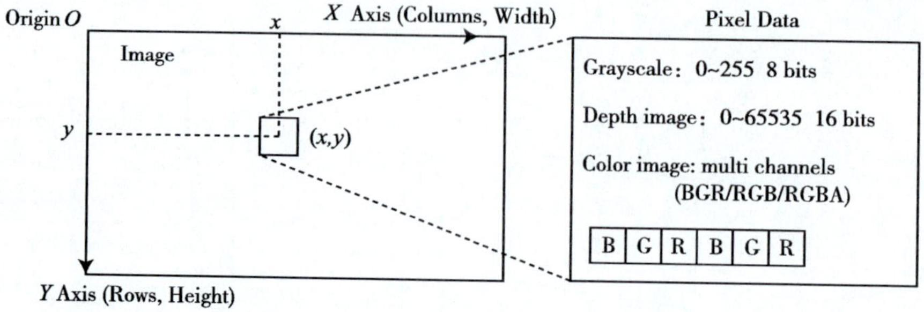

Fig. 4.8 Pixels in an image

$$(I)(x, y) : \mathbb{R}^2 \mapsto \mathbb{R},$$

where (x, y) is the coordinate of the pixel. However, computers cannot express real numbers, so we need to quantify the subscripts and image readings within a certain range. For example, x, y are usually integers starting with 0 to $w - 1, h - 1$. In common grayscale images, an integer of 0–255 (that is, an unsigned char in C++, 1 byte) is used to express the grayscale reading of the image. Then, a grayscale image with a width of 640 pixels and a height of 480 pixels can be expressed as:

Listing 4.1 Use 2D array to express an image

```
unsigned char image[480][640];
```

Why does the two-dimensional array here have the size of 480×640? Because in the program, the first index of the 2D array is the row, and the second index is the column. In an image, the number of rows (or the y axis) in the array corresponds to the height of the image, and the number of columns (or the x axis) corresponds to the width of the image.

Let's examine the content of this image. Images are naturally made up of pixels. When accessing a certain pixel, you need to specify its coordinates, as shown in Fig. 4.8. The left side of the figure shows how the traditional pixel coordinate system is defined. The origin is in the upper left corner of the image, the X axis is from left to right, and the Y is top-down. If it has a third axis, the Z axis, then according to the right-hand rule, the Z axis should be from outside to inside (or front in 3D space). This definition is consistent with the camera coordinate system. The width or number of columns of an image corresponds to the X axis; the number of rows or the height of an image corresponds to its Y axis.

According to this definition, if we discuss a pixel located at x, y, then the code of accessing its memory in the computer should be:

Listing 4.2 Accessing image pixels

```
unsigned char pixel = image[y][x];
```

It corresponds to the reading of the gray value $I(x, y)$. Please note the order of x and y here. Although we tirelessly discuss the problem of coordinate systems, errors like this index sequence will still be one of the most common errors encountered by novices during debugging. If you accidentally change the coordinates of x, y when writing a program, the compiler cannot provide any useful information at compile-time. All you can see is a segment fault in runtime.

A pixel's grayscale can be recorded as an 8-bit unsigned integer, which is a value of 0–255. If we have more information to record, one byte is probably not enough. For example, in the depth map of an RGB-D camera, each pixel's distance is recorded. This distance is usually measured in millimeters, and the range of RGB-D cameras is usually around a dozen meters, exceeding 255. At this time, people will use 16-bit integers (unsigned short in C++) to record the depth map information, that is, the value at 0–65535. When converted to meters, it can represent up to 65 m, which is enough for RGB-D cameras.

The representation of a color image requires the concept of a channel. In computers, we use three colors: red, green, and blue to express any color. Therefore, for each pixel, three R, G, and B values are recorded, and each value is called a channel. For example, the most common color image has three channels, represented by an 8-bit integer. Under this rule, one pixel occupies a 24-bit space.

The number and order of channels can be freely defined. In OpenCV color images, the default order of channels is B, G, R, which means when we get a 24-bit pixel, the first 8 bits represent the blue value, the middle 8 bits represent the green, and the last 8 bits represent the red. Similarly, the order of R, G, and B can also be used to describe a color image. If you want to express the image's transparency, use R, G, B, A four channels.

4.3 Practice: Images in Computer Vision

4.3.1 Basic Usage of OpenCV

The following is a demo program to help you understand how to access the image in OpenCV and how to visit its pixels.

4.3.1.1 Install OpenCV

OpenCV[7] provides many open-source image algorithms and is a very widely used image processing algorithm library in computer vision. This book also uses OpenCV for basic image processing. Before using, readers must install it from the pre-compiled library or from source code. Under Ubuntu, there are two options: *install from source code* or *only install binary library files*:

---

[7] Official homepage: http://opencv.org.

1. Install from source means to download all OpenCV source code from the OpenCV website, compile and install on the machine for usage. The advantage is that you can freely choose which version to install, and the source code is accessible, but it takes some compilation time.
2. Or, we can only install the binary library file, which means it was pre-compiled by the Ubuntu community, so there is no need to recompile it.

If we use a newer version of OpenCV than that in the apt source, we must install it from the source code. First, you can adjust some compilation options to match the programming environment (for example, to disable some unused modules or turn on the GPU acceleration, etc.). OpenCV currently maintains two major versions, divided into OpenCV 2.4 series and OpenCV 3 series.[8] This book uses the OpenCV *3* or higher.

Because the OpenCV project is relatively large, it will not be placed under 3rdparty in this book. Readers can download it from http://opencv.org/downloads.html and select the Linux version. You will get a compressed package like opencv-3.1.0.zip. Unzip it to any directory, we can found that OpenCV is also a CMake project.

Before compiling, first, install the dependencies of OpenCV:

Listing 4.3 Terminal input:

```
sudo apt-get install build-essential libgtk2.0-dev libvtk5-dev libjpeg-dev
    libtiff4-dev libjasper-dev libopenexr-dev libtbb-dev
```

In fact, OpenCV has many dependencies, and the lack of certain dependency items will affect some of its functions (but we will not use all the functions). OpenCV will check whether the dependencies will be installed during CMake and adjust its own configurations. If you have a GPU on your computer and the relevant dependencies are installed, OpenCV will also enable GPU acceleration. But for this book, the above dependencies are sufficient.

Subsequent compilation and installation are the same as ordinary CMake projects. After make, please call "sudo make install" to install OpenCV on your machine (instead of just compiling it). Depending on the machine configuration, this compilation process may take from 20 min to an hour. If your CPU is powerful, you can use commands like "make -j4" to call multiple threads to compile (the parameter after -j is the number of threads used). After installation, OpenCV is stored in the /usr/local directory by default. You can look for OpenCV header files and library files' installation location to see where they are. Besides, if you have installed the OpenCV 2 series before, it is recommended that you install OpenCV 3 elsewhere (think about how this should be done).

4.3.2 Basic OpenCV Images Operations

Now let's go through the basic image operations in OpenCV from a simple example.

[8] In 2020, we can also use version 4.0 or higher.

Listing 4.4 slambook/ch5/imageBasics/imageBasics.cpp

```
#include <iostream>
#include <chrono>

using namespace std;

#include <opencv2/core/core.hpp>
#include <opencv2/highgui/highgui.hpp>

int main(int argc, char **argv) {
  // Read the image in argv[1]
  cv::Mat image;
  image = cv::imread(argv[1]); // call cv::imread to read the image from
       file

  // check the data is correctly loaded
  if (image.data == nullptr) { // maybe the file does not exist
    cerr << "file" << argv[1] << " not exist." << endl;
    return 0;
  }

  // print some basic information
  cout << "Image cols: " << image.cols << ", rows: " << image.rows
    << ", channels: " << image.channels() << endl;
  cv::imshow("image", image); // use cv::imshow to show the image
  cv::waitKey(0); // display and wait for a keyboard input

  // check image type
  if (image.type() != CV_8UC1 && image.type() != CV_8UC3) {
    // we need grayscale image or RGB image
    cout << "image type incorrect." << endl;
    return 0;
  }

  // check hte pixels
  chrono::steady_clock::time_point t1 = chrono::steady_clock::now();
  for (size_t y = 0; y < image.rows; y++) {
    // use cv::Mat::ptr to get the pointer of each row
    unsigned char *row_ptr = image.ptr<unsigned char>(y); // row_ptr is the
         pointer to y-th row
    for (size_t x = 0; x < image.cols; x++) {
      // read the pixel on (x,y), x=column, y=row
      unsigned char *data_ptr = &row_ptr[x * image.channels()]; // data_ptr
           is the pointer to (x,y)
      // visit the pixel in each channel
      for (int c = 0; c != image.channels(); c++) {
        unsigned char data = data_ptr[c]; // data should be pixel of I(x,y)
             in c-th channel
      }
    }
  }
  chrono::steady_clock::time_point t2 = chrono::steady_clock::now();
  chrono::duration<double> time_used = chrono::duration_cast < chrono::
       duration < double >> (t2 - t1);
  cout << "time used: " << time_used.count() << " seconds." << endl;

  // copying cv::Mat
  // operator = will not copy the image data, but only the reference
  cv::Mat image_another = image;
  // changing image_another will also change image
  image_another(cv::Rect(0, 0, 100, 100)).setTo(0); // set top-left 100*100
       block to zero
  cv::imshow("image", image);
  cv::waitKey(0);

  // use cv::Mat::clone to actually clone the data
```

```
60    cv::Mat image_clone = image.clone();
61    image_clone(cv::Rect(0, 0, 100, 100)).setTo(255);
62    cv::imshow("image", image);
63    cv::imshow("image_clone", image_clone);
64    cv::waitKey(0);
65
66    // We are not going to copy the OpenCV's documentation here
67    // please take a look at it for other image operations like clipping,
          rotating and scaling.
68
69    cv::destroyAllWindows();
70    return 0;
71 }
```

In this example, we demonstrated the following operations: image reading, displaying, pixel vising, copying, assignment, etc. When compiling the program, you need to add the OpenCV header file in your "CMakeLists.txt", and then link the program to the OpenCV's library. At the same time, due to the use of C++ 11 standards (such as the nullptr and chrono), you also need to set up the c++ standard in the compiler flag:

Listing 4.5 slambook/ch5/imageBasics/CMakeLists.txt

```
# use c++11 standard
set( CMAKE_CXX_FLAGS "-std=c++11" )

# find OpenCV
find_package( OpenCV REQUIRED )
# include its headers
include_directories( ${OpenCV_INCLUDE_DIRS} )

add_executable( imageBasics imageBasics.cpp )

# link the exe to opencv's libs
target_link_libraries( imageBasics ${OpenCV_LIBS} )
```

Let's give some notes for the code:

1. The program reads the image position from argv[1], the first parameter on the command line. We prepared an image (ubuntu.png, an Ubuntu wallpaper, hope you like it) for readers to test. Therefore, after compilation, use the following command to call this program:

Listing 4.6 Terminal input:

```
./build/imageBasics ubuntu.png
```

If you call this program in the IDE, be sure to give it parameters at the same time. This can be configured in the launch configuration dialog if you are using Clion.
2. In line 10–18, we use the cv::imread function to read the image. And then, we display the image and its basic information.
3. In line 35–46, we iterate over all pixels in the image and calculates the time spent in the entire loop. Please note that the pixel visiting method is not unique, and the method given by the example is not the most efficient way. OpenCV provides an iterator of cv::Mat. You can traverse the pixels of the image through the iterator. Or, cv::Mat::data provides a raw pointer to the beginning of the image data. You

can also directly calculate the offset through this pointer, and then get the memory location of the pixel. The method used in the example is to facilitate the reader to understand the structure of the image.

4. OpenCV provides many functions for manipulating images. We will not list them one by one. Otherwise, this book will become an OpenCV operation manual. The example shows the most common things like image reading and displaying and the deep copy function in cv::Mat. During the programming process, readers will also encounter operations such as image rotation and interpolation. At this time, you should refer to the corresponding documentation of the function to understand their principles and usage.

It should be noted that OpenCV is not the only image library. It is just one of the more widely used ones. However, most image libraries have similar image operations. We hope that readers can understand the representation of images in other libraries after using OpenCV to quickly adjust to any other libraries. Since cv::Mat is also a matrix class, we can also use it to store matrix data such as rotation matrix and do some linear algebra operations. But it is generally believed that *Eigen* is more efficient for use with fixed-size matrices.

4.3.3 Image Undistortion

We've introduced the rad-tan distortion model in the previous section, now let write an example to show the implementation. OpenCV has provided the cv::Undistort function for us, but we will also give a hand-written undistortion function to show the principles.

Listing 4.7 slambook/ch5/imageBasics/undistortImage.cpp

```
#include <opencv2/opencv.hpp>
#include <string>
using namespace std;
string image_file = "./distorted.png"; // the distorted image

int main(int argc, char **argv) {
  // In this program we implement the undistortion by ourselves rather than using opencv
  // rad-tan model params
  double k1 = -0.28340811, k2 = 0.07395907, p1 = 0.00019359, p2 = 1.76187114e-05;
  // intrinsics
  double fx = 458.654, fy = 457.296, cx = 367.215, cy = 248.375;

  cv::Mat image = cv::imread(image_file, 0); // the image type is CV_8UC1
  int rows = image.rows, cols = image.cols;
  cv::Mat image_undistort = cv::Mat(rows, cols, CV_8UC1); // the undistorted image

  // computate the pixels in the undistorted one
  for (int v = 0; v < rows; v++) {
    for (int u = 0; u < cols; u++) {
      // note we are computing the pixel of (u,v) in the undistorted image
      // according to the rad-tan model, compute the coordinates in the distorted image
```

```
      double x = (u - cx) / fx, y = (v - cy) / fy;
      double r = sqrt(x * x + y * y);
      double x_distorted = x * (1 + k1 * r * r + k2 * r * r * r * r) + 2 * p1
          * x * y + p2 * (r * r + 2 * x * x);
      double y_distorted = y * (1 + k1 * r * r + k2 * r * r * r * r) + p1 * (
          r * r + 2 * y * y) + 2 * p2 * x * y;
      double u_distorted = fx * x_distorted + cx;
      double v_distorted = fy * y_distorted + cy;

      // check if the pixel is in the image boarder
      if (u_distorted >= 0 && v_distorted >= 0 && u_distorted < cols &&
          v_distorted < rows) {
        image_undistort.at<uchar>(v, u) = image.at<uchar>((int) v_distorted,
            (int) u_distorted);
      } else {
        image_undistort.at<uchar>(v, u) = 0;
      }
    }
  }

  // show the undistorted image
  cv::imshow("distorted", image);
  cv::imshow("undistorted", image_undistort);
  cv::waitKey();
  return 0;
}
```

Please check the difference between the two images by yourself.

4.4 Practice: 3D Vision

4.4.1 Stereo Vision

We have introduced the imaging principle of stereo vision. Now we start from the left and right images, calculate the disparity map corresponding to the left eye, and then calculate each pixel's coordinates in the camera coordinate system, which will form a *point cloud*. We have prepared left and right images for the readers, as shown in Fig. 4.9. The following code demonstrates the calculation of disparity map and point cloud:

Listing 4.8 slambook/ch5/stereoVision/stereoVision.cpp (Part)

```
int main(int argc, char **argv) {
  // intrinsics
  double fx = 718.856, fy = 718.856, cx = 607.1928, cy = 185.2157;
  // baseline
  double b = 0.573;

  cv::Mat left = cv::imread(left_file, 0);
  cv::Mat right = cv::imread(right_file, 0);
  cv::Ptr<cv::StereoSGBM> sgbm = cv::StereoSGBM::create(
    0, 96, 9, 8 * 9 * 9, 32 * 9 * 9, 1, 63, 10, 100, 32); // SGBM is
        sensitive to parameters
  cv::Mat disparity_sgbm, disparity;
  sgbm->compute(left, right, disparity_sgbm);
  disparity_sgbm.convertTo(disparity, CV_32F, 1.0 / 16.0f);

```

```
  // compute the point cloud
  vector<Vector4d, Eigen::aligned_allocator<Vector4d>> pointcloud;

  // change v++ and u++ to v+=2, u+=2 if your machine is slow to get a
      sparser cloud
  for (int v = 0; v < left.rows; v++)
  for (int u = 0; u < left.cols; u++) {
    if (disparity.at<float>(v, u) <= 10.0 || disparity.at<float>(v, u) >=
        96.0) continue;

    Vector4d point(0, 0, 0, left.at<uchar>(v, u) / 255.0); // the first
        three dimensions are xyz, the 4-th is the color

    // compute the depth from disparity
    double x = (u - cx) / fx;
    double y = (v - cy) / fy;
    double depth = fx * b / (disparity.at<float>(v, u));
    point[0] = x * depth;
    point[1] = y * depth;
    point[2] = depth;

    pointcloud.push_back(point);
  }

  cv::imshow("disparity", disparity / 96.0);
  cv::waitKey(0);

  // show the point cloud in pangolin
  showPointCloud(pointcloud);
  return 0;
}
```

In this example, we call the SGBM (Semi-global Batch Matching) [26] algorithm implemented by OpenCV to calculate the disparity of the left and right images and then transform it into the 3D space of the camera through the geometric model of the binocular camera. We use a classic parameter configuration from the internet, and we mainly adjust the maximum and minimum disparity. The disparity data combined with the camera's internal parameters and baseline can determine each point's position in three-dimensional space. We omit the code related to displaying the point cloud to save some space.

This book is not going to introduce the disparity calculation algorithm of the binocular camera. Interested readers can read the relevant Refs. [27, 28]. In addition to the binocular algorithm implemented by OpenCV, there are many other libraries focused on achieving efficient parallax calculations. It is still an active and complex subject today.

4.4.2 RGB-D Vision

Finally, we demonstrate an example of RGB-D vision. The convenience of RGB-D cameras is that they can obtain pixel depth information through physical methods. If the camera's intrinsic and extrinsic are known, we can calculate any pixel position in the world coordinate system, thereby creating a point cloud map. Now let's demonstrate how to do it.

Fig. 4.9 Stereo image example. Top-left: left image, top-right: right image, mid: SGBM disparity map, bottom: point cloud. Note that since some of the pixels in the left image is not seen in the right one, so the disparity map will have some empty values

We have prepared 5 pairs of images located in the slambook2/ch5/rgbd folder. There are 5 RGB images from 1.png to 5.png under the color/ directory and 5 corresponding depth images under the depth/. At the same time, the "pose.txt" file gives the camera poses of the 5 images (in the form of $\mathbf{T}_{wc}$). The format of the pose record is the same as before, with the translation vector plus a rotation quaternion:

$$[x, y, z, q_x, q_y, q_z, q_w],$$

where q_w is the real part of the quaternion. For example, the parameters of the first pair of image are:

$$[-0.228993, 0.00645704, 0.0287837, -0.0004327, -0.113131, -0.0326832, 0.993042].$$

Below we write a program to accomplish two things: (1) We calculate the point cloud corresponding to each pair of RGB-D images based on internal parameters; (2) According to the camera pose of each image, we put the points to a global cloud by the camera poses.

Listing 4.9 slambook/ch5/rgbd/jointMap.cpp (Part)

```
int main(int argc, char **argv) {
  vector<cv::Mat> colorImgs, depthImgs;
  TrajectoryType poses; // camera poses

  ifstream fin("./pose.txt");
  if (!fin) {
    cerr << "Please run the program in the directory that has pose.txt" <<
         endl;
    return 1;
  }

  for (int i = 0; i < 5; i++) {
    boost::format fmt("./%s/%d.%s"); // the image format
    colorImgs.push_back(cv::imread((fmt % "color" % (i + 1) % "png").str()))
         ;
    depthImgs.push_back(cv::imread((fmt % "depth" % (i + 1) % "pgm").str(),
         -1)); // use -1 flag to load the depth image

    double data[7] = {0};
    for (auto &d:data) fin >> d;
    Sophus::SE3d pose(Eigen::Quaterniond(data[6], data[3], data[4], data[5])
         ,
    Eigen::Vector3d(data[0], data[1], data[2]));
    poses.push_back(pose);
  }

  // compute the point cloud using camera intrinsics
  double cx = 325.5;
  double cy = 253.5;
  double fx = 518.0;
  double fy = 519.0;
  double depthScale = 1000.0;
  vector<Vector6d, Eigen::aligned_allocator<Vector6d>> pointcloud;
  pointcloud.reserve(1000000);

  for (int i = 0; i < 5; i++) {
    cout << "Converting RGBD images " << i + 1 << endl;
    cv::Mat color = colorImgs[i];
    cv::Mat depth = depthImgs[i];
    Sophus::SE3d T = poses[i];
    for (int v = 0; v < color.rows; v++)
    for (int u = 0; u < color.cols; u++) {
      unsigned int d = depth.ptr<unsigned short>(v)[u]; // depth value is
           16-bit
      if (d == 0) continue; // 0 means no valid value
      Eigen::Vector3d point;
      point[2] = double(d) / depthScale;
      point[0] = (u - cx) * point[2] / fx;
      point[1] = (v - cy) * point[2] / fy;
      Eigen::Vector3d pointWorld = T * point;

      Vector6d p;
      p.head<3>() = pointWorld;
      p[5] = color.data[v * color.step + u * color.channels()]; // blue
      p[4] = color.data[v * color.step + u * color.channels() + 1]; // green
      p[3] = color.data[v * color.step + u * color.channels() + 2]; // red
      pointcloud.push_back(p);
    }
  }

  cout << "global point cloud has " << pointcloud.size() << " points." <<
       endl;
  showPointCloud(pointcloud);
  return 0;
}
```

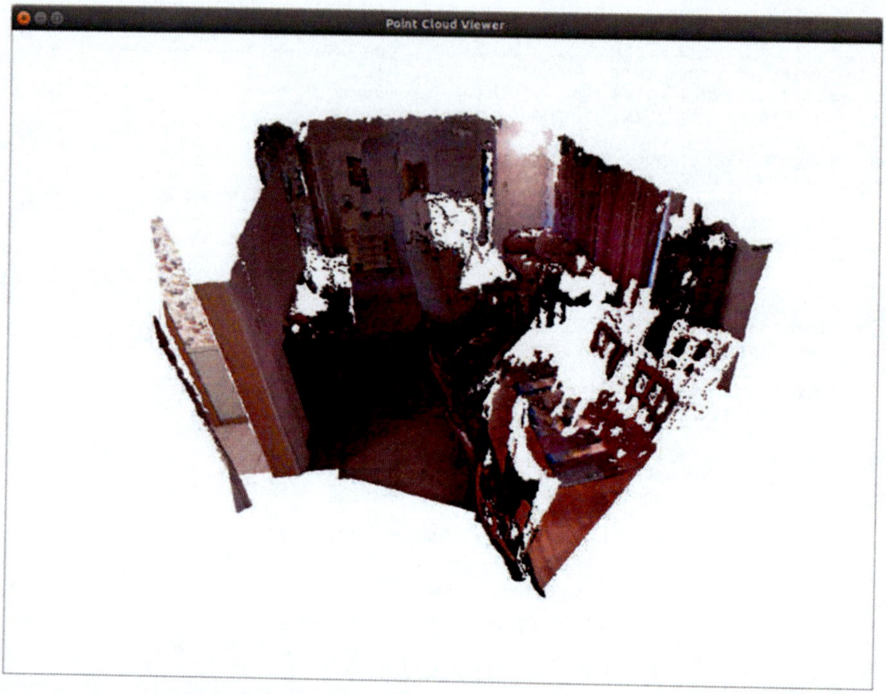

Fig. 4.10 The global point cloud from 5 RGBD image pairs

We can see the point cloud in *Pangolin* after building it (see Fig. 4.10).

We demonstrated some common monocular, binocular, and rgbd camera algorithms in computer vision through these examples. We hope readers can understand the meaning of the intrinsics, extrinsics, and distortion model through them.

Exercise

1. *Find a camera (use the camera of your mobile phone or laptop if you don't have one) and calibrate its internal parameters. You may use a calibration board or print out a checkerboard for calibration.
2. Describes the physical meaning of the camera's intrinsics. If the resolution of a camera is doubled and the rest is unchanged, how does its intrinsic change?
3. Search for the calibration method of special cameras (fisheye or panoramic cameras). Where are the differences between them and the pinhole models?
4. Investigate the similarities and differences between a global shutter camera and a rolling shutter camera. What are their advantages and disadvantages in SLAM?
5. How are RGB-D cameras calibrated? Taking Kinect as an example, what parameters need to be calibrated? (Refer to https://github.com/code-iai/iai_kinect2.)
6. In addition to the way of traversing the image demonstrated by the sample program, what other methods can you give to traverse the image?
7. *Read the official OpenCV tutorial to learn its basic usage.

Chapter 5
Nonlinear Optimization

Goal of Study

1. Understand how to form the batch state estimation problem into a least-square and how to solve the least-square problem.
2. Understand the Gauss-Newton and Levenburg-Marquardt method and implement them.
3. Learn how to use the Google Ceres and *g2o* library to solve a least-square problem.

In the previous lectures, we introduced the motion and observation equations of the classic SLAM model. Now we know that the pose in the equation can be described by the transformation matrix and then optimized by Lie algebra. The observation equation is given by the camera imaging model, in which the internal parameter is fixed with the camera, and the external parameter is the pose of the camera. So, we have figured out the concrete expression of the classic visual SLAM model.

However, due to the presence of noise, the equations of motion and observation can not be exactly met. Although the camera can fit the pinhole model very well, unfortunately, the data we get is usually affected by various unknown noises. Even if we have a high-precision camera and controller, the motion and observations equations can only be approximated. Therefore, instead of assuming that the data must conform to the equation precisely, it is better to find an estimation approach to get the state from the noisy data.

Solving the state estimation problem requires a certain degree of optimization background knowledge. This section will introduce the basic unconstrained nonlinear optimization method and introduce optimization libraries *g2o* and Ceres.

X. Gao and T. Zhang, *Introduction to Visual SLAM*,
https://doi.org/10.1007/978-981-16-4939-4_5

5.1 State Estimation

5.1.1 From Batch State Estimation to Least-Square

According to the previous sections, the SLAM process can be described by a discrete-time motion and observation equations like (1.5):

$$\begin{cases} \mathbf{x}_k = \mathbf{f}\left(\mathbf{x}_{k-1}, \mathbf{u}_k\right) + \mathbf{w}_k \\ \mathbf{z}_{k,j} = \mathbf{h}\left(\mathbf{y}_j, \mathbf{x}_k\right) + \mathbf{v}_{k,j} \end{cases}. \tag{5.1}$$

Through the knowledge in lecture 3, we learned that $\mathbf{x}_k$ is the pose of the camera, which can be described by SE(3). As for the observation equation, we have already explained in lecture 4 that it is just the pinhole camera model. To give readers a deeper impression of them, we may wish to discuss their specific parameterized form. First, the pose variable $\mathbf{x}_k$ can be expressed by $\mathbf{T}_k \in \mathrm{SE}(3)$. Second, the motion is related to the specific form of the input, but there is no particularity in visual SLAM (should be the same as ordinary robots and vehicles). We will not talk about it for now. The observation equation is given by the pinhole model. Assuming an observation of the road sign $\mathbf{y}_j$ at $\mathbf{x}_k$, corresponding to the pixel position on the image $\mathbf{z}_{k,j}$, then, the observation equation can be expressed as:

$$s\mathbf{z}_{k,j} = \mathbf{K}(\mathbf{R}_k\mathbf{y}_j + \mathbf{t}_k), \tag{5.2}$$

where $\mathbf{K}$ is the intrinsic matrix of the camera, and s is the distance of pixels, which is also the third element of $(\mathbf{R}_k\mathbf{y}_j + \mathbf{t}_k)$. If we use transformation matrix $\mathbf{T}_k$ to describe the pose, then the points $\mathbf{y}_j$ must be described in homogeneous coordinates, and then converted to non-homogeneous coordinates afterwards. If you are not familiar with this process, please go back to the previous lectures.

Now, consider what happens when the data is affected by noise. In the motion and observation equations, we *usually* assume that the two noise terms $\mathbf{w}_k, \mathbf{v}_{k,j}$ satisfy a Gaussian distribution with zero mean, like this:

$$\mathbf{w}_k \sim \mathcal{N}\left(\mathbf{0}, \mathbf{R}_k\right), \mathbf{v}_k \sim \mathcal{N}\left(\mathbf{0}, \mathbf{Q}_{k,j}\right), \tag{5.3}$$

where $\mathcal{N}$ means Gaussian distribution, $\mathbf{0}$ means zero mean, and $\mathbf{R}_k, \mathbf{Q}_{k,j}$ is the covariance matrix. Under the influence of these noises, we hope to infer the pose $\mathbf{x}$ and the map $\mathbf{y}$ from the noisy data $\mathbf{z}$ and $\mathbf{u}$ (and their probability density distribution), which constitutes a state estimation problem.

Generally, there are two ways to deal with this state estimation problem. Since these data come gradually over time in the SLAM process, we should intuitively hold an estimated state at the current moment and then update it with new data. This method is called the *incremental* method, or *filtering*. For a long time in history, researchers have used filters, especially the extended Kalman filter (EKF) and its derivatives, to solve it. The other way is to record the data into a file and looking for the best trajectory and map in all time. This method is called the *batch* estimation. In other words, we can collect all the input and observation data from time 0 to k together and ask, with such input and observation, how to estimate the entire trajectory and map from time 0 to k?

These two different processing methods lead to many estimation methods. In general, the incremental method only cares about the state estimation of the *current moment* $\mathbf{x}_k$ but does not consider much about the previous state. Conversely, the batch method can be used to get an optimized trajectory in a long time, which is considered superior to the traditional filters [13], and has become the mainstream method of current visual SLAM. In extreme cases, we can let robots or drones collect data and then bring it back to the computing center for unified processing, which is also the mainstream practice of SfM (structure from motion). Of course, in these cases, the method is obviously not *real-time*, which is not the most common application scenario of SLAM. So in SLAM, practical methods are usually some compromises. For example, we fix some historical trajectories and only optimize some trajectories close to the current moment, leading to the sliding window estimation method described later.

In theory, the batch method is easier to introduce. At the same time, understanding the batch method also makes it easier to understand the incremental method. Therefore, in this section, we focus on the batch optimization method based on nonlinear optimization. The Kalman filter and more in-depth knowledge will be discussed in the backend chapter. Since the batch method is discussed, we will consider all the moments from time 1 to N and assume M map points. Define the robot pose and map coordinates at all times as:

$$\mathbf{x} = \{\mathbf{x}_1, \ldots, \mathbf{x}_N\}, \quad \mathbf{y} = \{\mathbf{y}_1, \ldots, \mathbf{y}_M\}.$$

Similarly, $\mathbf{u}$ without subscript is used for input at all times, and $\mathbf{z}$ is used for observation data at all times. Then we say that the state estimation of the robot, from a probabilistic point of view, is to find the state $\mathbf{x}, \mathbf{y}$ under the condition that the input data $\mathbf{u}$ and the observation data $\mathbf{z}$. Or, the conditional probability distribution of:

$$P(\mathbf{x}, \mathbf{y}|\mathbf{z}, \mathbf{u}). \tag{5.4}$$

In particular, when we do not know the control input and only have one image, that is, only considering the data brought by the observation equation, it is equivalent to estimate the conditional probability distribution $P(\mathbf{x}, \mathbf{y}|\mathbf{z})$. Such a problem is also called Structure from Motion (SfM), that is, how to reconstruct the three-dimensional spatial structure from images only [29].

To estimation the conditional pdf, we use the Bayes equation to switch the variables:

$$P\left(\mathbf{x}, \mathbf{y}|\mathbf{z}, \mathbf{u}\right) = \frac{P\left(\mathbf{z}, \mathbf{u}|\mathbf{x}, \mathbf{y}\right) P\left(\mathbf{x}, \mathbf{y}\right)}{P\left(\mathbf{z}, \mathbf{u}\right)} \propto \underbrace{P\left(\mathbf{z}, \mathbf{u}|\mathbf{x}, \mathbf{y}\right)}_{\text{likelihood}} \underbrace{P\left(\mathbf{x}, \mathbf{y}\right)}_{\text{prior}}. \tag{5.5}$$

The left side is called *posterior probability*, and $P(\mathbf{z}|\mathbf{x})$ on the right is called *likelihood* (or likehood), and the other part is $P(\mathbf{x})$ is called *prior*. *It is normally difficult to find the posterior distribution directly (in nonlinear systems), but it is feasible to find an optimal point which maximize the posterior* (Maximize a Posterior, MAP):

$$(\mathbf{x}, \mathbf{y})^*{}_{\mathrm{MAP}} = \arg\max P\left(\mathbf{x}, \mathbf{y}|\mathbf{z}, \mathbf{u}\right) = \arg\max\ P(\mathbf{z}, \mathbf{u}|\mathbf{x}, \mathbf{y})P(\mathbf{x}, \mathbf{y}). \tag{5.6}$$

Please note that the denominator part of Bayes' rule has nothing to do with the state $\mathbf{x}$, $\mathbf{y}$ to be estimated, so it can be just ignored. Bayes' rule tells us that solving the maximum posterior probability is *equivalent to the estimate the product of maximum likelihood and a priori*. Further, we can also say, "I'm sorry, I don't know in prior where the robot pose or the map points are", then there is no *prior*. In this case, we can solve the *Maximize Likelihood Estimation* (MLE):

$$(\mathbf{x}, \mathbf{y})^*{}_{\mathrm{MLE}} = \arg\max\ P(\mathbf{z}, \mathbf{u}|\mathbf{x}, \mathbf{y}). \tag{5.7}$$

Intuitively speaking, likelihood refers to "what observation data may be generated in the current pose". Since we know the observation data, the maximum likelihood estimation can be understood as: "under what state, it is most likely to produce the data currently observed".

5.1.2 Introduction to Least-Squares

Now we have formulated the state estimation problem into a MAP/MLE problem, and the next question is how to solve it. Under the assumption of Gaussian distribution, we can have a simpler form of the maximum likelihood problem. Looking back at the observation model, for a certain kind of observation, we have:

$$\mathbf{z}_{k,j} = h\left(\mathbf{y}_j, \mathbf{x}_k\right) + \mathbf{v}_{k,j},$$

Since we assume the noise item is Gaussian, which means $\mathbf{v}_k \sim \mathcal{N}\left(\mathbf{0}, \mathbf{Q}_{k,j}\right)$, so the conditional probability of the observation data is:

$$P(\mathbf{z}_{j,k}|\mathbf{x}_k, \mathbf{y}_j) = N\left(h(\mathbf{y}_j, \mathbf{x}_k), \mathbf{Q}_{k,j}\right),$$

which is, of course, still a Gaussian distribution. Now let's consider solving the maximum likelihood estimation under this single observation.

We can rewrite this maximum problem into a *minimum of negative logarithm* one. Gaussian distributions have better mathematical forms under negative logarithm. Consider an arbitrary multi-dimensional Gaussian distribution $\mathbf{x} \sim \mathcal{N}(\bar{\ }, \mathbf{\Sigma})$, its probability density function expansion form is:

$$P(\mathbf{x}) = \frac{1}{\sqrt{(2\pi)^N \det(\mathbf{\Sigma})}} \exp\left(-\frac{1}{2}(\mathbf{x}-\boldsymbol{\mu})^T \mathbf{\Sigma}^{-1}(\mathbf{x}-\boldsymbol{\mu})\right). \tag{5.8}$$

Take the negative logarithm of both sides:

$$-\ln(P(\mathbf{x})) = \underbrace{\frac{1}{2}\ln\left((2\pi)^N \det(\mathbf{\Sigma})\right)}_{\text{discarded}} + \frac{1}{2}(\mathbf{x}-\boldsymbol{\mu})^T \mathbf{\Sigma}^{-1}(\mathbf{x}-\boldsymbol{\mu}). \tag{5.9}$$

Because the logarithm function is monotonically increasing, maximizing the original function is equivalent to minimizing the negative logarithm. When minimizing $\mathbf{x}$ in the above formula, the first term has nothing to do with $\mathbf{x}$ and can be omitted. Therefore, as long as the quadratic term on the right is minimized, the state's maximum likelihood estimate is obtained. Substituting into the SLAM observation model, it is equivalent to find such a solution:

$$\begin{aligned}(\mathbf{x}_k, \mathbf{y}_j)^* &= \arg\max \mathcal{N}(h(\mathbf{y}_j, \mathbf{x}_k), \mathbf{Q}_{k,j}) \\ &= \arg\min\left(\left(\mathbf{z}_{k,j} - h\left(\mathbf{x}_k, \mathbf{y}_j\right)\right)^T \mathbf{Q}_{k,j}^{-1}\left(\mathbf{z}_{k,j} - h\left(\mathbf{x}_k, \mathbf{y}_j\right)\right)\right).\end{aligned} \tag{5.10}$$

We found that this equation is equivalent to a quadratic form that minimizes the noise term (i.e., the error). This quadratic form is called *Mahalanobis distance*. It can also be regarded as the Euclidean distance ($\mathcal{L}_2$-norm) weighted by $\mathbf{Q}_{k,j}^{-1}$, where $\mathbf{Q}_{k,j}^{-1}$ is also called the *information matrix*, which is exactly the *inverse* of the Gaussian covariance matrix.

Now we put all the observations together. It is usually assumed that the inputs and observations are independent of each other, which means that each input is independent, each observation is independent, and input and observation are still independent. So we can factorize the joint distribution like this:

$$P(\mathbf{z}, \mathbf{u}|\mathbf{x}, \mathbf{y}) = \prod_k P(\mathbf{u}_k|\mathbf{x}_{k-1}, \mathbf{x}_k) \prod_{k,j} P\left(\mathbf{z}_{k,j}|\mathbf{x}_k, \mathbf{y}_j\right), \tag{5.11}$$

It shows that we can handle the movement and observation at each moment independently. Let's define the error between the model and real data:

$$\begin{aligned}\mathbf{e}_{u,k} &= \mathbf{x}_k - f(\mathbf{x}_{k-1}, \mathbf{u}_k) \\ \mathbf{e}_{z,j,k} &= \mathbf{z}_{k,j} - h\left(\mathbf{x}_k, \mathbf{y}_j\right),\end{aligned} \tag{5.12}$$

Then, minimizing the Mahalanobis distance between the estimated value and the measurements from sensors is equivalent to finding the maximum likelihood estimation. The negative logarithm allows us to turn the product into a summation:

$$\min J(\mathbf{x}, \mathbf{y}) = \sum_k \mathbf{e}_{u,k}^T \mathbf{R}_k^{-1} \mathbf{e}_{u,k} + \sum_k \sum_j \mathbf{e}_{z,k,j}^T \mathbf{Q}_{k,j}^{-1} \mathbf{e}_{z,k,j}. \quad (5.13)$$

In this way, a *least-square problem* is obtained with the same solution as the MLE problem. Intuitively speaking, due to the presence of noise, when we substitute the estimated trajectory and map into the SLAM motion and observation models, they will not be perfectly fit. What shall we do at this time? We perform *fine-tuning* on the state's estimated value so that the overall error is reduced. Of course, finally, we will reach a (local) *minimum value*. This is a typical nonlinear optimization process.

Observing the formula (5.13) carefully, we find that the least-squares problem in SLAM has some specific structures:

- First, the objective function of the whole problem consists of many (weighted) error quadratic forms. Although the overall state variable's dimensionality is very high, each error term is simple and is only related to one or two state variables. For example, the motion error is only related to $\mathbf{x}_{k-1}, \mathbf{x}_k$, and the observation error is only related to $\mathbf{x}_k, \mathbf{y}_j$. This relationship will give a *sparse* least-square problem, which we will investigate further in the backend chapter.
- Secondly, if you use Lie algebra to represent the increment, the problem is the least-squares problem of *unconstrained*. However, if the rotation matrix/transformation matrix is used to describe the pose, the constraint of the rotation matrix itself will be introduced, that is, "s.t. $\mathbf{R}^T\mathbf{R} = \mathbf{I}$ and $\det(\mathbf{R}) = 1$" need to be added to the problem. Additional constraints can make optimization more difficult.
- Finally, we used a quadratic metric error, exactly the $\mathcal{L}_2$−norm. The information matrix is used as the weights of each element. For example, if an observation is very accurate, the covariance matrix will be "small" and the information matrix will be "large", so this error term will have a higher weight than the others in the whole problem. We will see some drawbacks of the $\mathcal{L}_2$ error later.

Next, we will introduce how to solve this least-squares problem, which requires some basic knowledge of nonlinear optimization. In particular, we want to discuss how to solve such a general unconstrained nonlinear least-squares problem. In the following lectures, we will make extensive use of this lecture's results and discuss its application in the SLAM's front and backends.

5.1.3 Example: Batch State Estimation

Maybe it's better to give an example here. Consider a very simple discrete-time system:

$$\begin{aligned} x_k &= x_{k-1} + u_k + w_k, \qquad & w_k &\sim \mathcal{N}(0, Q_k) \\ z_k &= x_k + n_k, \qquad & n_k &\sim \mathcal{N}(0, R_k) \end{aligned}, \quad (5.14)$$

which describes a car moving forward or backward along the x axis. The first formula is the motion model, where u_k is the input, and w_k is the noise; the second formula is the observation model, where z_k is the measurement of the position. We set the time $k = 1, \ldots, 3$, and want to estimate the states based on the existing v, y. Suppose the initial state x_0 is known. Let's derive the maximum likelihood estimation of the batch state estimation.

First, let the batch state variable be $\mathbf{x} = [x_0, x_1, x_2, x_3]^T$, and the batch observation be $\mathbf{z} = [z_1, z_2, z_3]^T$, define $\mathbf{u} = [u_1, u_2, u_3]^T$ in the same way. According to the previous derivation, we know that the maximum likelihood estimate is:

$$\begin{aligned}\mathbf{x}^*_{\text{map}} &= \arg\max P(\mathbf{x}|\mathbf{u}, \mathbf{z}) = \arg\max P(\mathbf{u}, \mathbf{z}|\mathbf{x}) \\ &= \prod_{k=1}^{3} P(u_k|x_{k-1}, x_k) \prod_{k=1}^{3} P\left(z_k|x_k\right),\end{aligned} \tag{5.15}$$

For each item, such as the motion equation, we have:

$$P(u_k|x_{k-1}, x_k) = \mathcal{N}(x_k - x_{k-1}, Q_k). \tag{5.16}$$

The observation equation is similar:

$$P\left(z_k|x_k\right) = \mathcal{N}\left(x_k, R_k\right). \tag{5.17}$$

According to the previous statements, the error variable can be constructed as:

$$e_{u,k} = x_k - x_{k-1} - u_k, \quad e_{z,k} = z_k - x_k, \tag{5.18}$$

Then the objective function of the least-squares is:

$$\min \sum_{k=1}^{3} e_{u,k}^T Q_k^{-1} e_{u,k} + \sum_{k=1}^{3} e_{z,k}^T R_k^{-1} e_{z,k}. \tag{5.19}$$

In addition, since this system is a linear one, we can easily write the equations in the vector/matrix form. Define the vector $\mathbf{y} = [\mathbf{u}, \mathbf{z}]^T$, then the error can be defined as:

$$\mathbf{y} - \mathbf{H}\mathbf{x} = \mathbf{e} \sim \mathcal{N}(\mathbf{0}, \mathbf{\Sigma}), \tag{5.20}$$

where

$$\mathbf{H} = \left[\begin{array}{cccc} 1 & -1 & 0 & 0 \\ 0 & 1 & -1 & 0 \\ 0 & 0 & 1 & -1 \\ \hline 0 & 1 & 0 & 0 \\ 0 & 0 & 1 & 0 \\ 0 & 0 & 0 & 1 \end{array}\right], \tag{5.21}$$

and $\mathbf{\Sigma} = \text{diag}(Q_1, Q_2, Q_3, R_1, R_2, R_3)$. The whole question can be written as:

$$\mathbf{x}^*_{\text{map}} = \arg\min \mathbf{e}^T \mathbf{\Sigma}^{-1} \mathbf{e}, \tag{5.22}$$

Later we will see that this problem has a unique solution:

$$\mathbf{x}^*_{\text{map}} = (\mathbf{H}^T \mathbf{\Sigma}^{-1} \mathbf{H})^{-1} \mathbf{H}^T \mathbf{\Sigma}^{-1} \mathbf{y}, \tag{5.23}$$

since $(\mathbf{H}^T \mathbf{\Sigma}^{-1} \mathbf{H})^{-1}$ is invertible.

5.2 Nonlinear Least-Square Problem

First, consider a simple least-square problem:

$$\min_{\mathbf{x}} F(\mathbf{x}) = \frac{1}{2} \| f(\mathbf{x}) \|_2^2, \tag{5.24}$$

where the state variable is $\mathbf{x} \in \mathbb{R}^n$, and f is any scalar nonlinear function $f(\mathbf{x}) : \mathbb{R}^n \mapsto \mathbb{R}$. Note that the coefficient $\frac{1}{2}$ here is not important. Some literature has this coefficient, and some not. It will not affect the subsequent conclusions. Obviously, if f is a mathematically simple function, then the problem can be solved in an analytical form. Let the derivative of the objective function be zero, and then find the optimal value of $\mathbf{x}$, just like finding the extreme value of a scalar function:

$$\frac{\mathrm{d}F}{\mathrm{d}\mathbf{x}} = \mathbf{0}. \tag{5.25}$$

We reach the minimum, maximum, or saddle points by solving this equation (or, intuitively, by letting the derivative be zero). But is this equation easy to solve? Well, it depends on the form of the derivative function of f. If f is just a simple linear function, then the problem is only a simple linear least-square problem. But, some derivative functions may be complicated in form, making the equation difficult to solve. Solving this equation requires us to know the *global property* of the objective function, which is usually not possible. For the least-square problem that is inconvenient to solve directly, we can use the *iterated methods* to start from an initial value and continuously update the current estimations to reduce the objective function. The specific steps can be listed as follows:

1. Give an initial value $\mathbf{x}_0$.
2. For k-th iteration, we find an incremental value of $\Delta\mathbf{x}_k$, such that the object function $\| f(\mathbf{x}_k + \Delta\mathbf{x}_k) \|_2^2$ reaches a smaller value.
3. If $\Delta\mathbf{x}_k$ is small enough, stop the algorithm.
4. Otherwise, let $\mathbf{x}_{k+1} = \mathbf{x}_k + \Delta\mathbf{x}_k$ and return to step 2.

Now things get much simpler. We turn the problem of solving the derivative function equals zero into a problem of looking for decreasing increments $\Delta\mathbf{x}_k$. We will see that since the objective function can be linearly approximated at the current estimation, the increment calculation will be more straightforward.[1] When the function decreases until the increment becomes very small, it is considered that the algorithm converges, and the objective function reaches a minimum value. In this process, the problem is how to find the increment at each iteration point, which is a local problem. We only need to be concerned about the local properties of f at the iteration value rather than the global properties. Such methods are widely used in optimization, machine learning, and other fields.

Next, we examine how to find this increment $\Delta \mathbf{x}_k$. This part of knowledge belongs to the field of numerical optimization. Let's take a quick look at some widely used results.

5.2.1 The First and Second-Order Method

Now consider the k-th iteration. Suppose we are at $\mathbf{x}_k$ and want to find the increment $\Delta \mathbf{x}_k$, then the most intuitive way is to make the Taylor expansion of the objective function in $\mathbf{x}_k$:

$$F(\mathbf{x}_k + \Delta \mathbf{x}_k) \approx F(\mathbf{x}_k) + \mathbf{J}(\mathbf{x}_k)^T \Delta \mathbf{x}_k + \frac{1}{2} \Delta \mathbf{x}_k^T \mathbf{H}(\mathbf{x}_k) \Delta \mathbf{x}_k, \tag{5.26}$$

where $\mathbf{J}(\mathbf{x}_k)$ is the first derivative of $F(\mathbf{x})$ with respect to $\mathbf{x}$ (also called gradient, *Jacobi* Matrix, or *Jacobian*),[2] $\mathbf{H}$ is the second-order derivative (or *Hessian*), which are all taken at $\mathbf{x}_k$. Readers should be familiar with them in the undergraduate course like multivariate calculus. We can choose to keep the first-order or second-order terms of the Taylor expansion, and the corresponding solution is called the first-order or the second-order method. In the simplest way, if we only keep the first-order one, then taking the increment at the minus gradient direction will ensure that the function decreases:

$$\Delta \mathbf{x}^* = -\mathbf{J}(\mathbf{x}_k). \tag{5.27}$$

Of course, this is only a direction. Usually, we have to compute another step length parameter, say, λ. The step length can be calculated according to certain conditions [30]. There are also some empirical methods in machine learning, but we will not discuss them. This method is called *steepest descent method*. Its intuitive meaning is very simple. As long as we move along the reverse gradient direction, the objective function must decrease if the first-order (linear) approximation still holds.

Note that the above discussion was carried out during the k-th iteration and did not involve any information about k. To simplify the notation, we will omit the subscript k later and think that these discussions are valid for each iteration.

---

[1] Linear cases are always the easiest ones.

[2] We write $\mathbf{J}(\mathbf{x})$ as a column vector, then it can be inner product with $\Delta \mathbf{x}$ to get a scalar.

On the other hand, we can choose to keep the second step information, and the increment equation is:

$$\Delta \mathbf{x}^* = \arg\min \left(F(\mathbf{x}) + \mathbf{J}(\mathbf{x})^T \Delta \mathbf{x} + \frac{1}{2} \Delta \mathbf{x}^T \mathbf{H} \Delta \mathbf{x} \right). \tag{5.28}$$

The right side only contains the zero-order, first-order and quadratic terms of $\Delta \mathbf{x}$. Finding the derivative of $\Delta \mathbf{x}$ on the right side of the equation and setting it to zero leads to[3]:

$$\mathbf{J} + \mathbf{H}\Delta \mathbf{x} = \mathbf{0} \Rightarrow \mathbf{H}\Delta \mathbf{x} = -\mathbf{J}. \tag{5.29}$$

We can also solve this linear equation and get the increment. This method is also called *Newton's method.*

We have seen that both the first-order and second-order methods are very intuitive, as long as we can calculate the Taylor expansion of f and find the increments. We say, hey, the function looks like a linear or quadratic one. We can use the approximated function's minimum value to guess the minimum value of the original function. As long as the original objective function really looks like a first-order or quadratic function locally, this type of algorithm is always valid (this is also the most cases in reality). However, these two methods also have their drawbacks. The steepest descent method is too greedy and easy to lead a zigzag way and increases the number of iterations. However, Newton's method needs to calculate the $\mathbf{H}$ matrix of the objective function, which is very time-expensive when the problem is large. We usually tend to avoid the calculation of $\mathbf{H}$. For general problems, some quasi-Newton methods can get better results. For least-square problems, there are several more practical methods: the *Gauss-Newton's method* and the (Levernburg-Marquardt's method).

5.2.2 *The Gauss-Newton Method*

The Gauss-Newton method is one of the simplest methods in optimization algorithms. Its idea is to carry out a first-order Taylor expansion of $f(\mathbf{x})$. Please note that this is not the objective function $F(\mathbf{x})$ but the lower case $f(\mathbf{x})$, otherwise it is the same as Newton's method.

$$f(\mathbf{x} + \Delta \mathbf{x}) \approx f(\mathbf{x}) + \mathbf{J}(\mathbf{x})^T \Delta \mathbf{x}. \tag{5.30}$$

Here $\mathbf{J}(\mathbf{x})$ is the derivative of $f(\mathbf{x})$ with respect to $\mathbf{x}$, which is a column vector of $n \times 1$. According to the previous framework, the current goal is to find the increment $\Delta \mathbf{x}$ such that $\| f(\mathbf{x} + \Delta \mathbf{x}) \|^2$ reached the minimum. In order to find $\Delta \mathbf{x}$, we need to solve a linear least-square problem[4]:

---

[3] For students who are not familiar with matrix derivation, please refer to Appendix B.

[4] We omit the information matrix to simplify the derivation here.

$$\Delta \mathbf{x}^* = \arg\min_{\Delta \mathbf{x}} \frac{1}{2} \left\| f(\mathbf{x}) + \mathbf{J}(\mathbf{x})^T \Delta \mathbf{x} \right\|^2. \tag{5.31}$$

What's the difference from before? According to the extreme conditions, we set the derivative with $\Delta \mathbf{x}$ to zero to reach the extreme value. To do this, let's first expand the square term of the objective function:

$$\begin{aligned} \frac{1}{2} \left\| f(\mathbf{x}) + \mathbf{J}(\mathbf{x})^T \Delta \mathbf{x} \right\|^2 &= \frac{1}{2} \left(f(\mathbf{x}) + \mathbf{J}(\mathbf{x})^T \Delta \mathbf{x} \right)^T \left(f(\mathbf{x}) + \mathbf{J}(\mathbf{x})^T \Delta \mathbf{x} \right) \\ &= \frac{1}{2} \left(\| f(\mathbf{x}) \|_2^2 + 2 f(\mathbf{x}) \mathbf{J}(\mathbf{x})^T \Delta \mathbf{x} + \Delta \mathbf{x}^T \mathbf{J}(\mathbf{x}) \mathbf{J}(\mathbf{x})^T \Delta \mathbf{x} \right). \end{aligned}$$

Find the derivative of the above formula with respect to $\Delta \mathbf{x}$ and set it to zero:

$$\mathbf{J}(\mathbf{x}) f(\mathbf{x}) + \mathbf{J}(\mathbf{x}) \mathbf{J}^T(\mathbf{x}) \Delta \mathbf{x} = \mathbf{0}.$$

The following equation can be obtained:

$$\underbrace{\mathbf{J}(\mathbf{x}) \mathbf{J}^T}_{\mathbf{H}(\mathbf{x})}(\mathbf{x}) \Delta \mathbf{x} = \underbrace{-\mathbf{J}(\mathbf{x}) f(\mathbf{x})}_{\mathbf{g}(\mathbf{x})}. \tag{5.32}$$

This equation is a *linear equation* of the variable $\Delta \mathbf{x}$. We call it *normal equation* or *Gauss-Newton equation*. We define the coefficients on the left as $\mathbf{H}$ and the coefficient on the right as $\mathbf{g}$, then the above formula becomes:

$$\mathbf{H} \Delta \mathbf{x} = \mathbf{g}. \tag{5.33}$$

It makes sense to mark the left side as $\mathbf{H}$ here. Compared with Newton's method (5.29), Gauss-Newton's method uses $\mathbf{J}\mathbf{J}^T$ *as the approximation* of the second-order Hessian matrix in Newton's method, thus omitting the calculation of $\mathbf{H}$. Please note that solving the normal equation is the core of the entire optimization problem. If we can get the $\Delta \mathbf{x}$ in each iteration, then the algorithm of the Gauss-Newton method can be written as:

1. Set it initial value as $\mathbf{x}_0$.
2. For k-th iteration, calculate the Jacobian $\mathbf{J}(\mathbf{x}_k)$ and residual $f(\mathbf{x}_k)$.
3. Solve the normal equation: $\mathbf{H} \Delta \mathbf{x}_k = \mathbf{g}$.
4. If $\Delta \mathbf{x}_k$ is small enough, stop the algorithm. Otherwise let $\mathbf{x}_{k+1} = \mathbf{x}_k + \Delta \mathbf{x}_k$ and return to step 2.

It can be seen from the algorithm steps that the solution of the incremental equation occupies the major position. As long as we can solve the increment smoothly, we can ensure that the objective function decreases correctly.

In order to solve the incremental equation, we need to solve $\mathbf{H}^{-1}$, which requires the $\mathbf{H}$ matrix to be invertible. However, the calculated $\mathbf{J}\mathbf{J}^T$ is only semi-positive definite. If $\mathbf{J}$ has null-spaces, which means we can find non-zero $\Delta \mathbf{x}$ such that $\mathbf{J} \Delta \mathbf{x} =$

0, then we can not determine which $\Delta \mathbf{x}$ can really makes the objective function decreases. In that case, the algorithm is probably not converging, and we may obtain erroneous results. Intuitively, the local approximation of the original function at this point is not like a quadratic function. Furthermore, even if we assume that $\mathbf{H}$ is not singular or ill-conditioned, but if we take a very large step size $\Delta \mathbf{x}$, the result can still be bad since linear approximation is not accurate enough at that point. So in practice, we can't guarantee that the Gauss-Newton method converges. Sometimes they can reach even larger objective function values than the initial one.

Although the Gauss-Newton method has these shortcomings, it is still a simple and effective method for nonlinear optimization, and it is worth learning. In nonlinear optimization, quite a lot of algorithms can be taken as the Gauss-Newton method's variants. These algorithms all use the idea of the Gauss-Newton method and correct their shortcomings through their own improvements. For example, some *line search method* adds an extra step size α. After determining $\Delta \mathbf{x}$, we may further find the α to make $\|f(\mathbf{x} + \alpha \Delta \mathbf{x})\|^2$ is minimized, instead of simply making $\alpha = 1$.

The Levenberg-Marquardt method corrects these problems to a certain extent. It is generally considered more robust than the Gauss-Newton method, but its convergence rate may be slower than the Gauss-Newton method. Such a kind of method is also called as *damped Newton Method*.

5.2.3 The Levernberg-Marquatdt Method

The approximate second-order Taylor expansion used in the Gauss-Newton method can only have a good approximation effect near the expansion point. So, we naturally thought a range should be added to $\Delta \mathbf{x}$, called *trust-region*. This range defines under what circumstances the second-order approximation is valid. This type of method is also called *trust-region method*. We think the approximation is valid only in the trusted region; otherwise, it may go wrong if the approximation goes outside.

So how to determine the scope of this trust-region? A good method is to determine it based on the difference between our approximate model and the real object function: if the difference is small, it means that the approximation is good, and we may expand the trust-region; conversely, if the difference is large, we will reduce the range of approximation. We define an indicator ρ to describe the degree of approximation:

$$\rho = \frac{f(\mathbf{x} + \Delta \mathbf{x}) - f(\mathbf{x})}{\mathbf{J}(\mathbf{x})^T \Delta \mathbf{x}}. \tag{5.34}$$

The numerator of ρ is the decreasing value of the real object function, and the denominator is the decreasing value of the approximate model. If ρ is close to 1, we may say the approximation is good. A small ρ is indicating that the actual reduced value is far less than the approximate reduced value. The approximation is poor in this case, and the trust-region needs to be reduced. Conversely, if ρ is relatively large, it means that the actual decline is greater than expected, and we can enlarge the approximate range.

Therefore, we build an improved version of the nonlinear optimization framework, which will have a better effect than the Gauss-Newton method:

1. Give the initial value $\mathbf{x}_0$ and initial trust-region radius μ.
2. For k-th iteration, we solve a linear problem based on Gauss-Newton's method added with a trust-region:

$$\min_{\Delta \mathbf{x}_k} \frac{1}{2} \left\| f(\mathbf{x}_k) + \mathbf{J}(\mathbf{x}_k)^T \Delta \mathbf{x}_k \right\|^2, \quad \text{s.t.} \quad \|\mathbf{D}\Delta \mathbf{x}_k\|^2 \leqslant \mu, \tag{5.35}$$

where μ is the radius and $\mathbf{D}$ is a coefficient matrix, which will be discussed in below.
3. Compute ρ using Eq. (5.34).
4. If $\rho > \frac{3}{4}$, set $\mu = 2\mu$.
5. Otherwise, if $\rho < \frac{1}{4}$, set $\mu = 0.5\mu$.
6. If ρ is larger than a given threshold, set $\mathbf{x}_{k+1} = \mathbf{x}_k + \Delta \mathbf{x}_k$.
7. Go back to step 2 if not converged, otherwise return the result.

Here, the range expansion's multiply factor and thresholds are empirical values and can be replaced with other values. In the formula (5.35), we limit the increment to a sphere with a radius of μ and think that it is valid only in this sphere. After bringing into the $\mathbf{D}$, this sphere can be seen as an ellipsoid. In the optimization method proposed by Levenberg, taking $\mathbf{D}$ into the unit matrix $\mathbf{I}$ is equivalent to directly constraining $\Delta \mathbf{x}_k$ in a sphere. Subsequently, Marquardt proposed to take $\mathbf{D}$ as a non-negative diagonal matrix. In practice, the square root of the diagonal elements of $\mathbf{J}^T\mathbf{J}$ is usually used as $\mathbf{D}$ so that the constraint range is larger on the dimensions with small gradient.

In any case, in Levenberg-Marquardt optimization, we need to solve a sub-problem like (5.35) to obtain the gradient. This sub-problem is an optimization problem with inequality constraints. We use Lagrangian multipliers to put the constraints in the objective function to form the Lagrangian function:

$$\mathcal{L}(\Delta \mathbf{x}_k, \lambda) = \frac{1}{2} \left\| f(\mathbf{x}_k) + \mathbf{J}(\mathbf{x}_k)^T \Delta \mathbf{x}_k \right\|^2 + \frac{\lambda}{2} \left(\|\mathbf{D}\Delta \mathbf{x}_k\|^2 - \mu \right), \tag{5.36}$$

where λ is the Lagrange multiplier. Similar to the method in the Gauss-Newton, let the derivative of the Lagrangian function with respect to $\Delta \mathbf{x}$ be zero, and we still need to solve the linear equation for calculating the increment:

$$\left(\mathbf{H} + \lambda \mathbf{D}^T \mathbf{D}\right) \Delta \mathbf{x}_k = \mathbf{g}. \tag{5.37}$$

It can be seen that the incremental equation has an extra $\lambda \mathbf{D}^T \mathbf{D}$ compared to the

Gauss-Newton method. If you consider its simplified form $\mathbf{D} = \mathbf{I}$, then it is equivalent to solving[5]:

$$(\mathbf{H} + \lambda \mathbf{I})\, \Delta \mathbf{x}_k = \mathbf{g}.$$

We can see that when the parameter λ is relatively small, $\mathbf{H}$ dominates the normal equation, which shows that the quadratic approximation model is better in this range. The Levenberg-Marquardt method is more close to the Gauss-Newton method. On the other hand, when λ is relatively large, $\lambda \mathbf{I}$ occupies a dominant position. The Levenberg-Marquardt method is closer to the one-step descent method (the steepest descent), which means that the quadratic approximation is not good enough. The solution method of the Levenberg-Marquardt method can avoid the non-singular and ill-conditioned problems of the coefficient matrix of linear equations to a certain extent and provide more stable and accurate increments $\Delta \mathbf{x}$.

In practice, there are also many other ways to solve the increment, such as Dog-Leg [31] and other methods. What we have introduced here is just the most common and basic method, and it is also the most widely-used method in visual SLAM. We usually choose one of the Gauss-Newton or Levenberg-Marquardt methods as the gradient descent strategy in practical problems. When the problem is well-formed, Gauss-Newton is used. Otherwise, in ill-formed problems, we use the Levenberg-Marquardt method.

5.2.4 *Conclusion*

Since I don't want this book to become a headache-inducing mathematics textbook, only two of the most common nonlinear optimization schemes are listed here: the Gauss-Newton method and the Levenberg-Marquardt method. We avoided many discussions of mathematical properties. If readers are interested in optimization, you can read other books dedicated to numerical optimization (this is a big topic) like [31]. The optimization methods represented by the Gauss-Newton method and the Levenberg-Marquardt method have been implemented and provided to users in many open-source optimization libraries. We will conduct experiments below. Optimization is an essential mathematical tool for dealing with many practical problems. It plays a central role in visual SLAM and one of the core methods for solving problems in other fields such as deep learning (the amount of deep learning data is substantial, where the first-order method is a major choice). We hope that readers can learn more about optimization algorithms based on their own capabilities.

---

[5] Strict readers may not be satisfied with the description here. In addition to the Lagrangian function derivation to zero, the KKT condition also has some other constraints: $\lambda > 0$, and $\lambda(\|\mathbf{D}\Delta\mathbf{x}\|^2 - \mu) = 0$. But in the L-M iteration, we might as well regard it as a penalty term (augmented Lagrangian) with λ as the weight on the objective function of the original problem. After each iteration, if the trust-region condition is not satisfied, or the objective function increases, the weight of λ is increased until the trust-region condition is finally satisfied. Therefore, there are different interpretations of the L-M algorithm in theory, but we only care about whether it works smoothly in practice.

Perhaps you have discovered that whether it is the Gauss-Newton method or the Levenberg-Marquardt method, the variable's initial value needs to be provided when doing the optimization calculation. You may ask, can this initial value be set arbitrarily? Of course not. In fact, iterative solutions of nonlinear optimization require users to provide a good initial value. Because the objective function is too complicated, it is difficult to predict the solution space change. Providing different initial values for the problem often leads to different calculation results. This situation is a common problem of nonlinear optimization: most algorithms easily fall into local minima. Therefore, no matter what kind of scientific problem, we should provide the initial value carefully. For example, in the visual SLAM problem, we will use ICP, PnP, and other algorithms to provide the optimized initial value. In short, a good initial value is significant for optimization problems!

Readers may also have questions about the optimization mentioned above: how to solve the incremental linear equations? We have only mentioned that the incremental equation is linear, but does it require many calculations to invert the coefficient matrix directly? Of course not. In the visual SLAM algorithm, the dimension of $\Delta\mathbf{x}$ is often as large as hundreds or thousands. If you are doing large-scale visual 3D reconstruction, you will often find that this dimension can be easily achieved at hundreds of thousands or even higher levels. Most processors cannot afford such a large matrix inversion computation, so there are many numerical solutions for linear equations. There are different solutions in different fields, but there is almost no way to directly find the general inverse coefficient matrix. We will use matrix decomposition to solve linear equations, such as QR, Cholesky, and other decomposition methods. These methods can usually be found in textbooks, such as the matrix theory, and we will not introduce them.

Fortunately, the visual SLAM matrix has a specific sparse form, which can be solved in real-time applications. We will introduce its principle in detail in lecture 8. Using the sparse form of elimination, decomposition, and finally solving increment will significantly improve the solution's efficiency. In many open source optimization libraries, variables with a dimension of more than 10,000 can be solved in a few seconds or less on a general PC. The reason is that more advanced mathematical tools are used. The visual SLAM algorithm can now be implemented in real-time, thanks to the coefficient matrix is sparse. If the matrix is dense, I am afraid that optimization of this kind of visual SLAM algorithm will not be widely adopted by the academic community [32–34].

5.3 Practice: Curve Fitting

5.3.1 Curve Fitting with Gauss-Newton

Next, we use a simple example to demonstrate how to solve the least-square problem. We will demonstrate how to write the Gauss-Newton method by hand and then introduce how to use the optimization library to solve this problem. For the same problem, these implementations will get the same result because their core algorithms are the same.

Consider a curve that satisfies the following equation:

$$y = \exp(ax^2 + bx + c) + w,$$

where a, b, c are the parameters of the curve, and w is the Gaussian noise, satisfying $w \sim (0, \sigma^2)$. We deliberately chose such a nonlinear model so that the problem is not too easy. Now, suppose we have N observation data points about $xandy$ and want to find the parameters of the curve based on these data points. Then, we solve the following least-square problem to estimate the curve parameters:

$$\min_{a,b,c} \frac{1}{2} \sum_{i=1}^{N} \left\| y_i - \exp\left(ax_i^2 + bx_i + c\right) \right\|^2. \tag{5.38}$$

Please note that the estimated variables are a, b, c, not x. In our program, the true value of x, y is generated according to the model, and then the Gaussian noise is added to the true value. Subsequently, the Gauss-Newton method was used to fit a parametric model from the noisy data. Define the error as:

$$e_i = y_i - \exp\left(ax_i^2 + bx_i + c\right), \tag{5.39}$$

Then we can find the derivative of each error term with respect to the state variable:

$$\begin{aligned} \frac{\partial e_i}{\partial a} &= -x_i^2 \exp\left(ax_i^2 + bx_i + c\right) \\ \frac{\partial e_i}{\partial b} &= -x_i \exp\left(ax_i^2 + bx_i + c\right) \\ \frac{\partial e_i}{\partial c} &= -\exp\left(ax_i^2 + bx_i + c\right) \end{aligned} \tag{5.40}$$

So $\mathbf{J}_i = \left[\frac{\partial e_i}{\partial a}, \frac{\partial e_i}{\partial b}, \frac{\partial e_i}{\partial c}\right]^T$, and the normal equation of the Gauss-Newton method is:

$$\left(\sum_{i=1}^{100} \mathbf{J}_i (\sigma^2)^{-1} \mathbf{J}_i^{T}\right) \Delta \mathbf{x}_k = \sum_{i=1}^{100} -\mathbf{J}_i (\sigma^2)^{-1} e_i, \tag{5.41}$$

Of course, we can also choose to arrange all $\mathbf{J}_i$ in a row and write this equation in matrix form, but its meaning is consistent with the summation form. The following code demonstrates how this process works.

Listing 5.1 slambook2/ch6/gaussNewton.cpp

```
#include <iostream>
#include <opencv2/opencv.hpp>
```

```
#include <Eigen/Core>
#include <Eigen/Dense>

using namespace std;
using namespace Eigen;

int main(int argc, char **argv) {
      double ar = 1.0, br = 2.0, cr = 1.0;          // ground-truth values
      double ae = 2.0, be = -1.0, ce = 5.0;        // initial estimation
      int N = 100;                                 // num of data points
      double w_sigma = 1.0;                        // sigma of the noise
      double inv_sigma = 1.0 / w_sigma;
      cv::RNG rng;                                 // Random number generator

      vector<double> x_data, y_data;      // the data
      for (int i = 0; i < N; i++) {
            double x = i / 100.0;
            x_data.push_back(x);
            y_data.push_back(exp(ar * x * x + br * x + cr) + rng.gaussian(w_sigma * w_sigma));
      }

      // start Gauss-Newton iterations
      int iterations = 100;
      double cost = 0, lastCost = 0;

      chrono::steady_clock::time_point t1 = chrono::steady_clock::now();
      for (int iter = 0; iter < iterations; iter++) {
        Matrix3d H = Matrix3d::Zero(); // Hessian = J^T W^{-1} J in Gauss-Newton
        Vector3d b = Vector3d::Zero(); // bias
        cost = 0;

        for (int i = 0; i < N; i++) {
              double xi = x_data[i], yi = y_data[i]; // the i-th data
            double error = yi - exp(ae * xi * xi + be * xi + ce);
            Vector3d J; // jacobian
            J[0] = -xi * xi * exp(ae * xi * xi + be * xi + ce); // de/da
            J[1] = -xi * exp(ae * xi * xi + be * xi + ce); // de/db
            J[2] = -exp(ae * xi * xi + be * xi + ce); // de/dc

            H += inv_sigma * inv_sigma * J * J.transpose();
            b += -inv_sigma * inv_sigma * error * J;

            cost += error * error;
        }

        // solve Hx=b
        Vector3d dx = H.ldlt().solve(b);
        if (isnan(dx[0])) {
            cout << "result is nan!" << endl;
            break;
        }

        if (iter > 0 && cost >= lastCost) {
            cout << "cost: " << cost << ">= last cost: " << lastCost << ", break." << endl;
            break;
        }

        ae += dx[0];
        be += dx[1];
        ce += dx[2];

        lastCost = cost;

```

```
        cout << "total cost: " << cost << ", \t\tupdate: " << dx.transpose() <<
            "\t\testimated params: " << ae << "," << be << "," << ce << endl;
    }

    chrono::steady_clock::time_point t2 = chrono::steady_clock::now();
    chrono::duration<double> time_used = chrono::duration_cast<chrono::duration<double>>(t2 - t1);
    cout << "solve time cost = " << time_used.count() << " seconds. " << endl;
    cout << "estimated abc = " << ae << ", " << be << ", " << ce << endl;
    return 0;
}
```

In this example, we demonstrate how to optimize a simple fitting problem iteratively. Through our handwritten code, it is easy to see the entire optimization process. The program outputs the objective function value and updates the amount of each iteration, as follows:

Listing 5.2 Terminal output:

```
/home/xiang/Code/slambook2/ch6/cmake-build-debug/gaussNewton
total cost: 3.19575e+06, update: 0.0455771 0.078164 -0.985329 estimated params: 2.04558,-0.921836,4.01467
total cost: 376785, update: 0.065762 0.224972 -0.962521 estimated params: 2.11134,-0.696864,3.05215
total cost: 35673.6, update: -0.0670241 0.617616 -0.907497 estimated params: 2.04432,-0.0792484,2.14465
total cost: 2195.01, update: -0.522767 1.19192 -0.756452 estimated params: 1.52155,1.11267,1.3882
total cost: 174.853, update: -0.537502 0.909933 -0.386395 estimated params: 0.984045,2.0226,1.00181
total cost: 102.78, update: -0.0919666 0.147331 -0.0573675 estimated params: 0.892079,2.16994,0.944438
total cost: 101.937, update: -0.00117081 0.00196749 -0.00081055 estimated params: 0.890908,2.1719,0.943628
total cost: 101.937, update: 3.4312e-06 -4.28555e-06 1.08348e-06 estimated params: 0.890912,2.1719,0.943629
total cost: 101.937, update: -2.01204e-08 2.68928e-08 -7.86602e-09 estimated params: 0.890912,2.1719,0.943629
cost: 101.937>= last cost: 101.937, break.
solve time cost = 0.000212903 seconds.
estimated abc = 0.890912, 2.1719, 0.943629
```

It is easy to see that the objective function of the whole problem approaches convergence after 9 iterations, and the updated amount approaches zero. The final estimated value is close to the true value, and the function image is shown in Fig. 5.1. On my machine (my CPU is i7-8700), the optimization takes about 0.2 milliseconds. Let's try to use the optimized library to accomplish the same task.

5.3.2 *Curve Fitting with Google Ceres*

In the next two sections we introduce two C++ optimization libraries: the Ceres library [35] from Google and the *g2o* library [36] based on graph optimization. Since in *g2o*, we still need some knowledge about graph optimizations, we first go through

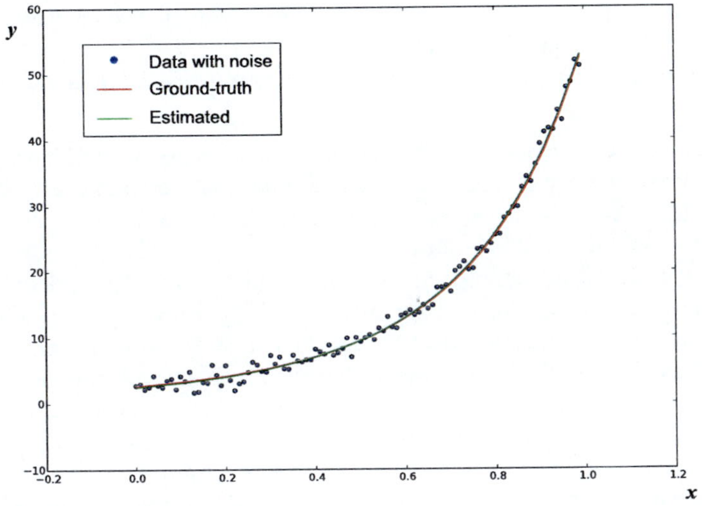

Fig. 5.1 Estimated curve when the noise $\sigma = 1$

the Ceres here, then introduce some graph optimization theories, and finally talk about *g2o*. Since the optimization algorithms will still appear in the later chapters, please make sure you understand the meaning of the optimization algorithm and the content of the program.

5.3.2.1 Introduction to Ceres

Google Ceres is a widely used optimization library for least-square problems. In Ceres, as users, we only need to define the optimization problem to be solved according to specific steps and then hand it over to the solver for calculation. The most general form of the least-square problem solved by Ceres is as follows (kernel function least-squares with boundary):

$$\begin{aligned} &\min_{x} \frac{1}{2} \sum_{i} \rho_i \left(\left\| f_i \left(x_{i_1}, \ldots x_{i_n} \right) \right\|^2 \right) \\ &\text{s.t.} \quad l_j \leqslant x_j \leqslant u_j. \end{aligned} \tag{5.42}$$

In this question, $x_1, \ldots, x_n$ are optimized variables, also called parameter blocks, f_i is called cost function, or residual blocks. l_j and u_j are the upper and lower limits of the j-th optimization variable. In the simplest case, we can set $l_j = -\infty, u_j = \infty$ (do not limit the boundary of the optimization variable). At this time, the objective function is composed of many square terms after the kernel function $\rho(\cdot)$ and then the sum.[6] Similarly, you can take ρ as the identity function. The objective function is the sum of the squares of many terms.

In order to tell Ceres the definition of the problem, we need to do the following things:

1. Defines each parameter block. The parameter block is usually a trivial vector, but it can also be defined as a particular structure such as quaternion and Lie algebra in SLAM. If it is a vector, we need to allocate a double array for each parameter block to store the variable's value.
2. Then, define the calculation method of the residual block. The residual block is usually associated with several parameter blocks, performs some custom calculations on them, and then returns the residual value. After that, Ceres will sum the squares residuals, which is used as the overall objective function.
3. In the residual blocks, we also need to define the Jacobian calculation method. In Ceres, we can use the "automatic derivative" function or manually specify the Jacobian calculation process. If you want to use automatic derivation, then the residual block must be written in a specific way: the residual calculation should be implemented as a bracketed operator with a template. We will illustrate this point through an example.
4. Finally, add all the parameter blocks and residual blocks to Ceres's Problem object and call the Solve function to solve it. Before solving, we can pass some configuration information, such as the number of iterations, termination conditions, etc., or use the default configuration.

Next, let's actually operate Ceres to solve the curve-fitting problem and understand the optimization process.

5.3.2.2 Install Ceres

In order to use Ceres, we first need to compile and install it. Ceres' GitHub address is https://github.com/ceres-solver/ceres-solver, and you can also directly use Ceres in our 3rdparty directory of the code repository so that you will use the same version as mine.

Like the libraries encountered before, Ceres is also a CMake project. Before compiling it, we need to install the dependencies first. You can install them with apt-get in Ubuntu, mainly some logging and testing tools used by Google:

Listing 5.3 Terminal input:

```
sudo apt-get install liblapack-dev libsuitesparse-dev libcxsparse3
     libgflags-dev libgoogle-glog-dev libgtest-dev
```

Then, enter the Ceres library directory, use cmake to compile and install it. We have done this process many times, so I won't repeat it here. After the installation is complete, we can find the Ceres header file under /usr/local/include/ceres, and find

[6] Kernel function is discussed in Chap. 8.

the library file named libceres.a under /usr/local/lib/. With these files, we are able to include the Ceres headers and do the optimization calculations.

5.3.2.3 Use Ceres for Curve Fitting

The following code demonstrates how to use Ceres to solve the same problem.

Listing 5.4 slambook/ch6/ceresCurveFitting.cpp

```
#include <iostream>
#include <opencv2/core/core.hpp>
#include <ceres/ceres.h>
#include <chrono>

using namespace std;

// residual
struct CURVE_FITTING_COST {
    CURVE_FITTING_COST(double x, double y) : _x(x), _y(y) {}

    // implement operator () to compute the error
    template<typename T>
   bool operator()(
        const T *const abc, // the estimated variables, 3D vector
        T *residual) const {
        // y-exp(ax^2+bx+c)
        residual[0] = T(_y) - ceres::exp(abc[0] * T(_x) * T(_x) + abc[1] *
            T(_x) + abc[2]);
        return true;
    }

    const double _x, _y; // x,y data
};

int main(int argc, char **argv) {
    // same as before
    double ar = 1.0, br = 2.0, cr = 1.0;                // ground-truth
        values
    double ae = 2.0, be = -1.0, ce = 5.0;               // initial estimation
    int N = 100;                                        // num of data points
    double w_sigma = 1.0;                               // sigma of the noise
    double inv_sigma = 1.0 / w_sigma;
    cv::RNG rng;                                        // Random number
        generator

    vector<double> x_data, y_data;          // the data
    for (int i = 0; i < N; i++) {
        double x = i / 100.0;
        x_data.push_back(x);
        y_data.push_back(exp(ar * x * x + br * x + cr) + rng.gaussian(
            w_sigma * w_sigma));
    }

    double abc[3] = {ae, be, ce};

    // construct the problem in ceres
    ceres::Problem problem;
    for (int i = 0; i < N; i++) {
        problem.AddResidualBlock( // add i-th residual into the problem
            // use auto-diff, template params: redisual type, output
                dimension, input dimension
            // shoule be same as the struct written before
```

```
                new ceres::AutoDiffCostFunction<CURVE_FITTING_COST, 1, 3>(
                    new CURVE_FITTING_COST(x_data[i], y_data[i])
                ),
                nullptr, // kernel function, don't use here
                abc      // estimated variables
        );
    }

    // set the solver options
    ceres::Solver::Options options; // actually there're lots of params can be adjusted
    options.linear_solver_type = ceres::DENSE_NORMAL_CHOLESKY; // use cholesky to solve the normal equation
    options.minimizer_progress_to_stdout = true; // print to cout

    ceres::Solver::Summary summary;
    chrono::steady_clock::time_point t1 = chrono::steady_clock::now();
    ceres::Solve(options, &problem, &summary); // do optimization!
    chrono::steady_clock::time_point t2 = chrono::steady_clock::now();
    chrono::duration<double> time_used = chrono::duration_cast<chrono::duration<double>>(t2 - t1);
    cout << "solve time cost = " << time_used.count() << " seconds. " << endl;

    // get the outpus
    cout << summary.BriefReport() << endl;
    cout << "estimated a,b,c = ";
    for (auto a:abc) cout << a << " ";
    cout << endl;

    return 0;
}
```

The code is quite self-explained because we write a lot of comments here. As you can see, we used OpenCV's noise generator to generate 100 data with Gaussian noise and then used Ceres for fitting. The Ceres usage demonstrated here has the following steps:

1. Defines the class of the residual block. The method is to write a class (or structure) and define the () operator with template parameters in the class to become a functor.[7] This definition allows Ceres to call the $< \text{double} >$ () method for an instance of this class like a function. In fact, Ceres will pass the Jacobian matrix as a type parameter to this function to realize automatic derivation (auto-diff, which is one of the best features of Ceres).
2. The double array abc[3] in the program is the parameter block. We construct a CURVE_FITTING_COST object for each data for the residual block and then call AddResidualBlock to add the error term to the objective function. Since the optimization requires gradients, we have several options: (1) Use Ceres's auto diff feature. (2) Use numeric diff[8]; (3) Derive the analytical derivative form by yourself and provide it to Ceres. Because automatic derivation is the most convenient in coding, we use automatic derivation here.

[7] The functor is a C++ term. Such a class can be called as if it were a function because the operator () is overloaded.

[8] Automatic derivation is also implemented with numerical derivatives, but since it is a template operation, it runs faster.

3. Automatic derivation requires specifying the dimensions of the error term and optimization variable. The error here is a scalar with a dimension of 1; the optimized three quantities are a, b, c with 3. Therefore, the variable dimensions are set to 1, 3 in the template parameter of the auto-derivation class AutoDiffCostFunction.
4. After setting the problem, call the Solve function to solve it. You can configure (very detailed) optimization options in Ceres. For example, you can choose to use line search or trust region, the number of iterations, step size, and so on. Readers can check the Ceres options to see what optimization methods are available. Of course, the default configuration can be used for a wide range of problems.

Finally, let's see the experimental results by calling build/ceresCurveFitting:

Listing 5.5 erminal output:

```
iter cost cost_change |gradient| |step| tr_ratio tr_radius ls_iter
    iter_time total_time
0 1.597873e+06 0.00e+00 3.52e+06 0.00e+00 0.00e+00 1.00e+04 0 2.10e-05
    7.92e-05
1 1.884440e+05 1.41e+06 4.86e+05 9.88e-01 8.82e-01 1.81e+04 1 5.60e-05
    1.05e-03
2 1.784821e+04 1.71e+05 6.78e+04 9.89e-01 9.06e-01 3.87e+04 1 2.00e-05
    1.09e-03
3 1.099631e+03 1.67e+04 8.58e+03 1.10e+00 9.41e-01 1.16e+05 1 6.70e-05
    1.16e-03
4 8.784938e+01 1.01e+03 6.53e+02 1.51e+00 9.67e-01 3.48e+05 1 1.88e-05
    1.19e-03
5 5.141230e+01 3.64e+01 2.72e+01 1.13e+00 9.90e-01 1.05e+06 1 1.81e-05
    1.22e-03
6 5.096862e+01 4.44e-01 4.27e-01 1.89e-01 9.98e-01 3.14e+06 1 1.79e-05
    1.25e-03
7 5.096851e+01 1.10e-04 9.53e-04 2.84e-03 9.99e-01 9.41e+06 1 1.81e-05
    1.28e-03
solve time cost = 0.00130755 seconds.
Ceres Solver Report: Iterations: 8, Initial cost: 1.597873e+06, Final
    cost: 5.096851e+01, Termination: CONVERGENCE
estimated a,b,c = 0.890908 2.1719 0.943628
```

The final optimized value is basically the same as our experimental result in the previous section, but Ceres is relatively slow in running speed. Ceres used about 1.3 milliseconds on my machine, which is about six times slower than the handwritten Gauss-Newton method.

I hope readers have a general understanding of how to use Ceres through this simple example. Its advantage is that it provides an automatic derivation tool, making it unnecessary to calculate the troublesome Jacobian matrix. Ceres's automatic derivation is realized through template elements, and the automatic derivation can be completed at compile-time, but it is still a numerical derivative. Most of the time in this book, I will still introduce the Jacobian matrix calculation because it helps understand the problem. Besides, Ceres' optimization process configuration is also very rich, making it suitable for a wide range of least-squares optimization problems, including various problems other than SLAM.

5.3.3 *Curve Fitting with g2o*

The second practice part of this lecture is about another optimization library (widely used mainly in the SLAM field): *g2o* (general graphic optimization, g^2o). It is a library based on *graphic optimization*. Graph optimization is a theory that combines nonlinear optimization with graph theory, so before using it, let's spend a little space to introduce graph optimization theory.

5.3.3.1 Introduction to Graph Optimization Theory

We have introduced the solution of nonlinear least-squares. The least-square problem in SLAM is normally composed of the sum of many small error terms. However, if we only treat them as variables and residuals, it would be hard to tell the *relationship* between them. For example, how many error terms are related to a certain optimization variable x_j? If we adjust some variables, does the overall cost function change or keep the same? Furthermore, we hope to visually see what optimization problem *looks like*. Therefore, graph optimization is involved.

Graph optimization is a way to express the optimization problem as a graph. A graph consists of a number of vertices and edges connecting these vertices. A vertex is used to represent an optimization variable, and an edge is used to represent an error term. Therefore, for any of the above-mentioned nonlinear least-squares problems, we can construct a corresponding graph. We can simply call it a graph or use the probability graph definition, call it a Bayesian graph or a factor graph. Sometimes it is also called a hypergraph because an edge can be connected to more than two variables, e.g., where an error term is related to more than two variables.

Figure 5.2 is a simple graph optimization example. We use triangles to represent the camera poses and circles to represent landmark points, which constitute the vertices. At the same time, the solid line represents the camera's motion model, and the dashed line represents the observation model, which forms the edges of the graph optimization. At this point, although the mathematical form of the entire problem is still like (5.13), now we can intuitively see the structure of the problem. If you want, you can also make improvements like removing isolated vertices or preferentially optimizing vertices with more edges (or, in graph theory terms, greater degrees). But the most basic idea is just to use a graph model to express a nonlinear least-squares optimization problem. And we can use some properties of the graph model to do better optimization.

The *g2o* is a *general* graph optimization library. "General" means that you can solve any least-squares problem that can be expressed as graph optimization, obviously including the curve-fitting problem discussed above. Let us demonstrate how to do that.

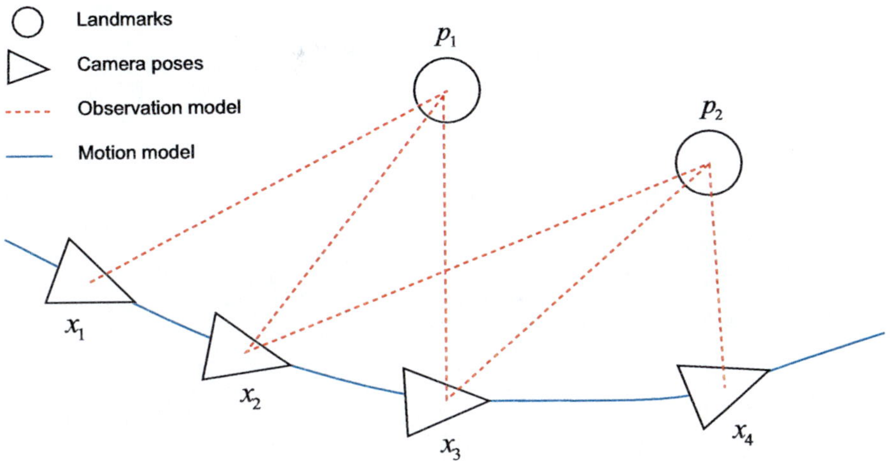

Fig. 5.2 An example of the graph optimization

5.3.3.2 Compilation and Installation of *g2o*

Just like the Ceres section, we should first compile and install the *g2o* library. Readers should have experienced this process many times, and they are basically the same. Regarding *g2o*, readers can download it from GitHub: https://github.com/RainerKuemmerle/g2o, or obtain it from the third-party code library provided in this book. Since *g2o* is still being updated, I suggest using *g2o* under 3rdparty to ensure that the version is the same as mine.

g2o is also a cmake project. Let's install its dependencies first (some dependencies overlap with Ceres):

Listing 5.6 Terminal input:

```
sudo apt-get install qt5-qmake qt5-default libqglviewer-dev-qt5
    libsuitesparse-dev libcxsparse3 libcholmod3
```

Then, compile and install *g2o* according to the cmake method. The description of the process is omitted here. After the installation is complete, the header files of *g2o* will be located under "/usr/local/g2o", and the library files will be located under /usr/local/lib/. We reconsider the curve fitting experiment in the Ceres routine and do that experiment again in *g2o*.

5.3.3.3 Curve Fitting with *g2o*

In order to use *g2o*, the curve fitting problem must first be constructed into graph optimization. In this process, just remember that nodes are the optimization variables, and edges are the error terms. The graph optimization problem of curve fitting can be drawn in the form of Fig. 5.3.

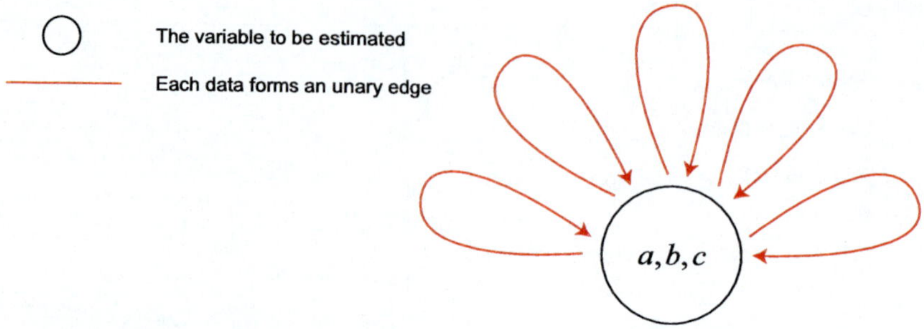

Fig. 5.3 Graph model of the curve fitting problem

The entire problem has only one vertex: the parameters a, b, c of the curve model. Each noisy data point constitutes an error term, which is the edge of the graph optimization. But the edges here are not the same as we usually think. They are unary edges, which means that the edges connect only one vertex. Because the entire graph has only one vertex. So in Fig. 5.3, we can only draw it as if it is connected to itself. An edge in graph optimization can connect one, two, or more vertices, reflecting how many optimization variables each error is related to. In a slightly mysterious way, we call it a hyperedge. The whole graph is called hypergraph.[9]

After clarifying the graph model, the next step is to build the model in *g2o* for optimization. As a user of *g2o*, what we need to do mainly includes the following steps:

1. Define the type of vertices and edges.
2. Build the graph.
3. Select the optimization algorithm.
4. Call *g2o* to optimize and get the result.

This part is very similar to Ceres. Of course there will be some differences in coding. Let's demonstrate the program.

Listing 5.7 slambook/ch6/g2oCurveFitting.cpp

```
#include <iostream>
#include <g2o/core/g2o_core_api.h>
#include <g2o/core/base_vertex.h>
#include <g2o/core/base_unary_edge.h>
#include <g2o/core/block_solver.h>
#include <g2o/core/optimization_algorithm_levenberg.h>
#include <g2o/core/optimization_algorithm_gauss_newton.h>
#include <g2o/core/optimization_algorithm_dogleg.h>
#include <g2o/solvers/dense/linear_solver_dense.h>
#include <Eigen/Core>
#include <opencv2/core/core.hpp>
#include <cmath>
#include <chrono>

using namespace std;
```

[9] I personally would better call it just as a graph to avoid being mysterious.

```
// vertex: 3d vector
class CurveFittingVertex : public g2o::BaseVertex<3, Eigen::Vector3d> {
public:
        EIGEN_MAKE_ALIGNED_OPERATOR_NEW

        // override the reset function
        virtual void setToOriginImpl() override {
                _estimate << 0, 0, 0;
        }

        // override the plus operator, just plain vector addition
        virtual void oplusImpl(const double *update) override {
                _estimate += Eigen::Vector3d(update);
        }

        // the dummy read/write function
        virtual bool read(istream &in) {}
        virtual bool write(ostream &out) const {}
};

// edge: 1D error term, connected to exactly one vertex
class CurveFittingEdge : public g2o::BaseUnaryEdge<1, double,
     CurveFittingVertex> {
public:
        EIGEN_MAKE_ALIGNED_OPERATOR_NEW

        CurveFittingEdge(double x) : BaseUnaryEdge(), _x(x) {}

        // define the error term computation
        virtual void computeError() override {
                const CurveFittingVertex *v = static_cast<const
                     CurveFittingVertex *> (_vertices[0]);
                const Eigen::Vector3d abc = v->estimate();
                _error(0, 0) = _measurement - std::exp(abc(0, 0) * _x * _x +
                     abc(1, 0) * _x + abc(2, 0));
        }

        // the jacobian
        virtual void linearizeOplus() override {
                const CurveFittingVertex *v = static_cast<const
                     CurveFittingVertex *> (_vertices[0]);
                const Eigen::Vector3d abc = v->estimate();
                double y = exp(abc[0] * _x * _x + abc[1] * _x + abc[2]);
                _jacobianOplusXi[0] = -_x * _x * y;
                _jacobianOplusXi[1] = -_x * y;
                _jacobianOplusXi[2] = -y;
        }

        virtual bool read(istream &in) {}
        virtual bool write(ostream &out) const {}
public:
        double _x; // x data, note y is given in _measurement
};

int main(int argc, char **argv) {
        // ... we omit the data sampling code, same as before
        typedef g2o::BlockSolver<g2o::BlockSolverTraits<3, 1>>
             BlockSolverType; // bolck solver
        typedef g2o::LinearSolverDense<BlockSolverType::PoseMatrixType>
             LinearSolverType; // linear solver

        // choose the optimizatoin method from GN, LM, DogLeg
        auto solver = new g2o::OptimizationAlgorithmGaussNewton(
                g2o::make_unique<BlockSolverType>(g2o::make_unique<
                     LinearSolverType>()));
        g2o::SparseOptimizer optimizer; // graph optimizer
```

```
    optimizer.setAlgorithm(solver); // set the algorithm
    optimizer.setVerbose(true);        // print the results

    // add vertex
    CurveFittingVertex *v = new CurveFittingVertex();
    v->setEstimate(Eigen::Vector3d(ae, be, ce));
    v->setId(0);
    optimizer.addVertex(v);

    // add edges
    for (int i = 0; i < N; i++) {
        CurveFittingEdge *edge = new CurveFittingEdge(x_data[i]);
        edge->setId(i);
        edge->setVertex(0, v);                // connect to the vertex
        edge->setMeasurement(y_data[i]); // measurement
        edge->setInformation(Eigen::Matrix<double, 1, 1>::Identity() * 1 / (w_sigma * w_sigma)); // set the information matrix
        optimizer.addEdge(edge);
    }

    // carry out the optimization
    cout << "start optimization" << endl;
    chrono::steady_clock::time_point t1 = chrono::steady_clock::now();
    optimizer.initializeOptimization();
    optimizer.optimize(10);
    chrono::steady_clock::time_point t2 = chrono::steady_clock::now();
    chrono::duration<double> time_used = chrono::duration_cast<chrono::duration<double>>(t2 - t1);
    cout << "solve time cost = " << time_used.count() << " seconds. " << endl;

    // print the results
    Eigen::Vector3d abc_estimate = v->estimate();
    cout << "estimated model: " << abc_estimate.transpose() << endl;

    return 0;
}
```

In this program, we derive graph optimization vertices and edges for curve fitting from *g2o* built-in classes: CurveFittingVertex and CurveFittingEdge, which essentially expands the use of *g2o*. These two classes are derived from BaseVertex and BaseUnaryEdge, respectively. In the derived classes, we have rewritten some important virtual functions:

1. Vertex update function: *oplusImpl*. We know that the most important thing in the optimization process is the calculation of incremental $\Delta\mathbf{x}$, and this function deals with $\mathbf{x}_{k+1} = \mathbf{x}_k + \Delta\mathbf{x}$ process.
   Readers may think this is not something worth mentioning because it is just simple addition. Why doesn't *g2o* help us complete it? In the curve fitting process, since the optimization variables (curve parameters) are located in the plain vector space, this update calculation is indeed nothing but just simple addition. However, when the optimization variable is not in the vector space, for example, for $\mathbf{x}$ is the camera pose, it does not necessarily have an addition operation. At this time, it is necessary to redefine the behavior of how the increment is added to the existing estimate. According to the explanation in lecture 3, we may use the left-multiplication update or the right-multiplication update instead of direct addition.

2. Vertex reset function: *setToOriginImpl*. This is trivial, and we just set the estimate to zero.
3. The error calculation function: computeError. According to the curve model, this function needs to take out the current estimated value of the vertex connected by the edge and compare it with its observed value. This is consistent with the error model in the least-squares problem.
4. The Jacobian calculation function: *linearizeOplus*. In this function, we calculate the Jacobian of each edge relative to the vertex.
5. Save and read functions: read, write. Since we do not want to perform read/write operations, leave it blank.

After defining the vertices and edges, we declare a graph model in the main function, then add vertices and edges to the graph model according to the generated noise data, and finally call the optimization function for optimization. *g2o* will give the optimized result:

Listing 5.8 Terminal output:

```
start optimization
iteration= 0 chi2= 376785.128234 time= 3.3299e-05 cumTime= 3.3299e-05 edges
    = 100 schur= 0
iteration= 1 chi2= 35673.566018 time= 1.3789e-05 cumTime= 4.7088e-05 edges=
    100 schur= 0
iteration= 2 chi2= 2195.012304 time= 1.2323e-05 cumTime= 5.9411e-05 edges=
    100 schur= 0
iteration= 3 chi2= 174.853126 time= 1.3302e-05 cumTime= 7.2713e-05 edges=
    100 schur= 0
iteration= 4 chi2= 102.779695 time= 1.2424e-05 cumTime= 8.5137e-05 edges=
    100 schur= 0
iteration= 5 chi2= 101.937194 time= 1.2523e-05 cumTime= 9.766e-05 edges=
    100 schur= 0
iteration= 6 chi2= 101.937020 time= 1.2268e-05 cumTime= 0.000109928 edges=
    100 schur= 0
iteration= 7 chi2= 101.937020 time= 1.2612e-05 cumTime= 0.00012254 edges=
    100 schur= 0
iteration= 8 chi2= 101.937020 time= 1.2159e-05 cumTime= 0.000134699 edges=
    100 schur= 0
iteration= 9 chi2= 101.937020 time= 1.2688e-05 cumTime= 0.000147387 edges=
    100 schur= 0
solve time cost = 0.000919301 seconds.
estimated model: 0.890912 2.1719 0.943629
```

We use the Gauss-Newton method for gradient descent, and after 9 iterations, the optimization result is obtained, which is almost the same as the Ceres and handwritten Gauss-Newton method. From the running speed perspective, our experimental conclusion is that handwriting is faster than *g2o*, and *g2o* is faster than Ceres. This is a generally intuitive experience that versatility and efficiency are often contradictory. However, in this experiment, Ceres uses automatic derivation, and the solver configuration is not completely consistent with Gauss-Newton, so it seems slower.

5.4 Summary

This section introduces a nonlinear optimization problem often encountered in SLAM: the least-squares problem consisting of the sum of squares of many error terms. We introduced its definition and solution and discussed two main gradient descent methods: the Gauss-Newton method and the Levenberg-Marquardt method. In the practice part, two optimization libraries of the handwritten Gauss-Newton method, Ceres and *g2o*, were used to solve the same curve-fitting problem and found that they gave similar results.

Since the bundle adjustment has not been discussed in detail, we have chosen a simple but representative example of curve fitting in the practice part to demonstrate the general nonlinear least-squares solution method. In particular, if you use *g2o* to fit a curve, you must first convert the problem to graph optimization and define new vertices and edges. This approach is somewhat roundabout—the main purpose of *g2o* is not here. In contrast, Ceres' usage is much more natural because it is designed to be a general optimization library. However, the SLAM problem is about solving an optimization problem with many camera poses and many spatial points. In particular, when the camera pose is represented by Lie algebra, calculating the derivative of the camera pose in the error term will be worthy of detailed discussion. We will find in the follow-up content that *g2o* provides a large number of ready-to-use vertices and edges, which is very convenient for the camera pose estimation problem. In Ceres, we have to implement each cost function by ourselves, which has some inconveniences.

In the two programs of the practical part, we did not calculate the derivative of the curve model with respect to the three parameters but used the numerical derivative of the optimization library, which made the theory and code simpler. The Ceres library provides automatic derivation based on template elements and numerical derivation at runtime, while *g2o* only provides a numerical derivation method. However, for most problems, if you can derive the analytical form of the Jacobian matrix and tell the optimization library, you can avoid many problems in numerical derivation.

Finally, I hope readers can adapt to Ceres and *g2o*'s extensive use of template programming. It may seem scary at first (especially the parenthesis operator for Ceres to set the residual block and the code for the *g2o* initialization part). But once you are familiar with it, you will feel that this method is natural and easy to extend. We will continue to discuss issues such as sparsity, kernel functions, and pose graphs in the SLAM backend lecture.

Exercises

1. Prove that in the linear equation $\mathbf{A}\mathbf{x} = \mathbf{b}$, when the coefficient matrix $\mathbf{A}$ is over-determined, the least-square solution is $\mathbf{x} = (\mathbf{A}^T\mathbf{A})^{-1}\mathbf{A}^T\mathbf{b}$.
2. Investigate the advantages and disadvantages of the steepest descent method, Newton method, Gauss-Newton method, and Levenberg-Marquardt method. In

addition to the Ceres library and *g2o* library we cited, are there any commonly used optimization libraries? (You may find some libraries on MATLAB.)

3. Why does the coefficient matrix of the Gauss-Newton method's incremental equation may not be positive definite? What is the geometric meaning of indefinite? Why is the solution unstable in this case?
4. What is DogLeg? What are the similarities and differences between it and the Gauss-Newton method and the Levenberg-Marquardt method? Please search for relevant materials.[10]
5. Read Ceres' tutorials (http://ceres-solver.org/tutorial.html) to better understand its usage.
6. Read the documentation that comes with *g2o*, can you understand it? If you still can't fully understand it, please come back after lectures 8 and 9.
7.* Please change the curve model in the curve fitting experiment, and use Ceres and *g2o* to optimize it. For example, write an example with more parameters and more complex models.

[10] http://www.numerical.rl.ac.uk/people/nimg/course/lectures/raphael/lectures/lec7slides.pdf.

Part II
SLAM Technologies

Chapter 6
Visual Odometry: Part I

Goal of Study

1. Study how to extract feature points from images and match feature points in multiple images.
2. Learn the principle of epipolar geometry and use epipolar constraints to recover the camera's 3D motion between two images.
3. Study how to solve the PNP problem and use the known correspondence between the 3D structure and the 2D image to solve the camera's 3D motion.
4. Study the ICP algorithm and use the point clouds matching to estimate 3D motion.
5. Study how to obtain the 3D structure of corresponding points on the 2D image through triangulation.

In the previous chapters, we introduced the details of motion and observation equations and explained how nonlinear optimization is used to solve those equations. From this chapter, we finish introducing the fundamental knowledge and moving on to the next topic. Starting from this chapter, we will investigate visual odometry, backend optimization, loop detection, and map reconstruction. This chapter and the next chapter mainly focus on two commonly used visual odometry methods: the feature and direct methods. This chapter will introduce what feature points are, how to extract and match them, and how to estimate camera motion based on matched feature points.

6.1 Feature Method

In Chap. 1, we said that a SLAM system can be divided into frontend and backend, where the frontend is also called visual odometry (VO). VO estimates the rough camera movement based on the consecutive images' information and provides a good initial value for the backend. VO algorithms are mainly divided into two categories: feature method and direct method. The frontend based on the features has been

X. Gao and T. Zhang, *Introduction to Visual SLAM*,
https://doi.org/10.1007/978-981-16-4939-4_6

considered the mainstream of VO for a long time (even until now). It performs well thanks to its stability and insensitivity to lighting and dynamic objects. The feature method is relatively mature at present. This chapter will start with the feature method, learn how to extract and match image feature points, and then estimate the camera motion and scene structure between two frames to realize visual odometry between two frames. This type of algorithm is sometimes called two-view geometry.

The core problem of VO is how to estimate camera motion based on adjacent images. However, the image itself is a numerical matrix encoding brightness and color. The numbers in the matrix are abstract, and it is very difficult to compute the motion directly from the pixel level. Therefore, it is more convenient to do this: First, select some representative features from the image. These points will remain the same after a small change in the camera's angle of view. So we are able to find the same points in each image. Then, based on these points, we can investigate the problem of camera pose estimation and the 3D positions of these points. In the classic SLAM model, we call these points the landmarks. In visual SLAM, they are referred to as image features.

On Wikipedia, image features are defined as a set of information related to computing tasks. The computing tasks depend on the specific application [37]. Briefly speaking, the feature is another digital expression of image information. A good set of features is crucial to the final performance on the specified task, so researchers have comprehensive work on the features for many years. Digital images are stored in a computer as a gray value matrix, so at the simplest, a single image pixel can also be considered a feature. However, in visual odometry, we hope that feature points remain stable after the camera moves, and the gray value is severely affected by illumination, deformation, and object material. It varies significantly between different images and is not stable enough. Ideally, when the scene and camera angle of view changes slightly, the algorithm can also determine from the images which places refer to the same point. Therefore, the gray value alone is not feasible. We need to extract better features from the image.

We can say that the feature points are some special places in the image. Taking Fig. 6.1 as an example, we can see the corners, edges, and blocks as representative places in the image. It is easy for us to correctly point out that the same corner point appears in two images, whereas pointing out the same edge is slightly more difficult because the image patches are similar when moving along the edges. It is even harder for the blocks. We found that the corners and edges in the image are more *special* than pixels, and they are more distinguished between different images. Therefore, an intuitive way to extract features is to identify corners between different images and determine their correspondence. In this approach, the corners are the so-called features. There are many corner extraction algorithms, such as Harris corner [38], FAST corner [39], GFTT corner [40], etc. Most of them were proposed before 2000.

However, in the majority of applications, a single corner still cannot meet our needs. For example, a place that appears to be a corner from a long distance may not be viewed as a corner when the camera steps in. When the camera rotates, the appearance of the corner points will change, and it is not easy for us to recognize that they are the same corner point. For this reason, researchers in the field of computer

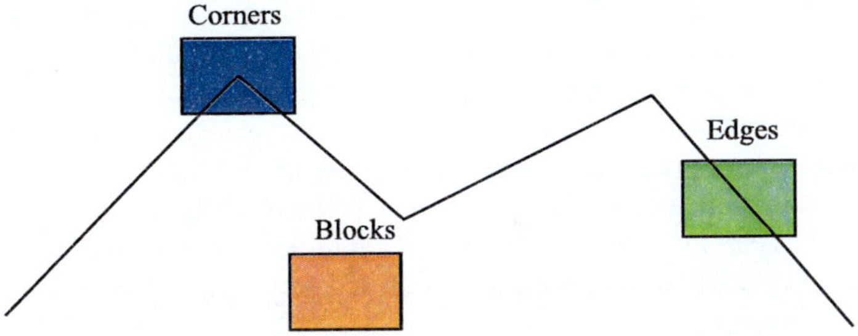

Fig. 6.1 Corners, edges and blocks can be used as image features

vision have designed many more stable local image features during years of research, such as the SIFT [41], SURF [42], ORB [43], etc. Compared with simple corner points, these handcrafted features should have the following properties:

1. **Repeatability**: The same feature can be found in different images.
2. **Distinctiveness**: Different features have different expressions.
3. **Efficiency**: In the same image, the number of feature points should be far smaller than the number of pixels.
4. **Locality**: The feature is only related to a small image area.

A feature point is composed of two parts: *key point* and *descriptor*. For example, when we say "calculate SIFT feature points in an image", we mean "extract SIFT key points and calculate the SIFT descriptors". The key point refers to the 2D position of the feature point. Some types of key points also hold other information, such as the orientation and size. According to some handcrafted rules, the descriptor is usually a vector, describing the information of the pixels around the key point. The descriptor should be designed according to the principle that features with similar appearance should have similar descriptors. Therefore, as long as the two features' descriptors are close in vector space, they can be considered the same feature.

Historically, researchers put forward many image features. Some of them are very accurate and robust. They still have similar expressions under camera movement and lighting changes, and consequentially they might require a large amount of calculation. Among them, SIFT (Scale-Invariant Feature Transform) is one of the most classic. To fully consider the changes in illumination, scale, and rotation during the image transformation, the SIFT comes with a considerable amount of calculation. The extraction and matching of image features is only a part compared to the entire SLAM process. Until now (2016), CPUs equipped on PCs cannot achieve real-time to calculate the SIFT features for localization and mapping.[1] So, we rarely use this luxury image feature in SLAM.

Some other features exchange accuracy and robustness for the calculation speed increase. For example, the FAST keypoint is a key point that is extremely fast to calculate (note the expression of keypoint here, which means that it has no descriptor),

[1] Real-time means the speed of 30 Hz.

while the ORB (Oriented FAST and Rotated BRIEF) feature is currently widely used for real-time image feature extraction. It solves the problem that the FAST detector [39] does not have descriptors and uses the extremely fast binary descriptor BRIEF [44] to make the whole image feature extraction process accelerate greatly. According to the author's experiment in the paper, extracting about 1000 feature points in the same image takes about 15.3ms for ORB, 217.3ms for SURF, and 5228.7ms for SIFT. It can be seen that ORB made a significant boost in speed while maintaining the features of rotation and scale invariance. It is a good choice for SLAM with high real-time requirements.

Most feature extractions have good parallelism and can be accelerated by GPU and other devices. SIFT accelerated by GPU can meet real-time requirements. However, the inclusion of a GPU will increase the cost of the entire SLAM system.[2] Whether the resulting performance improvement is sufficient to offset the computational cost requires careful consideration by the system designer.

Obviously, there are a large number of feature points in the field of computer vision, and we cannot introduce them one by one in the book. In the current SLAM scheme, ORB is a fair trade-off between quality and performance. Therefore, we take ORB as an example to introduce the entire process of extracting features. If readers are interested in feature extraction and matching algorithms, we recommend reading related books in this area [45].

6.1.1 ORB Feature

ORB features are also composed of two parts: *ORB key points* and *ORB descriptor*. Its key point is called "oriented FAST", which is an improved version of the FAST. We will introduce what the FAST corner below is. Its descriptor is called BRIEF (Binary Robust Independent Elementary Feature). Therefore, the extraction of ORB features is divided into the following two steps:

1. FAST corner point extraction: find the corner point in the image. Compared with the original FAST, the main direction of the feature points is calculated in ORB, making the subsequent BRIEF descriptor rotation-invariant.
2. BRIEF descriptor: describe the surrounding image area where the feature points were extracted in the previous step. ORB has made some improvements to BRIEF, mainly referring to utilizing the previously calculated direction.

Next, we will introduce FAST and BRIEF, respectively.

[2] But now we have many cheap embedded GPU chips, so it may not be an issue anymore.

6.1.1.1 The FAST Key Point

FAST is a kind of corner point, which mainly detects the local grayscale changes and is known for its fast speed. Its main idea is: if a pixel is very different from the neighboring pixels (too bright or too dark), it is more likely to be a corner point. Compared with other corner detection algorithms, FAST only needs to compare the pixels' brightness. Its entire procedure is as follows (see Fig. 6.2):

1. Select pixel p in the image, assuming its brightness as I_p
2. Set a threshold T (for example, 20% of I_p).
3. Take the pixel p as the center, and select the 16 pixels on a circle with a radius of 3.
4. If there are consecutive N points on the selected circle whose brightness is greater than $I_p + T$ or less than $I_p - T$, then the central pixel p can be considered a feature point. N usually takes 12, which is FAST-12. Other commonly used N values are 9 and 11, called FAST-9 and FAST-11, respectively).
5. Iterate through the above four steps on each pixel.

In the FAST-12 algorithm, to speed up, we can check the brightness of the 1, 5, 9, and 13 pixels on the circle to quickly exclude many pixels that are not corner points. Only when three of these four pixels are all greater than $I_p + T$ or less than $I_p - T$, the current pixel may potentially be a corner point. Otherwise, it should be directly excluded. Such a pre-processing operation greatly accelerates FAST corner detection. Also, the original FAST corners are often clustered, meaning a lot of FAST corners may present in the same area. Therefore, after the initial detection, non-maximal suppression is required. Only corner points with the maximum response in a specific location will be retained to avoid the corners concentrating.

The calculation of FAST feature points only compares the brightness difference between pixels, thus the speed is very fast, but it suffers from lousy repeatability and uneven distribution. Moreover, FAST corner points do not include direction

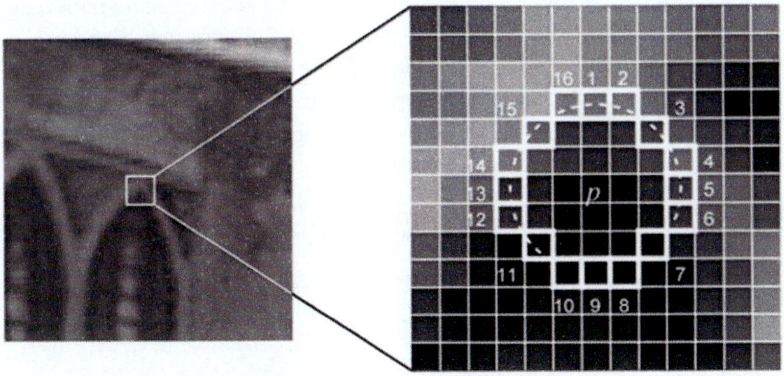

Fig. 6.2 FAST key points [39]. This figure is from OpenCV's document

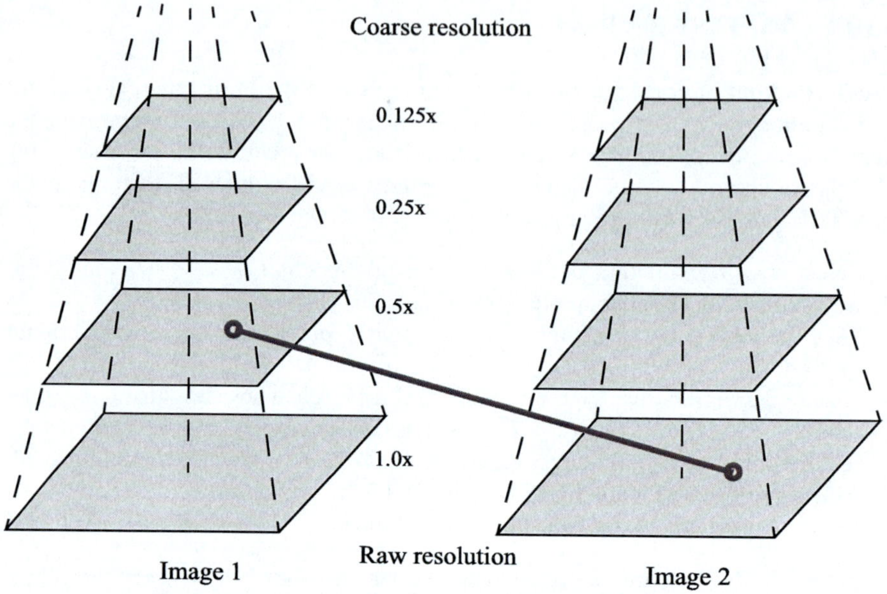

Fig. 6.3 Use pyramids to match images at different resolutions

information. Because it fixed the radius of the circle as 3, there is also a scaling problem: a place that looks like a corner from a distance may not be a corner when it comes close. To solve those, ORB adds the description of scale and rotation. The scale invariance is achieved by the image pyramid[3] and detect corner points on each layer. The rotation of features is realized by the intensity centroid method.

An image pyramid is a common approach in computer vision. For a schema, see Fig. 6.3. The bottom of the pyramid is the original image. For each layer up, the image is scaled with a fixed ratio so that we have images of different resolutions. The smaller image can be seen as a scene viewed from a distance. In the feature matching algorithm, we can match images on different layers to achieve scale invariance. For example, if the camera moves backward, we should find a match in the upper layer of the previous image pyramid and the lower layer of the next image.

In terms of rotation, we calculate the gray centroid of the image near the feature point. The so-called centroid refers to the gray value of the image block as the center of weight. The specific steps are as follows [46]:

1. In a small image block B, define the moment of the image block as:

$$m_{pq} = \sum_{x,y \in B} x^p y^q I(x, y), \quad p, q = \{0, 1\}.$$

[3] Pyramid refers to the downsampling of images at different levels to obtain images with different resolutions.

2. Calculate the centroid of the image block by the moment:

$$C = \left(\frac{m_{10}}{m_{00}}, \frac{m_{01}}{m_{00}}\right).$$

3. Connect the geometric center O and the centroid C of the image block to get a direction vector $\overrightarrow{OC}$, so the direction of the feature point can be defined as:

$$\theta = \arctan(m_{01}/m_{10}).$$

FAST corner points have a description of scale and rotation, which significantly improves the robustness of their representation between different images. This improved FAST is called oriented FAST in ORB.

6.1.1.2 BRIEF Descriptor

After extracting the Oriented FAST key points, we calculate the descriptor for each point. ORB uses an improved BRIEF feature description. Let's first introduce what BRIEF is.

BRIEF is a binary descriptor. Its description vector consists of many zeros and ones, which encode the size relationship between two random pixels near the key point (such as p and q): If p is greater than q, then take 1, otherwise take 0. If we take 128 such p, q pairs, we will finally get a 128-dimensional vector [44] consisting of 0s and 1s. The BRIEF implements the comparison of randomly selected points, which is very fast. Since it expresses in binary, it is also very convenient to store and suitable for real-time image matching. The original BRIEF descriptor does not have rotation invariance, so it is easy to get lost when the image is rotated. The ORB calculates the direction of the key points in the FAST feature point extraction stage. The direction information can be used to calculate the Steer BRIEF feature after the rotation so that the ORB descriptor has better rotation invariance.

Due to the consideration of rotation and scaling, the ORB performs well under translation, rotation, and scaling. Meanwhile, the combination of FAST and BRIEF is very efficient, which makes ORB features very popular in real-time SLAM. In Fig. 6.4, we show the result of extracting ORB from an image. In the following, we will move on to feature matching between different images.

6.1.2 Feature Matching

Feature matching is a critical step in visual SLAM (Fig. 6.5). Broadly speaking, feature matching solves the data association problem in SLAM, that is, to determine the view correspondence between the landmarks (feature points) and the landmarks

Fig. 6.4 ORB feature points detected by OpenCV

Fig. 6.5 Feature matching between two images

(feature points) seen before. By accurately matching the descriptors between images or between images and maps, we can reduce a lot of work for subsequent pose estimation and optimization operations. However, due to the locality of image features, mismatches are common and have not been effectively resolved for a long time. It has become a significant bottleneck to improve the performance of visual SLAM. It is somehow due to repeated textures in the scene, making the feature descriptions very similar. Under this circumstance, it is hard to resolve the mismatch by using local features only.

However, let's first assume the matching is correct and then go back to discuss the mismatch problem. Consider images at two moments. Assume features x_t^m, $m = 1, 2, \ldots, M$ are extracted in the image I_t, and x_{t+1}^n, $n = 1, 2, \ldots, N$ in I_{t+1}. How do we find the correspondence between these two sets? The simplest feature matching method is the brute-force matcher, which measures the distance between each pair of the features x_t^m and all x_{t+1}^n descriptors. Then sort the matching distance, and take the closest one as the matching point. The descriptor distance indicates the degree of similarity of two features. In practice, different distance metric norms can be used. For descriptors of floating-point type, using Euclidean distance to measure is a good choice. For binary descriptors (such as BRIEF), we often use Hamming distance as a metric. The Hamming distance between two binary vectors refers to the number of different digits.

When the number of feature points is large, the brute force matching's computational complexity will become too expensive, especially when you want to match a frame to a map. This does not meet our real-time requirements in SLAM. The Fast Approximate Nearest Neighbor (FLANN) algorithm is more suitable in this case. Since these matching algorithms' theory is mature and the implementation has already been integrated into OpenCV, the technical details will not be described here. Interested readers can refer to the literature [47]. In engineering, we can also limit the search range of the brute-force method to achieve real-time performance.

6.2 Practice: Feature Extraction and Matching

OpenCV has integrated most of the image features, and we can easily use them by function calls. Let's demonstrate two examples here. In the first one, we show the use of OpenCV for feature matching of ORB; in the second, we explain how to write an ORB feature from scratch based on the principles. Through the practice, readers will have a deeper understanding of the ORB calculation process. Then other features can be done analogously.

Fig. 6.6 Two images used in the practice

6.2.1 ORB Features in OpenCV

First, we call OpenCV to extract and match ORB. We prepared two images for this practice, 1.png and 2.png under "slambook2/ch7/", as shown in Fig. 6.6. They are two images from the public dataset [23]. We can see a slight movement of the camera. The procedure in this section demonstrates how to extract ORB features and perform matching. In the next section, we will show how to use matching results to estimate camera motion.

The following program demonstrates how to use ORB:

Listing 6.1 slambook2/ch7/orb_cv.cpp

```
#include <iostream>
#include <opencv2/core/core.hpp>
#include <opencv2/features2d/features2d.hpp>
#include <opencv2/highgui/highgui.hpp>
#include <chrono>

using namespace std;
using namespace cv;

int main(int argc, char **argv) {
  if (argc != 3) {
    cout << "usage: feature_extraction img1 img2" << endl;
    return 1;
  }
  //-- read images
  Mat img_1 = imread(argv[1], CV_LOAD_IMAGE_COLOR);
  Mat img_2 = imread(argv[2], CV_LOAD_IMAGE_COLOR);
  assert(img_1.data != nullptr && img_2.data != nullptr);

  //-- initialization
  std::vector<KeyPoint> keypoints_1, keypoints_2;
  Mat descriptors_1, descriptors_2;
  Ptr<FeatureDetector> detector = ORB::create();
  Ptr<DescriptorExtractor> descriptor = ORB::create();
  Ptr<DescriptorMatcher> matcher = DescriptorMatcher::create("BruteForce-
      Hamming");

  //-- detect Oriented FAST
  chrono::steady_clock::time_point t1 = chrono::steady_clock::now();
  detector->detect(img_1, keypoints_1);
  detector->detect(img_2, keypoints_2);
```

```cpp

  //-- compute BRIEF descriptor
  descriptor->compute(img_1, keypoints_1, descriptors_1);
  descriptor->compute(img_2, keypoints_2, descriptors_2);
  chrono::steady_clock::time_point t2 = chrono::steady_clock::now();
  chrono::duration<double> time_used = chrono::duration_cast<chrono::duration<double>>(t2 - t1);
  cout << "extract ORB cost = " << time_used.count() << " seconds. " << endl;

  Mat outimg1;
  drawKeypoints(img_1, keypoints_1, outimg1, Scalar::all(-1), DrawMatchesFlags::DEFAULT);
  imshow("ORB features", outimg1);

  //-- use Hamming distance to match the features
  vector<DMatch> matches;
  t1 = chrono::steady_clock::now();
  matcher->match(descriptors_1, descriptors_2, matches);
  t2 = chrono::steady_clock::now();
  time_used = chrono::duration_cast<chrono::duration<double>>(t2 - t1);
  cout << "match ORB cost = " << time_used.count() << " seconds. " << endl;

  //-- sort and remove the outliers
  // min and max distance
  auto min_max = minmax_element(matches.begin(), matches.end(),
    [](const DMatch &m1, const DMatch &m2) { return m1.distance < m2.distance; });
  double min_dist = min_max.first->distance;
  double max_dist = min_max.second->distance;

  printf("-- Max dist : %f \n", max_dist);
  printf("-- Min dist : %f \n", min_dist);

  // remove the bad matching
  std::vector<DMatch> good_matches;
  for (int i = 0; i < descriptors_1.rows; i++) {
    if (matches[i].distance <= max(2 * min_dist, 30.0)) {
      good_matches.push_back(matches[i]);
    }
  }

  //-- draw the results
  Mat img_match;
  Mat img_goodmatch;
  drawMatches(img_1, keypoints_1, img_2, keypoints_2, matches, img_match);
  drawMatches(img_1, keypoints_1, img_2, keypoints_2, good_matches, img_goodmatch);
  imshow("all matches", img_match);
  imshow("good matches", img_goodmatch);
  waitKey(0);

  return 0;
}
```

Run this program (you need to enter the paths of two images manually). The screen output should be like this:

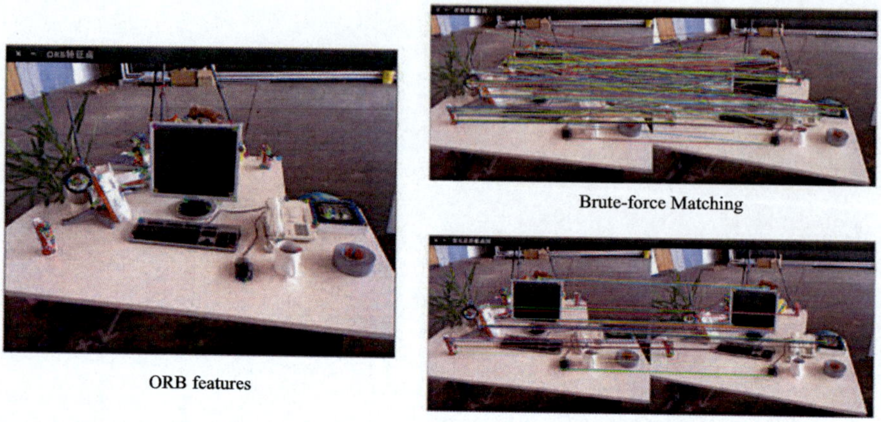

ORB features

Brute-force Matching

Good matches after sorting

Fig. 6.7 Feature extraction and matching results

Listing 6.2 Terminal input:

```
% build/orb_cv 1.png 2.png
extract ORB cost = 0.0229183 seconds.
match ORB cost = 0.000751868 seconds.
-- Max dist : 95.000000
-- Min dist : 4.000000
```

Figure 6.7 shows the results of the example. We see a large number of false matches before the filtering. After removing the bad matches, most of the remaining matches are correct. Here, we followed an empirical rule in engineering that good matches are selected from where the Hamming distance is less than twice the minimum distance. It may not have a theoretical explanation but just an engineering trick. However, although we can select out the correct matches in the example image, we still cannot guarantee that the remaining matches are correct. Therefore, during the motion estimation step, it is necessary to consider those mismatches. ORB extraction took 22.9 milliseconds (two images) on my machine, and matching took 0.75 milliseconds. It can be seen that most of the calculation is spent on feature extraction.

6.2.2 ORB Features from Scratch

Below we show the method of coding ORB features from scratch. Since the code is a bit long, only a snippet is shown here. Readers are recommended to obtain the rest from the GitHub codebase.

Listing 6.3 slambook2/ch7/orb_self.cpp (part)

```
typedef vector<uint32_t> DescType;
// ... omit some image loading and testing code
// compute the descriptor
```

```
void ComputeORB(const cv::Mat &img, vector<cv::key point> &key points, vector<DescType> &descriptors) {
      const int half_patch_size = 8;
      const int half_boundary = 16;
      int bad_points = 0;
      for (auto &kp: key points) {
            if (kp.pt.x < half_boundary || kp.pt.y < half_boundary || kp.pt.x >= img.cols - half_boundary || kp.pt.y >= img.rows - half_boundary) {
                  // outside
                  bad_points++;
                  descriptors.push_back({});
                  continue;
            }

            float m01 = 0, m10 = 0;
            for (int dx = -half_patch_size; dx < half_patch_size; ++dx) {
                  for (int dy = -half_patch_size; dy < half_patch_size; ++dy) {
                        uchar pixel = img.at<uchar>(kp.pt.y + dy, kp.pt.x + dx);
                        m01 += dx * pixel;
                        m10 += dy * pixel;
                  }
            }

            // angle should be arc tan(m01/m10);
            float m_sqrt = sqrt(m01 * m01 + m10 * m10);
            float sin_theta = m01 / m_sqrt;
            float cos_theta = m10 / m_sqrt;

            // compute the angle of this point
            DescType desc(8, 0);
            for (int i = 0; i < 8; i++) {
                  uint32_t d = 0;
                  for (int k = 0; k < 32; k++) {
                        int idx_pq = i * 8 + k;
                        cv::Point2f p(ORB_pattern[idx_pq * 4], ORB_pattern[idx_pq * 4 + 1]);
                        cv::Point2f q(ORB_pattern[idx_pq * 4 + 2], ORB_pattern[idx_pq * 4 + 3]);

                        // rotate with theta
                        cv::Point2f pp = cv::Point2f(cos_theta * p.x - sin_theta * p.y, sin_theta * p.x + cos_theta * p.y) + kp.pt;
                        cv::Point2f qq = cv::Point2f(cos_theta * q.x - sin_theta * q.y, sin_theta * q.x + cos_theta * q.y) + kp.pt;

                        // compare pp and qq
                        if (img.at<uchar>(pp.y, pp.x) < img.at<uchar>(qq.y, qq.x)) {
                              d |= 1 << k;
                        }
                  }
                  desc[i] = d;
            }
            descriptors.push_back(desc);
      }

      cout << "bad/total: " << bad_points << "/" << key points.size() << endl;
}

// brute-force matching
void BfMatch(
      const vector<DescType> &desc1, const vector<DescType> &desc2, vector<cv::DMatch> &matches) {
      const int d_max = 40;

      for (size_t i1 = 0; i1 < desc1.size(); ++i1) {
```

```
            if (desc1[i1].empty()) continue;
            cv::DMatch m{i1, 0, 256};
            for (size_t i2 = 0; i2 < desc2.size(); ++i2) {
                if (desc2[i2].empty()) continue;
                int distance = 0;
                for (int k = 0; k < 8; k++) {
                    distance += _mm_popcnt_u32(desc1[i1][k] ^desc2[i2][k]);
                }
                if (distance < d_max && distance < m.distance) {
                    m.distance = distance;
                    m.trainIdx = i2;
            }
        }
        if (m.distance < d_max) {
            matches.push_back(m);
        }
    }
}
```

We only show the ORB calculation and matching code. In the calculation, we use 256-bit binary description, which corresponds to 8 32-bit unsigned int data expressed as DescType with typedef. Then, we calculate the FAST feature point's angle according to the principle introduced above and then use the angle to calculate the descriptor. To accelerate, some complicated calculations, such as arctan, sin, and cos, are worked around by the principle of trigonometric functions. In the BfMatch function, we also use the _mm_popcnt_u32 function in the SSE instruction set to calculate the number of 1s in an unsigned int, which is used to achieve the effect of calculating the Hamming distance. The result of this program is as follows, and the matching result is shown in Fig. 6.8:

Listing 6.4 Terminal output:

```
bad/total: 43/638
bad/total: 8/595
extract ORB cost = 0.00390721 seconds.
match ORB cost = 0.000862984 seconds.
matches: 51
```

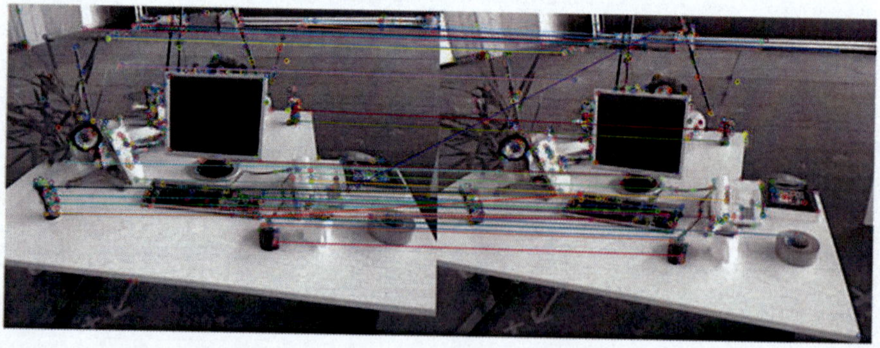

Fig. 6.8 Matching Result

In this program, ORB extraction takes 3.9 milliseconds, and the matching takes only 0.86 milliseconds. Through some simple modifications in implementation, we accelerated the extraction of ORB by 5.8 times. Note that compiling this program requires your CPU to support the SSE instruction set, which should be already supported on most modern PC CPUs. If we can further parallelize the feature extraction, the algorithm can be further accelerated.

6.2.3 Calculate the Camera Motion

Now, we have key points matching. In the next step, we need to estimate the camera's motion based on the matching. It is totally different for different camera settings (or the information available for the calculation):

1. When the camera is monocular, we only know the 2D pixel coordinates, so the problem is to estimate the motion according to two sets of 2D points. This problem is solved by *epipolar geometry*.
2. When the camera is binocular, RGB-D, or the distance is obtained by some method, then the problem is to estimate the motion according to two sets of 3D points. This problem is usually solved by ICP.
3. If one set is 3D and one set is 2D, we get some 3D points and their projection positions on the camera, and we can also estimate the camera's pose. This problem is solved by *PnP*.

The following sections will introduce camera motion estimation in these three situations. We will start from the 2D-2D case with the least information and see how it is dealt with and troublesome problems.

6.3 2D–2D: Epipolar Geometry

6.3.1 Epipolar Constraints

Suppose we have a pair of matched feature points from two images, as shown in Fig. 6.9. If there are several pairs of such matching points, the camera motion between the two frames can be recovered through the correspondence between these two-dimensional image points. How many pairs do we need? We will see later. Let's first take a look at the geometric relationship between the matching points in the two images.

Take Fig. 6.9 as an example. Our goal is to find the motion between two frames I_1, I_2. Define the motion from the first frame to the second frame as $\mathbf{R}, \mathbf{t}$, and the centers of the two cameras as O_1, O_2. Now, consider that there is a feature point p_1 in I_1, which corresponds to the feature point p_2 in I_2, obtained through feature

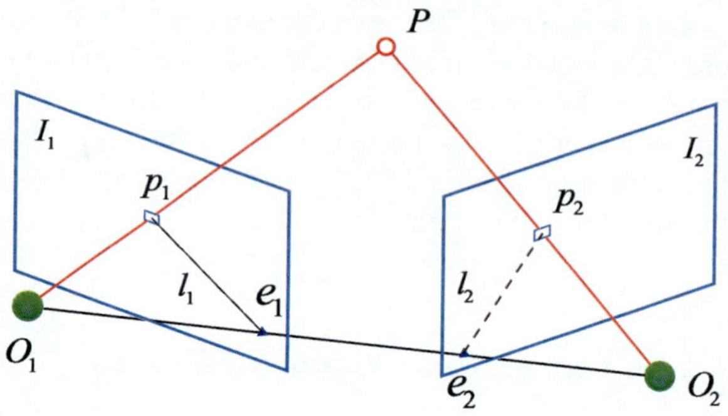

Fig. 6.9 The epipolar constraints

matching. If the matching is correct, it means that they are indeed the projection of the same point. Now we need some terms to describe the geometric relationship between them. First, the line $\overrightarrow{O_1 p_1}$ and the line $\overrightarrow{O_2 p_2}$ will intersect at the point P in the 3D space. The three points $O_1, O_2, and P$ can determine a plane, and it is called the epipolar plane. The intersection of the line of $O_1 O_2$ and the image plane I_1, I_2 is e_1, e_2, respectively. The e_1, e_2 is called epipoles, and $O_1 O_2$ is called the baseline. We call the intersecting line l_1, l_2 between the polar plane and the two image planes I_1, I_2 as the epipolar line.

From the first frame, the ray $\overrightarrow{O_1 p_1}$ represents the possible spatial locations where a pixel may appear since all points on the ray will be projected to the same pixel. Meanwhile, suppose we don't know the location of P. When we look at the second image, the connection $\overrightarrow{e_2 p_2}$ (i.e., the epipolar line in the second image) is the possible projected positions of the point P, as well as the projection of the ray $\overrightarrow{O_1 p_1}$ in the second image. Now, since we have determined the pixel location of p_2 through feature matching, we can infer the spatial location of P and the camera movement, as long as the feature matching is correct. If there is no feature matching, we can't determine where the p_2 is on the epipolar line. At that time, we must search on the epipolar line l_2 to get the correct match, which will be discussed in lecture 11.

Now, let's look at the geometric relationship algebraically. Define the spatial position of P in the first frame to be:

$$\mathbf{P} = [X, Y, Z]^T.$$

According to the pinhole camera model introduced in lecture 4, we know that the pixel positions of the two pixels $\mathbf{p}_1, \mathbf{p}_2$ are:

$$s_1 \mathbf{p}_1 = \mathbf{KP}, \quad s_2 \mathbf{p}_2 = \mathbf{K} \left(\mathbf{RP} + \mathbf{t} \right), \tag{6.1}$$

where $\mathbf{K}$ is the camera intrinsic matrix, and $\mathbf{R}, \mathbf{t}$ are the camera motions between two frames. Specifically, they are $\mathbf{R}_{21}$ and $\mathbf{t}_{21}$, i.e. transformation from the first frame to the second. We can also write them in Lie algebra form.

Sometimes, we use homogeneous coordinates to represent pixels. When using homogeneous coordinates, a vector will be equal to itself multiplied by any non-zero constant. This is usually used to express a projection relationship. For example, $s_1\mathbf{p}_1$ and $\mathbf{p}_1$ form a projection relationship, and they are equal in the sense of homogeneous coordinates. We call this equal *up to a scale*, denoted as:

$$s\mathbf{p} \simeq \mathbf{p}. \tag{6.2}$$

Then, the relationship between two projections can be written as:

$$\mathbf{p}_1 \simeq \mathbf{K}\mathbf{P}, \quad \mathbf{p}_2 \simeq \mathbf{K}(\mathbf{R}\mathbf{P} + \mathbf{t}). \tag{6.3}$$

Now, let

$$\mathbf{x}_1 = \mathbf{K}^{-1}\mathbf{p}_1, \quad \mathbf{x}_2 = \mathbf{K}^{-1}\mathbf{p}_2. \tag{6.4}$$

Here, $\mathbf{x}_1, \mathbf{x}_2$ are the coordinates on the normalized plane of two pixels. Substituting to the above equation, we get:

$$\mathbf{x}_2 \simeq \mathbf{R}\mathbf{x}_1 + \mathbf{t}. \tag{6.5}$$

Left multiply both sides by $\mathbf{t}^\wedge$. Recalling the definition of $^\wedge$, this is equivalent to the outer product of both sides with $\mathbf{t}$:

$$\mathbf{t}^\wedge\mathbf{x}_2 \simeq \mathbf{t}^\wedge\mathbf{R}\mathbf{x}_1. \tag{6.6}$$

Then, left multiply $\mathbf{x}_2^T$ on both sides:

$$\mathbf{x}_2^T\mathbf{t}^\wedge\mathbf{x}_2 \simeq \mathbf{x}_2^T\mathbf{t}^\wedge\mathbf{R}\mathbf{x}_1. \tag{6.7}$$

On the left side, $\mathbf{t}^\wedge\mathbf{x}_2$ is a vector orthogonal to both $\mathbf{t}$ and $\mathbf{x}_2$. So its inner product with $\mathbf{x}_2$ will get 0. Since the left side of the equation is strictly zero, it is also zero after multiplying by any non-zero constant, so we can turn $\simeq$ back to the usual equal sign. Therefore, we have a concise equation:

$$\mathbf{x}_2^T\mathbf{t}^\wedge\mathbf{R}\mathbf{x}_1 = 0. \tag{6.8}$$

Substituting the $\mathbf{p}_1, \mathbf{p}_2$ again, we have:

$$\mathbf{p}_2^T\mathbf{K}^{-T}\mathbf{t}^\wedge\mathbf{R}\mathbf{K}^{-1}\mathbf{p}_1 = 0. \tag{6.9}$$

Both equations are called **epipolar constraint**, which is famous for its conciseness. Geometrically, it means O_1, P, O_2 are coplanar. The epipolar constraint encodes both translation and rotation. We define two matrices: *Fundamental* matrix $\mathbf{F}$ and *essential* Matrix $\mathbf{E}$ from this equation:

$$\mathbf{E} = \mathbf{t}^\wedge\mathbf{R}, \quad \mathbf{F} = \mathbf{K}^{-T}\mathbf{E}\mathbf{K}^{-1}, \quad \mathbf{x}_2^T\mathbf{E}\mathbf{x}_1 = \mathbf{p}_2^T\mathbf{F}\mathbf{p}_1 = 0. \tag{6.10}$$

The epipolar constraint gives the spatial relationship of two matching points concisely. Therefore, the camera pose estimation problem can be summarized as the following two steps:

1. Find **E** or **F** based on the pixel positions of the matched points.
2. Find **R**, **t** based on **E** or **F**.

Since **E** and **F** only differ from the camera internal parameters, and the internal parameters are assumed to be known in SLAM problem,[4] so the simpler form **E** is often used in practice. Let's take **E** as an example to introduce how to solve the above two problems.

6.3.2 *Essential Matrix*

By definition, the essential matrix $\mathbf{E} = \mathbf{t}^\wedge \mathbf{R}$. It is a matrix of 3×3 with 9 unknown variables. So, can any matrix of the size 3×3 be an essential matrix? From the structure of **E**, there are the following points worth noting:

- The essential matrix is defined by the epipolar constraint. Since the epipolar constraint is the constraint of an *equal-to-zero* equation, after multiplying **E** by any non-zero constant, the constraint is still satisfied. We call this **E** 's equivalence under different scales.
- According to $\mathbf{E} = \mathbf{t}^\wedge \mathbf{R}$, it can be proved that [2], the singular value of the essential matrix **E** must be in the form of $[\sigma, \sigma, 0]^T$. This is called internal properties of essential matrix.
- On the other hand, since translation and rotation each have 3 degrees of freedom, $\mathbf{t}^\wedge \mathbf{R}$ has 6 degrees of freedom. But due to the equivalence of scales, **E** actually has 5 degrees of freedom.

The fact that **E** has 5 degrees of freedom indicates that we can use at least 5 pairs of points to solve **E**. However, the internal property of **E** is nonlinear, which could cause trouble in the estimation. Therefore, it is also possible to consider only its scale equivalence and use 8 pairs of matched points to estimate **E**. This is the classic *eight-point-algorithm* [48, 49]. The eight-point method only uses the linear properties of **E**, so it can be solved under the framework of linear algebra. Let's take a look at how the eight-point method works.

Consider a pair of matched points, their normalized coordinates are $\mathbf{x}_1 = [u_1, v_1, 1]^T$, $\mathbf{x}_2 = [u_2, v_2, 1]^T$. According to the polar constraints, we have:

$$(u_2, v_2, 1) \begin{pmatrix} e_1 & e_2 & e_3 \\ e_4 & e_5 & e_6 \\ e_7 & e_8 & e_9 \end{pmatrix} \begin{pmatrix} u_1 \\ v_1 \\ 1 \end{pmatrix} = 0. \tag{6.11}$$

---

[4] However, in SfM research, it may be unknown and need to be estimated.

We rewrite the matrix $\mathbf{E}$ in the vector form:

$$\mathbf{e} = [e_1, e_2, e_3, e_4, e_5, e_6, e_7, e_8, e_9]^T,$$

Then the epipolar constraint can be written in a linear form related to $\mathbf{e}$:

$$[u_2u_1, u_2v_1, u_2, v_2u_1, v_2v_1, v_2, u_1, v_1, 1] \cdot \mathbf{e} = 0. \tag{6.12}$$

Analogously, we stack all the points into one equation and obtain a linear equation system (where u^i, v^i represent the i-th feature point):

$$\begin{pmatrix} u_2^1u_1^1 & u_2^1v_1^1 & u_2^1 & v_2^1u_1^1 & v_2^1v_1^1 & v_2^1 & u_1^1 & v_1^1 & 1 \\ u_2^2u_1^2 & u_2^2v_1^2 & u_2^2 & v_2^2u_1^2 & v_2^2v_1^2 & v_2^2 & u_1^2 & v_1^2 & 1 \\ \vdots & \vdots & \vdots & \vdots & \vdots & \vdots & \vdots & \vdots & \vdots \\ u_2^8u_1^8 & u_2^8v_1^8 & u_2^8 & v_2^8u_1^8 & v_2^8v_1^8 & v_2^8 & u_1^8 & v_1^8 & 1 \end{pmatrix} \begin{pmatrix} e_1 \\ e_2 \\ e_3 \\ e_4 \\ e_5 \\ e_6 \\ e_7 \\ e_8 \\ e_9 \end{pmatrix} = 0. \tag{6.13}$$

These eight equations form a linear equation system. Its coefficient matrix is composed of 2D feature point positions, and its size is 8×9. $\mathbf{e}$ is located in the null space of this matrix. If the coefficient matrix is of full rank (i.e., 8), then its null space dimension is 1, meaning that $\mathbf{e}$ forms a line. This is consistent with the scale equivalence of $\mathbf{e}$. If the matrix composed of 8 pairs of matching points meets the condition of rank 8, then the elements of $\mathbf{E}$ can be solved uniquely by the above equation.

The next question is how to recover the movement of the camera $\mathbf{R}, \mathbf{t}$ according to the estimated essential matrix $\mathbf{E}$. This process is obtained by singular value decomposition (SVD). Let the SVD decomposition of $\mathbf{E}$ be:

$$\mathbf{E} = \mathbf{U}\boldsymbol{\Sigma}\mathbf{V}^T, \tag{6.14}$$

where $\mathbf{U}, \mathbf{V}$ are orthogonal matrices, and $\boldsymbol{\Sigma}$ is the singular value matrice. According to the internal properties of $\mathbf{E}$, we know that $\boldsymbol{\Sigma} = \mathrm{diag}(\sigma, \sigma, 0)$. In SVD decomposition, for any $\mathbf{E}$, there are two possible $\mathbf{t}, \mathbf{R}$:

$$\begin{aligned} \mathbf{t}_1^\wedge &= \mathbf{U}\mathbf{R}_Z(\tfrac{\pi}{2})\boldsymbol{\Sigma}\mathbf{U}^T, \quad \mathbf{R}_1 = \mathbf{U}\mathbf{R}_Z^T(\tfrac{\pi}{2})\mathbf{V}^T \\ \mathbf{t}_2^\wedge &= \mathbf{U}\mathbf{R}_Z(-\tfrac{\pi}{2})\boldsymbol{\Sigma}\mathbf{U}^T, \quad \mathbf{R}_2 = \mathbf{U}\mathbf{R}_Z^T(-\tfrac{\pi}{2})\mathbf{V}^T. \end{aligned} \tag{6.15}$$

Among them, $\mathbf{R}_Z\left(\frac{\pi}{2}\right)$ represents the rotation matrix obtained by rotating 90° along the Z axis. Since $-\mathbf{E}$ is equivalent to $\mathbf{E}$, taking the minus sign for any $\mathbf{t}$ will also get the same result. Therefore, when decomposing from $\mathbf{E}$ to $\mathbf{t}, \mathbf{R}$, there are a total of **four** possible solutions.

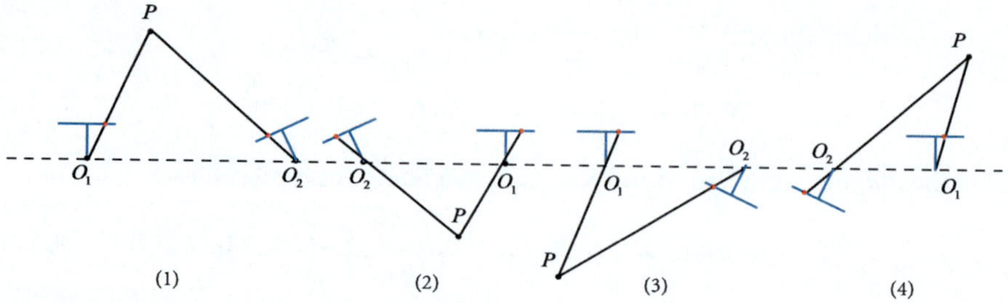

Fig. 6.10 We get four solutions when decomposing the essential matrix

Figure 6.10 shows the four solutions obtained by decomposing the essential matrix. We know the projection (red) of the space point on the camera (blue line) and want to solve the camera's motion. In the case of keeping the redpoint unchanged, four possible situations can be drawn. Fortunately, only in the first solution, P has positive depths in both cameras. Therefore, we can substitute any points into the four solutions and check the depth's sign to determine which solution is correct.

If we use the internal properties of $\mathbf{E}$, then it has only five degrees of freedom. So at least five pairs of matched points can be used to solve the camera motion [50, 51]. However, this approach is more complicated. Since there are usually dozens or even hundreds of matching points for engineering realization, it is often not helpful to reduce from 8 pairs to 5 pairs. To keep it simple, we only introduce the basic eight-point method here.

There is one remaining problem: $\mathbf{E}$ solved by linear equations may not satisfy the internal properties, i.e. its singular value is not necessarily in the form of $\sigma, \sigma, 0$. At this time, we will deliberately adjust the $\mathbf{\Sigma}$ matrix to look like the above. The usual procedure is to perform SVD decomposition on the $\mathbf{E}$ obtained by the eight-point method. Assume the singular value matrix is $\mathbf{\Sigma} = \mathrm{diag}(\sigma_1, \sigma_2, \sigma_3)$, and $\sigma_1 \geqslant \sigma_2 \geqslant \sigma_3$. We may take:

$$\mathbf{E} = \mathbf{U}\mathrm{diag}(\frac{\sigma_1 + \sigma_2}{2}, \frac{\sigma_1 + \sigma_2}{2}, 0)\mathbf{V}^T. \tag{6.16}$$

This is equivalent to projecting the calculated essential matrix onto the manifold where $\mathbf{E}$ is located. A simpler approach is to take the singular value matrix as $\mathrm{diag}(1, 1, 0)$, due to $\mathbf{E}$'s scale equivalence, so it is also reasonable.

6.3.3 Homography

In addition to the fundamental matrix and the essential matrix, there is another common matrix in two-view geometry: the homography matrix $\mathbf{H}$, which describes the mapping relationship between two planes. If the feature points in the scene all fall on the same plane (such as walls, ground, etc.), we can estimate the motion through the homography matrix. This situation is more common in top-view cameras carried by drones or sweepers.

The homography matrix usually describes the transformation between some points on a common plane between two images. Consider that there is a pair of matched feature points p_1 and p_2 in the images I_1 and I_2. These feature points fall on the plane P. Let this plane satisfy the equation of:

$$\mathbf{n}^T\mathbf{P} + d = 0. \tag{6.17}$$

Rearrange it:

$$-\frac{\mathbf{n}^T\mathbf{P}}{d} = 1. \tag{6.18}$$

Recalling (6.1), we have:

$$\begin{aligned}
\mathbf{p}_2 &\simeq \mathbf{K}(\mathbf{R}\mathbf{P} + \mathbf{t}) \\
&\simeq \mathbf{K}\left(\mathbf{R}\mathbf{P} + \mathbf{t} \cdot (-\frac{\mathbf{n}^T\mathbf{P}}{d})\right) \\
&\simeq \mathbf{K}\left(\mathbf{R} - \frac{\mathbf{t}\mathbf{n}^T}{d}\right)\mathbf{P} \\
&\simeq \underbrace{\mathbf{K}\left(\mathbf{R} - \frac{\mathbf{t}\mathbf{n}^T}{d}\right)\mathbf{K}^{-1}}_{\mathbf{H}}\mathbf{p}_1.
\end{aligned}$$

So, we get a direct description of the transformation between $\mathbf{p}_1$ and $\mathbf{p}_2$, denoting the middle part as $\mathbf{H}$, so:

$$\mathbf{p}_2 \simeq \mathbf{H}\mathbf{p}_1. \tag{6.19}$$

Its definition is related to the parameters of rotation, translation, and plane coefficients. Similar to the fundamental matrix $\mathbf{F}$, the homography matrix $\mathbf{H}$ is also a matrix of 3×3. Solving this is similar to that of $\mathbf{F}$. Calculate $\mathbf{H}$ based on the matching points, and then decompose it to find rotation and translation. Expand the above formula, get:

$$\begin{pmatrix} u_2 \\ v_2 \\ 1 \end{pmatrix} \simeq \begin{pmatrix} h_1 & h_2 & h_3 \\ h_4 & h_5 & h_6 \\ h_7 & h_8 & h_9 \end{pmatrix} \begin{pmatrix} u_1 \\ v_1 \\ 1 \end{pmatrix}. \tag{6.20}$$

Note that the equal sign here is still $\simeq$, not the ordinary equal sign, so the $\mathbf{H}$ matrix can also be multiplied by any non-zero constant. We can make $h_9 = 1$ in practice (when it takes a non-zero value). Then according to the third row, remove this non-zero factor. So we have:

$$u_2 = \frac{h_1 u_1 + h_2 v_1 + h_3}{h_7 u_1 + h_8 v_1 + h_9}$$
$$v_2 = \frac{h_4 u_1 + h_5 v_1 + h_6}{h_7 u_1 + h_8 v_1 + h_9}.$$

Rearrange it:

$$h_1 u_1 + h_2 v_1 + h_3 - h_7 u_1 u_2 - h_8 v_1 u_2 = u_2$$
$$h_4 u_1 + h_5 v_1 + h_6 - h_7 u_1 v_2 - h_8 v_1 v_2 = v_2.$$

A pair of matching points can construct two constraints, so the homography matrix with 8 degrees of freedom can be estimated by 4 pairs of matched features[5] (in the case of non-degenerate, which means these feature points cannot have three collinear points). We solve the following linear equations to compute the $\mathbf{H}$: (when $h_9 = 0$, the right side is zero):

$$\begin{pmatrix} u_1^1 & v_1^1 & 1 & 0 & 0 & 0 & -u_1^1 u_2^1 & -v_1^1 u_2^1 \\ 0 & 0 & 0 & u_1^1 & v_1^1 & 1 & -u_1^1 v_2^1 & -v_1^1 v_2^1 \\ u_1^2 & v_1^2 & 1 & 0 & 0 & 0 & -u_1^2 u_2^2 & -v_1^2 u_2^2 \\ 0 & 0 & 0 & u_1^2 & v_1^2 & 1 & -u_1^2 v_2^2 & -v_1^2 v_2^2 \\ u_1^3 & v_1^3 & 1 & 0 & 0 & 0 & -u_1^3 u_2^3 & -v_1^3 u_2^3 \\ 0 & 0 & 0 & u_1^3 & v_1^3 & 1 & -u_1^3 v_2^3 & -v_1^3 v_2^3 \\ u_1^4 & v_1^4 & 1 & 0 & 0 & 0 & -u_1^4 u_2^4 & -v_1^4 u_2^4 \\ 0 & 0 & 0 & u_1^4 & v_1^4 & 1 & -u_1^4 v_2^4 & -v_1^4 v_2^4 \end{pmatrix} \begin{pmatrix} h_1 \\ h_2 \\ h_3 \\ h_4 \\ h_5 \\ h_6 \\ h_7 \\ h_8 \end{pmatrix} = \begin{pmatrix} u_2^1 \\ v_2^1 \\ u_2^2 \\ v_2^2 \\ u_2^3 \\ v_2^3 \\ u_2^4 \\ v_2^4 \end{pmatrix}. \quad (6.21)$$

This approach regards the $\mathbf{H}$ matrix as a vector and estimates the $\mathbf{H}$ by solving the linear equations, also known as the direct linear transform (DLT). Like the essential matrix, the homography matrix needs to be decomposed after to get the corresponding rotation matrix $\mathbf{R}$ and translation vector $\mathbf{t}$. The decomposition method includes numerical method [52, 53] and analytical method [54]. Similar to the decomposition of the essential matrix, the decomposition of the homography matrix will also return four sets of rotation matrices and translation vectors as well as the plane coefficients. If the depths of the projected map points are all positive (in front of the camera), then two sets of solutions can be excluded. In the end, only two sets of solutions are left, and more prior information is needed to make a final decision. Usually, we can solve it by assuming the normal vector of the known scene plane. If the scene plane is parallel to the camera plane, then the theoretical value of the normal vector $\mathbf{n}$ is $\mathbf{1}^T$.

The homography is of great significance in SLAM. When the feature points are coplanar, or the camera motion is pure rotation, the degree of freedom of the fundamental matrix decreases, which causes degeneration. The real data is always noisy. If you stick to the eight-point method to solve the fundamental matrix, the fundamental matrix's redundant freedom will be mainly determined by the noise. To avoid the

[5] In fact, there are three constraints, but the third one is linear dependent on the former two, so only the first two are taken.

effects of degradation, we usually estimate the fundamental matrix **F** and the homography matrix **H** simultaneously and then choose one with the smaller reprojection error as the final result.

6.4 Practice: Solving Camera Motion with Epipolar Constraints

In the following section, let's solve the camera motion through the essential matrix. The program in the previous section provides feature matching results, and here we use them to calculate **E** or **F** and then decompose **E** to get **R** and **t**. The whole program is solved using the algorithm provided by OpenCV. We encapsulate the feature extraction in the previous section into functions for later use. This section only shows the code for the pose estimation.

Listing 6.5 slambook2/ch7/pose_estimation_2d2d.cpp (part)

```
void pose_estimation_2d2d(std::vector<key point> key points_1,
      std::vector<key point> key points_2,
      std::vector<DMatch> matches,
      Mat &R, Mat &t) {
      // Camera Intrinsics,TUM Freiburg2
      Mat K = (Mat_<double>(3, 3) << 520.9, 0, 325.1, 0, 521.0, 249.7, 0, 0, 1)
           ;

      //-- Convert the matching point to the form of vector<Point2f>
      vector<Point2f> points1;
      vector<Point2f> points2;

      for (int i = 0; i < (int) matches.size(); i++) {
           points1.push_back(key points_1[matches[i].queryIdx].pt);
           points2.push_back(key points_2[matches[i].trainIdx].pt);
      }

      //-- Calculate fundamental matrix
      Mat fundamental_matrix;
      fundamental_matrix = findFundamentalMat(points1, points2, CV_FM_8POINT);
      cout << "fundamental_matrix is " << endl << fundamental_matrix << endl;

      //-- Calculate essential matrix
      Point2d principal_point(325.1, 249.7); // camera principal point,
           calibrated in TUM dataset
      double focal_length = 521; // camera focal length, calibrated in TUM
           dataset
      Mat essential_matrix;
      essential_matrix = findEssentialMat(points1, points2, focal_length,
           principal_point);
      cout << "essential_matrix is " << endl << essential_matrix << endl;

      //-- Calculate homography matrix
      //-- But the scene is not planar, and calculating the homography matrix
           here is of little significance
      Mat homography_matrix;
      homography_matrix = findHomography(points1, points2, RANSAC, 3);
      cout << "homography_matrix is " << endl << homography_matrix << endl;

      //-- Recover rotation and translation from the essential matrix.
      recoverPose(essential_matrix, points1, points2, R, t, focal_length,
           principal_point);
```

```
    cout << "R is " << endl << R << endl;
    cout << "t is " << endl << t << endl;
}
```

This function shows how to solve the camera motion from the 2D feature points. Then, we call it in the main function to get the camera motion:

Listing 6.6 slambook2/ch7/pose_estimation_2d2d.cpp (part)

```
int main( int argc, char** argv ){
    if (argc != 3) {
        cout << "usage: pose_estimation_2d2d img1 img2" << endl;
        return 1;
    }
    //-- Fetch images
    Mat img_1 = imread(argv[1], CV_LOAD_IMAGE_COLOR);
    Mat img_2 = imread(argv[2], CV_LOAD_IMAGE_COLOR);
    assert(img_1.data && img_2.data && "Can not load images!");

    vector<key point> key points_1, key points_2;
    vector<DMatch> matches;
    find_feature_matches(img_1, img_2, key points_1, key points_2, matches)
        ;
    cout << "In total, we get " << matches.size() << " set of feature
        points" << endl;

    //-- Estimate the motion between two frames
    Mat R, t;
    pose_estimation_2d2d(key points_1, key points_2, matches, R, t);

    //-- Check E=t^R*scale
    Mat t_x =
        (Mat_<double>(3, 3) << 0, -t.at<double>(2, 0), t.at<double>(1, 0)
        ,
        t.at<double>(2, 0), 0, -t.at<double>(0, 0),
        -t.at<double>(1, 0), t.at<double>(0, 0), 0);
    cout << "t^R=" << endl << t_x * R << endl;

    //-- Check epipolar constraints
    Mat K = (Mat_<double>(3, 3) << 520.9, 0, 325.1, 0, 521.0, 249.7, 0, 0,
        1);
    for (DMatch m: matches) {
        Point2d pt1 = pixel2cam(key points_1[m.queryIdx].pt, K);
        Mat y1 = (Mat_<double>(3, 1) << pt1.x, pt1.y, 1);
        Point2d pt2 = pixel2cam(key points_2[m.trainIdx].pt, K);
        Mat y2 = (Mat_<double>(3, 1) << pt2.x, pt2.y, 1);
        Mat d = y2.t() * t_x * R * y1;
        cout << "epipolar constraint = " << d << endl;
    }
    return 0;
}
```

We get the values of **E**, **F** and **H** in the function, and then verify whether the epipolar constraint is satisfied, and **t**^**R** and **E** are equivalent to non-zero numbers. Now, execute this program to see the output result:

Listing 6.7 Terminal Input

```
% build/pose_estimation_2d2d 1.png 2.png
-- Max dist : 95.000000
-- Min dist : 4.000000
In total, we get 79 set of feature points
fundamental_matrix is
[4.844484382466111e-06, 0.0001222601840188731, -0.01786737827487386;
-0.0001174326832719333, 2.122888800459598e-05, -0.01775877156212593;
```

```
0.01799658210895528, 0.008143605989020664, 1]
essential_matrix is
[-0.0203618550523477, -0.4007110038118445, -0.03324074249824097;
0.3939270778216369, -0.03506401846698079, 0.5857110303721015;
-0.006788487241438284, -0.5815434272915686, -0.01438258684486258]
homography_matrix is
[0.9497129583105288, -0.143556453147626, 31.20121878625771;
0.04154536627445031, 0.9715568969832015, 5.306887618807696;
-2.81813676978796e-05, 4.353702039810921e-05, 1]
R is
[0.9985961798781875, -0.05169917220143662, 0.01152671359827873;
0.05139607508976055, 0.9983603445075083, 0.02520051547522442;
-0.01281065954813571, -0.02457271064688495, 0.9996159607036126]
t is
[-0.8220841067933337;
-0.03269742706405412;
0.5684264241053522]

t^R=
[0.02879601157010516, 0.5666909361828478, 0.04700950886436416;
-0.5570970160413605, 0.0495880104673049, -0.8283204827837456;
0.009600370724838804, 0.8224266019846683, 0.02034004937801349]
epipolar constraint = [0.002528128704106625]
epipolar constraint = [-0.001663727901710724]
epipolar constraint = [-0.0008009088410884102]
......
```

It can be seen from the program's output that the accuracy of the epipolar constraints is around the order of 10^{-3}. According to the previous discussion, there are 4 possibilities for $\mathbf{R}, \mathbf{t}$ obtained by decomposition. In this program, OpenCV will use triangulation to detect whether the detected point's depth is positive to select the correct solution.

6.4.1 Discussion

From the demo, we can see that the result **E** and **F** are different by a camera intrinsic matrix. Although it is not numerically intuitive, their mathematical relationship can be verified by the program. Motion can be decomposed from **E**, **F** and **H**, but **H** needs to assume that the feature points are on the same plane. For this experiment's data, this assumption is not true, so we mainly use **E** to find the motion.

It is worth mentioning that since **E** itself has scale equivalence, the **t**, **R** also have scale equivalence. And $\mathbf{R} \in \mathrm{SO}(3)$ has its own constraints, so we think that **t** has a **scale**. In other words, in the decomposition process, if **t** is multiplied by any non-zero constant, the decomposition is still valid. Therefore, we usually normalize **t** to make its length equal to 1.

6.4.1.1 Scale Ambiguity

The normalization of **t** directly leads to *scale ambiguity in monocular vision*. For example, the first dimension of **t** output in the program is about 0.822. We are unsure that it refers to 0.822 meters or 0.822 centimeters. Because after multiplying **t** by any constant, the epipolar constraint is still valid. In other words, in monocular SLAM, the trajectory and map are simultaneously zoomed at any multiple, and the image we get is still the same. This has already been introduced in Chap. 1.

In the monocular vision, the normalization of $\mathbf{t}$ for two images is equivalent to fixing the scale. Although we don't know its real length, we use this $\mathbf{t}$ as unit 1 to calculate the camera motion and the 3D position of the feature points. This is called the initialization phrase of monocular SLAM. After initialization, 3D-2D can be used to calculate camera motion. The latter trajectory and map will use the fixed scale during initialization. Therefore, monocular SLAM has an inevitable initialization step. The two initialized images must have a certain amount of translation, and then the unit of trajectory and map will be determined by this translation.

In addition to normalizing $\mathbf{t}$, another method is to set the average depth of all the feature points during initialization to 1, or a fixed scale. Compared to set the length of $\mathbf{t}$ to 1, normalizing the depth of the feature points can control the scale of the scene and make the calculation more numerically stable. But there is no theoretical difference.

6.4.1.2 The Problem of Pure Rotation

In the decomposition of $\mathbf{E}$ to get $\mathbf{R}, \mathbf{t}$, if the camera is purely rotated, causing $\mathbf{t}$ to be zero, then $\mathbf{E}$ will also be zero, which will make it impossible for us to solve $\mathbf{R}$. However, we can rely on $\mathbf{H}$ to find the rotation at this time, but we cannot generally assume the scene is a plane. Also, in the rotation-only case, we cannot use triangulation to estimate the feature points' spatial position (this will be introduced later). So we can conclude that the monocular initialization cannot be performed in the pure rotation case. It must have a certain amount of translation. If there is no translation, the monocular SLAM system can not be initialized. In practice, if the translation is too small during initialization, it will also make the pose estimation and triangulation results unstable and even failed. In contrast, if the camera is moved left and right instead of rotating, it will be easy to initialize the monocular SLAM. Therefore, experienced SLAM researchers often choose to move the camera left and right to smoothly initialize in monocular SLAM, while novices tend to rotate the camera sitting in the seat. You can tell if one is really an experienced SLAM engineer or not from this.

6.4.1.3 More Than Eight Pairs of Features

When we get more than eight pairs of matched points (for example, we found 79 pairs of matches), we can calculate a least-square solution. Recalling the linearized epipolar constraint in (6.13), we denote the coefficient matrix on the left as $\mathbf{A}$:

$$\mathbf{A}\mathbf{e} = \mathbf{0}. \tag{6.22}$$

For the eight-point method, the size of **A** is 8×9. If the given features are more than 8, the equation constitutes an overdetermined equation, that is, **e** does not necessarily exist to make the above formula perfectly true. Therefore, it can be solved by minimizing in a quadratic form:

$$\min_{\mathbf{e}} \|\mathbf{A}\mathbf{e}\|_2^2 = \min_{\mathbf{e}} \mathbf{e}^T \mathbf{A}^T \mathbf{A}\mathbf{e}. \tag{6.23}$$

So the **E** matrix in the sense of least-squares is obtained. However, we prefer to use random sample consensus (RANSAC) instead of least-squares to solve the above problem due to potential mismatches. RANSAC is general, applicable to many cases with incorrect data, and can handle data with incorrect matching.

6.5 Triangulation

In the previous two sections, we introduced using the epipolar constraint to estimate the camera motion and discussed its limitations. After estimating the motion, the next step is to use the camera motion to estimate the spatial positions of the feature points. In monocular SLAM, the depths of the pixels cannot be obtained by a single image. We need to estimate the depths of the map points by the method of triangulation, shown in Fig. 6.11.

Triangulation refers to observing the same landmark point at different locations and determining the distance of the landmark point from the observed locations. Triangulation was first proposed by Gauss and used in metrology. It can also be applied to astronomy and geography. For example, we can estimate the distance to

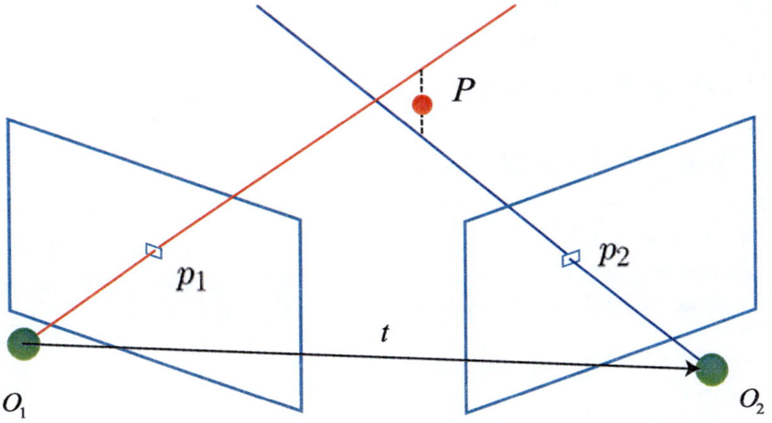

Fig. 6.11 Use triangulation to estimate the depth

us from the angle of the star observed in different seasons. In SLAM, we mainly use triangulation to estimate the distance of pixels.

Similar to the previous section, consider the images I_1 and I_2, with the left image as a reference. The transform matrix to the right image is $\mathbf{T}$. The principal points of the camera are O_1 and O_2. The feature p_1 is in I_1, which corresponds to a feature p_2 in I_2. In theory, the straight line $O_1 p_1$ and $O_2 p_2$ will intersect at a point P in the scene, which is the 3D map point. However, due to the noise, these two lines often fail to exactly intersect. Therefore, it can be solved in the sense of least-square.

According to the definition in epipolar geometry, let $\mathbf{x}_1, \mathbf{x}_2$ be the normalized coordinates of two feature points, then they satisfy:

$$s_2 \mathbf{x}_2 = s_1 \mathbf{R} \mathbf{x}_1 + \mathbf{t}. \tag{6.24}$$

Now we know the $\mathbf{R}$ and $\mathbf{t}$, we want to find the depth s_1 and s_2 of two feature points. Geometrically, you can find a 3D point on the ray $O_1 p_1$ to make its projection close to $\mathbf{p}_2$. Similarly, you can also find it on $O_2 p_2$ or in the middle of two lines. Different strategies correspond to different calculation methods, but the results are similar. For example, if we want to calculate s_1, we first multiply both sides of the above formula by $\mathbf{x}_2^\wedge$ to get:

$$s_2 \mathbf{x}_2^\wedge \mathbf{x}_2 = 0 = s_1 \mathbf{x}_2^\wedge \mathbf{R} \mathbf{x}_1 + \mathbf{x}_2^\wedge \mathbf{t}. \tag{6.25}$$

The left side of this equation is zero, and the right side can be regarded as an equation of s_1, and s_1 can be obtained directly from it. With s_1, s_2 is also very easy to calculate. Thus, we get the depth of the points under the two frames and determine their spatial coordinates. Definitely, due to the existence of noise, our estimated $\mathbf{R}, \mathbf{t}$ may not exactly make the Eq. (6.25) zero, so a more common approach is to find the least-square solution rather than a direct solution.

6.6 Practice: Triangulation

6.6.1 Triangulation with OpenCV

We demonstrate how to obtain the feature point's spatial position in the previous section through triangulation based on the camera pose solved by epipolar geometry. We will use the triangulation function provided by OpenCV here.

Listing 6.8 slambook2/ch7/triangulation.cpp (part)

```
void triangulation(
  const vector<key point> &key point_1,
  const vector<key point> &key point_2,
  const std::vector<DMatch> &matches,
  const Mat &R, const Mat &t,
  vector<Point3d> &points) {
```

```
  Mat T1 = (Mat_<float>(3, 4) <<
    1, 0, 0, 0,
    0, 1, 0, 0,
    0, 0, 1, 0);
  Mat T2 = (Mat_<float>(3, 4) <<
    R.at<double>(0, 0), R.at<double>(0, 1), R.at<double>(0, 2), t.at<double>(0, 0),
    R.at<double>(1, 0), R.at<double>(1, 1), R.at<double>(1, 2), t.at<double>(1, 0),
    R.at<double>(2, 0), R.at<double>(2, 1), R.at<double>(2, 2), t.at<double>(2, 0)
  );

  Mat K = (Mat_<double>(3, 3) << 520.9, 0, 325.1, 0, 521.0, 249.7, 0, 0, 1);
  vector<Point2f> pts_1, pts_2;
  for (DMatch m:matches) {
    // Convert pixel coordinates to camera coordinates
    pts_1.push_back(pixel2cam(key point_1[m.queryIdx].pt, K));
    pts_2.push_back(pixel2cam(key point_2[m.trainIdx].pt, K));
  }

  Mat pts_4d;
  cv::triangulatePoints(T1, T2, pts_1, pts_2, pts_4d);

  // Convert to non-homogeneous coordinates
  for (int i = 0; i < pts_4d.cols; i++) {
    Mat x = pts_4d.col(i);
    x /= x.at<float>(3, 0); //
    Point3d p(
      x.at<float>(0, 0),
      x.at<float>(1, 0),
      x.at<float>(2, 0)
    );
    points.push_back(p);
  }
}
```

Meanwhile, we can add the triangulation part to the main function and then draw each point's depth. Readers can run this program to view the triangulation results.

6.6.2 *Discussion*

Regarding triangulation, there is one more thing that must be noted.

Triangulation is caused by the *translation*. Only when there is enough amount of translation, triangles in the epipolar geometry can be formed, and then the triangulation can be implemented. Therefore, triangulation cannot be used for pure rotation because the triangle does not exist in this case. Definitely, the real data is often not wholly equal to zero. In the presence of translation, we should also concern about the uncertainty of triangulation, which will lead to the triangulation contradiction.

As shown in Fig. 6.12, when the translation is small, the pixel's uncertainty will result in a more enormous depth uncertainty. That is to say, if the feature point moves by one pixel δx, the sight angle changes by an angle $\delta\theta$, then the measured depth will experience a difference of δd. It can be seen from the geometric relationship that when $\mathbf{t}$ is larger, δd will be significantly smaller, meaning that when the translation is larger, the triangulation measurement will be more accurate at the same camera

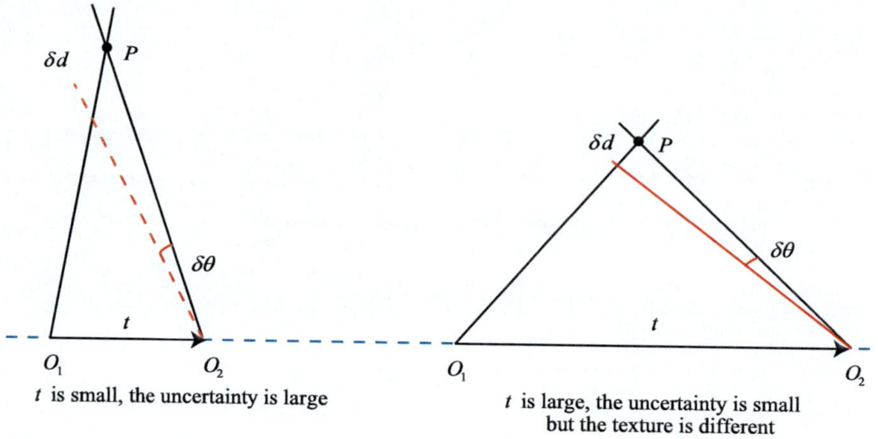

Fig. 6.12 The contradiction of triangulation

resolution. The quantitative analysis of the process can be proceeded by using the law of sine.

Therefore, to improve the accuracy of triangulation, one is to improve the accuracy of feature points extraction, which means to increase the image resolution, but this will cause the image to become larger and increase the computational cost. Another way is to increase the amount of translation. However, this will cause obvious changes in the image's appearance. For example, the box's side view initially blocked may be exposed, the lighting of the object changes, etc. Changes in appearance will make feature extraction and matching more difficult. In a nutshell, increasing the translation may lead to failure of matching; and if the translation is too small, the triangulation's accuracy is insufficient. This is the contradiction of triangulation.

In the monocular vision, since the image has no depth information, we have to wait for the feature points to be tracked for a few frames and then use triangulation to determine the new feature points' depth. This is also called delayed triangulation [55]. However, if the camera rotates in place, causing the parallax to be small, it is difficult to estimate the newly observed feature points' depth. This situation is common in robotics, as rotation is also a standard command for robots. In this case, monocular vision may suffer from tracking failures and incorrect scales.

Although this section only introduces the depth estimation of triangulation, we can also quantitatively calculate each feature point's location and uncertainty. Therefore, if we assume that the feature point obeys the Gaussian distribution and continuously observes it with the correct information, we can expect its variance will continue to decrease and even converge. This can be referred to as the depth filter. However, since its principle is more complicated, we will discuss it in later Chap. 11. Next, we will discuss the estimation of camera motion from 3D-2D matching points and 3D-3D estimation methods.

6.7 3D–2D PnP

PnP (Perspective-n-Point) is a method to solve 3D to 2D motion estimation. It describes how to estimate the camera's pose when the n 3D space points and their projection positions are known. As mentioned earlier, the 2D-2D epipolar geometry method requires eight or more point pairs (take the eight-point method as an example), and there have problems with initialization, pure rotation, and scale. However, if the 3D position of one of the feature points in the two images is known, then we need at least three pairs (and at least one additional point to verify the result) to estimate the camera motion. The 3D position of the feature point can be determined by triangulation or the depth map of an RGB-D camera. Therefore, in binocular or RGB-D visual odometry, we can directly use PnP to estimate camera motion. While in the monocular case, initialization must be conducted before using PnP. The 3D-2D method does not require epipolar constraints and can obtain better motion estimation in a few matching points. It is the most important pose estimation method.

There are many ways to solve PnP problems, for example, P3P [56], direct linear transformation (DLT), EPnP (Efficient PnP) [57], UPnP [58], etc. In addition, non-linear optimization can be used to construct a least-square problem and iteratively solve it, which is commonly called the bundle adjustment. Let's look at DLT first, and then we will explain the bundle adjustment approach.

6.7.1 *Direct Linear Transformation*

Consider such a problem: we know the 3D positions of a point set and their projections in the camera, now we want to find the camera's pose. This problem can be used to solve the camera pose when a given map and image are given. If the 3D point is regarded as a point in another camera coordinate system, it can also solve the two cameras' relative motion problem. We will start with simple questions.

Consider a 3D spatial point P, its homogeneous coordinates are $\mathbf{P}=(X, Y, Z, 1)^T$. In the image I_1, it is projected to the feature point $\mathbf{x}_1 = (u_1, v_1, 1)^T$ (expressed as the normalized homogeneous coordinates). At this time, the pose of the camera $\mathbf{R}, \mathbf{t}$ is unknown. Similar to the solution of the homography matrix, we define the 3×4 augmented matrix $[\mathbf{R}|\mathbf{t}]$, encoding rotation and translation information.[6] We will write its expanded form as follows:

$$s \begin{pmatrix} u_1 \\ v_1 \\ 1 \end{pmatrix} = \underbrace{\begin{pmatrix} t_1 & t_2 & t_3 & t_4 \\ t_5 & t_6 & t_7 & t_8 \\ t_9 & t_{10} & t_{11} & t_{12} \end{pmatrix}}_{[\mathbf{R}|\mathbf{t}]} \begin{pmatrix} X \\ Y \\ Z \\ 1 \end{pmatrix}. \tag{6.26}$$

[6] This is different from the transformation matrix $\mathbf{T}$ in SE(3).

Eliminate s with the last row to get two constraints:

$$u_1 = \frac{t_1 X + t_2 Y + t_3 Z + t_4}{t_9 X + t_{10} Y + t_{11} Z + t_{12}}, \quad v_1 = \frac{t_5 X + t_6 Y + t_7 Z + t_8}{t_9 X + t_{10} Y + t_{11} Z + t_{12}}.$$

To simplify the representation, define $\mathbf{T}$ as a row vector:

$$\mathbf{t}_1 = (t_1, t_2, t_3, t_4)^T, \mathbf{t}_2 = (t_5, t_6, t_7, t_8)^T, \mathbf{t}_3 = (t_9, t_{10}, t_{11}, t_{12})^T.$$

Now we have:

$$\mathbf{t}_1^T \mathbf{P} - \mathbf{t}_3^T \mathbf{P} u_1 = 0,$$

and

$$\mathbf{t}_2^T \mathbf{P} - \mathbf{t}_3^T \mathbf{P} v_1 = 0.$$

Please note that $\mathbf{t}$ is the variable to be determined. As you can see, each feature point provides two linear constraints on $\mathbf{t}$. Assuming there are a total of N feature points, the following linear equations can be constructed:

$$\begin{pmatrix} \mathbf{P}_1^T & 0 & -u_1 \mathbf{P}_1^T \\ 0 & \mathbf{P}_1^T & -v_1 \mathbf{P}_1^T \\ \vdots & \vdots & \vdots \\ \mathbf{P}_N^T & 0 & -u_N \mathbf{P}_N^T \\ 0 & \mathbf{P}_N^T & -v_N \mathbf{P}_N^T \end{pmatrix} \begin{pmatrix} \mathbf{t}_1 \\ \mathbf{t}_2 \\ \mathbf{t}_3 \end{pmatrix} = 0. \tag{6.27}$$

Since $\mathbf{t}$ has a total dimension of 12, the linear solution of the matrix $\mathbf{T}$ can be achieved by at least six pairs of matching points. This method is called Direct Linear Transform (DLT). When the matching points are greater than 6 pairs, methods such as SVD can also be used to find the least-square solution of the overdetermined equation.

In the DLT solution, we directly regard the $\mathbf{T}$ matrix as 12 unknowns, ignoring the correlation between them. Because the rotation matrix $\mathbf{R} \in \mathrm{SO}(3)$, the solution obtained by DLT does not necessarily satisfy the SE(3) constraint. It is just a general matrix. The translation vector is easier to handle. It belongs to the vector space. For the rotation matrix $\mathbf{R}$, we must look for the best rotation matrix to approximate the matrix block of 3×3 on the left of $\mathbf{T}$ estimated by DLT. This can be done by QR decomposition [2, 59], or it can be calculated like [5, 60]:

$$\mathbf{R} \leftarrow \left(\mathbf{R}\mathbf{R}^T\right)^{-\frac{1}{2}} \mathbf{R}. \tag{6.28}$$

This can be seen as reprojecting the result from the matrix space onto the SE(3) manifold and converting it into two parts: rotation and translation.

What needs to be mentioned is that our $\mathbf{x}_1$ here uses normalized plane coordinates and neglects the influence of the intrinsic matrix $\mathbf{K}$. This is because $\mathbf{K}$ is usually

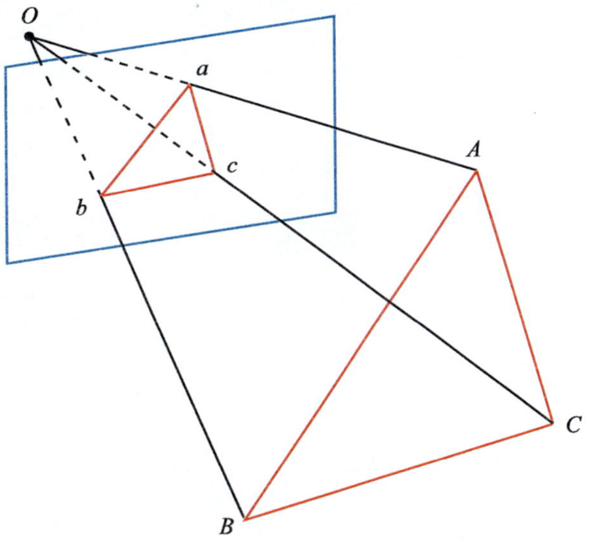

Fig. 6.13 The P3P problem

assumed to be known in SLAM. Even if the intrinsic parameters are unknown, PnP can still be used to estimate the three quantities $\mathbf{K}$, $\mathbf{R}$, $\mathbf{t}$. However, due to the increase in the number of unknown variables, the result's quantity may be worse.

6.7.2 P3P

P3P is another way to solve PnP. It only uses 3 pairs of matching points and requires less data (this part refers to the [61]).

P3P requires establishing geometric relationships of the given 3 points. Its input data is 3 pairs of 3D-2D matching points. Define 3D points as A, B, C, 2D points as a, b, c, where the point represented by the lowercase letter is the projection of the point on the camera image plane represented by the corresponding uppercase letter, as shown in Fig. 6.13. Also, P3P needs a pair of verification points to select the correct one from the possible solutions (similar to epipolar geometry). Denote the verification point pair as $D - d$ and the principal camera point as O. Suppose that A, B, C are in the world coordinate frame, not the camera coordinate. Once the coordinates of the 3D point in the camera coordinate system can be calculated, we get the 3D–3D corresponding point and convert the PnP problem to the ICP problem.

Obviously, there is a relationship between triangles:

$$\Delta Oab - \Delta OAB, \quad \Delta Obc - \Delta OBC, \quad \Delta Oac - \Delta OAC. \tag{6.29}$$

Consider the relationship between Oab and OAB. Using the law of cosines, there are:

$$\overline{OA}^2 + \overline{OB}^2 - 2.\overline{OA}.\overline{OB}.\cos\langle a, b\rangle = \overline{AB}^2. \tag{6.30}$$

The other two triangles have similar properties, so we further get:

$$\begin{aligned} \overline{OA}^2 + \overline{OB}^2 - 2.\overline{OA}.\overline{OB}.\cos\langle a, b\rangle &= \overline{AB}^2 \\ \overline{OB}^2 + \overline{OC}^2 - 2.\overline{OB}.\overline{OC}.\cos\langle b, c\rangle &= \overline{BC}^2 \\ \overline{OA}^2 + \overline{OC}^2 - 2.\overline{OA}.\overline{OC}.\cos\langle a, c\rangle &= \overline{AC}^2. \end{aligned} \tag{6.31}$$

Divide all the above three equations by $\overline{OC}^2$ on both sides, and consider $x = \overline{OA}/\overline{OC}$, $y = \overline{OB}/\overline{OC}$, we get:

$$\begin{aligned} x^2 + y^2 - 2.x.y.\cos\langle a, b\rangle &= \overline{AB}^2/\overline{OC}^2 \\ y^2 + 1^2 - 2.y.\cos\langle b, c\rangle &= \overline{BC}^2/\overline{OC}^2 \\ x^2 + 1^2 - 2.x.\cos\langle a, c\rangle &= \overline{AC}^2/\overline{OC}^2. \end{aligned} \tag{6.32}$$

Let $v = \overline{AB}^2/\overline{OC}^2$, $u.v = \overline{BC}^2/\overline{OC}^2$, $w.v = \overline{AC}^2/\overline{OC}^2$, then we have:

$$\begin{aligned} x^2 + y^2 - 2.x.y.\cos\langle a, b\rangle - v &= 0 \\ y^2 + 1^2 - 2.y.\cos\langle b, c\rangle - u.v &= 0 \\ x^2 + 1^2 - 2.x.\cos\langle a, c\rangle - w.v &= 0. \end{aligned} \tag{6.33}$$

Move v in the first equation to the right side, and combine it with the other two equations, we have:

$$\begin{aligned} (1-u)\,y^2 - ux^2 - 2\cos\langle b, c\rangle\, y + 2uxy\cos\langle a, b\rangle + 1 &= 0 \\ (1-w)\,x^2 - wy^2 - 2\cos\langle a, c\rangle\, x + 2wxy\cos\langle a, b\rangle + 1 &= 0. \end{aligned} \tag{6.34}$$

Please distinguish the known from the unknown quantities in these equations. Since we know the positions of the 2D points in the image, the 3 cosine angles $\cos\langle a, b\rangle$, $\cos\langle b, c\rangle$, $\cos\langle a, c\rangle$ can be calculated. Meanwhile, $u = \overline{BC}^2/\overline{AB}^2$, $w = \overline{AC}^2/\overline{AB}^2$ can be calculated by the coordinates of A, B, C in the world frame. Transforming to the camera frame does not change the ratio. The x and y are unknown and will change as the camera moves. Therefore, the equations are quadratic equations about two unknowns x, y. Analytically solving the equations is complicated and requires Wu's elimination method. It will not be introduced here. Interested readers could refer to the literature [56]. Analogous to the case of decomposing **E**, this equation may get 4 solutions at most. Still, we can use the verification points to select the most probable solution and get the 3D of A, B, C in the camera frame. Then, based on the 3D–3D point pair, the camera movement $\mathbf{R}, \mathbf{t}$ can be calculated, which will be introduced in Sect. 6.9.

From the principle of P3P to solve PnP, we use the triangle relationships to solve the 3D coordinates of the projection points a, b, c in the camera frame, and finally convert the problem into a 3D to 3D pose estimation problem. As we will see later, it is easy to solve the 3D-3D pose with matching information, so this idea is very effective. Some other methods, such as EPnP, also adopted this idea. However, P3P also has some deficiencies:

1. P3P only includes the information of 3 points. When the given matched points are more than 3, it is difficult to use more information.
2. If the 3D point or 2D point is affected by noise or a mismatch, the algorithm goes into trouble.

People also proposed many other methods, such as EPnP, UPnP, and so on. They use more information and iteratively optimize the camera pose to eliminate noise as much as possible. However, compared to P3P, the principles are more complicated, so we recommend that readers read the original papers or understand the PnP process by practice. In SLAM, the usual approach is to first estimate the camera pose using P3P/EPnP and then construct a least-squares optimization problem to adjust the estimated values (bundle adjustment). When the camera motion is sufficiently continuous, or you can also assume that the camera does not move or move at a constant speed, you can use the estimated values as the initial values for optimization. Next, we look at the PnP problem from the perspective of nonlinear optimization.

6.7.3 Solve PnP by Minimizing the Reprojection Error

Other than the linear method, we can also construct the PnP problem as a nonlinear least-square problem about reprojection errors. This will use the knowledge from the Chaps. 3 and 4 of this book. The linear method mentioned above is often divided into many steps, such as estimating the camera pose first and then the point's position. While nonlinear optimization treats them as optimization variables and optimizes them together. This is a very general solution method. We can use it to optimize the results given by PnP or ICP. This type of problem, putting the camera and 3D points together to minimize, is generally referred to as bundle adjustment.[7]

We can build a bundle adjustment problem in PnP to optimize the camera pose. If the camera is moving continuously (as in most SLAM processes), we can also use BA directly to solve the camera pose. In this section, we will give the basic two-view form of this problem and then discuss the larger-scale BA problem in lecture 8.

Suppose we have n 3D space points P and their projection p, we want to calculate the pose $\mathbf{R}, \mathbf{t}$ of the camera. Suppose the coordinates of a point are $\mathbf{P}_i = [X_i, Y_i, Z_i]^T$, and their projected pixel coordinates are $\mathbf{u}_i = [u_i, v_i]^{\mathrm{T}}$. According to Chap. 4, the relationship between the 2D pixel position and the 3D spatial position is:

$$s_i \begin{bmatrix} u_i \\ v_i \\ 1 \end{bmatrix} = \mathbf{KT} \begin{bmatrix} X_i \\ Y_i \\ Z_i \\ 1 \end{bmatrix}. \tag{6.35}$$

---

[7] The meaning of BA in different documents and contexts are not exactly the same. Some people only refer to problems that minimize the reprojection errors as BA, while others have a broader definition of BA. Even if the BA has only one camera, it can be called BA. I personally prefer the broader definition, so the method of calculating PnP here is also called BA.

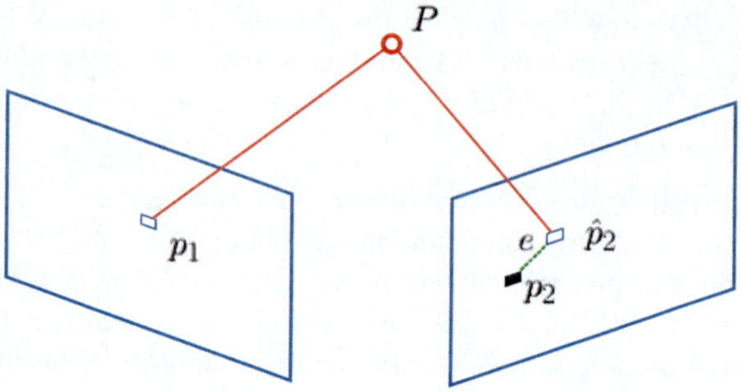

Fig. 6.14 The reprojection error

Or in the matrix form:

$$s_i \mathbf{u}_i = \mathbf{KTP}_i.$$

This equation includes a conversion from homogeneous coordinates to non-homogeneous coordinates implicitly. Or we can also use $\mathbf{RP} + \mathbf{t}$. Now, due to the unknown camera pose and the noise of the observation points, there is a residual in the equation. Therefore, we sum up the residuals, construct a least-square problem, and then minimize it to find the most possible camera pose:

$$\mathbf{T}^* = \arg\min_{\mathbf{T}} \frac{1}{2} \sum_{i=1}^{n} \left\| \mathbf{u}_i - \frac{1}{s_i} \mathbf{KTP}_i \right\|_2^2. \tag{6.36}$$

The residual term is the error of the projected position and the observed position, which is called the reprojection error. With homogeneous coordinates, this error has three dimensions. However, since the last dimension of $\mathbf{u}$ is 1, the error of this dimension is always zero, so we normally use non-homogeneous coordinates. Therefore, the error has only 2 dimensions. As shown in Fig. 6.14, we know that p_1 and p_2 are projections of the same space point P through feature matching, but we don't know the pose of the camera. In the initial value, there is a certain distance between the projection of P $\hat{p}_2$ and the observed p_2. So we adjusted the pose of the camera to make this distance smaller. However, since this adjustment needs to consider many points, the goal is to reduce the overall error, and the error of each point usually can not be exactly zero.

We have already discussed the least-square optimization problem in Chap. 5. Using Lie algebra, we can construct an unconstrained optimization problem on the manifold, easily solved using optimization algorithms such as the Gauss-Newton method and Levenberg-Marquardt method. However, we need to calculate the derivative of each error term with respect to the optimization variable, which is also the linearization:

$$\mathbf{e}(\mathbf{x} + \Delta\mathbf{x}) \approx \mathbf{e}(\mathbf{x}) + \mathbf{J}^T \Delta\mathbf{x}. \tag{6.37}$$

The form of $\mathbf{J}^T$ is worth discussing. Definitely, we can use numerical derivatives, but if we can derive an analytical form, we will prefer the analytical derivatives. Now, $\mathbf{e}$ is the pixel coordinate error (2-d) and $\mathbf{x}$ is the camera pose (6-d), $\mathbf{J}^T$ is a matrix of 2×6. Let's derive the form of $\mathbf{J}^T$.

We have introduced how to use the perturbation model to find the pose variable's derivative (Chap. 3). First, define the coordinates of the space point in the camera frame as $\mathbf{P}'$, and take out the first 3 dimensions:

$$\mathbf{P}' = (\mathbf{TP})_{1:3} = [X', Y', Z']^T. \tag{6.38}$$

Then, the camera projection model with respect to $\mathbf{P}'$ is:

$$s\mathbf{u} = \mathbf{KP}'. \tag{6.39}$$

Expand it:

$$\begin{bmatrix} su \\ sv \\ s \end{bmatrix} = \begin{bmatrix} f_x & 0 & c_x \\ 0 & f_y & c_y \\ 0 & 0 & 1 \end{bmatrix} \begin{bmatrix} X' \\ Y' \\ Z' \end{bmatrix}. \tag{6.40}$$

Use the third row to eliminate s (actually it is the distance of $\mathbf{P}'$), we get:

$$u = f_x \frac{X'}{Z'} + c_x, \quad v = f_y \frac{Y'}{Z'} + c_y. \tag{6.41}$$

This is consistent with the camera model described in Chap. 4. When we find the error, we can compare the u, v here with the measured value to find the difference. After defining the intermediate variables, we left multiply $\mathbf{T}$ by a disturbance quantity $\delta\boldsymbol{\xi}$, and then consider the derivative of the change of $\mathbf{e}$ with respect to the disturbance quantity. Using the chain rule, it is:

$$\frac{\partial \mathbf{e}}{\partial \delta\boldsymbol{\xi}} = \lim_{\delta\boldsymbol{\xi} \to 0} \frac{\mathbf{e}\left(\delta\boldsymbol{\xi} \oplus \boldsymbol{\xi}\right) - \mathbf{e}(\boldsymbol{\xi})}{\delta\boldsymbol{\xi}} = \frac{\partial \mathbf{e}}{\partial \mathbf{P}'} \frac{\partial \mathbf{P}'}{\partial \delta\boldsymbol{\xi}}. \tag{6.42}$$

Here $\oplus$ refers to the disturbance left multiplication in Lie algebra. The first item is the derivative of the error with respect to the projection point. The relationship between the variables is in the Eq. (6.41), and it is easy to get:

$$\frac{\partial \mathbf{e}}{\partial \mathbf{P}'} = -\begin{bmatrix} \frac{\partial u}{\partial X'} & \frac{\partial u}{\partial Y'} & \frac{\partial u}{\partial Z'} \\ \frac{\partial v}{\partial X'} & \frac{\partial v}{\partial Y'} & \frac{\partial v}{\partial Z'} \end{bmatrix} = -\begin{bmatrix} \frac{f_x}{Z'} & 0 & -\frac{f_x X'}{Z'^2} \\ 0 & \frac{f_y}{Z'} & -\frac{f_y Y'}{Z'^2} \end{bmatrix}. \tag{6.43}$$

The second term is the derivative of the transformed point with respect to the Lie algebra. According to the Sect. 3.3.5, we get:

$$\frac{\partial \left(\mathbf{TP}\right)}{\partial \delta\boldsymbol{\xi}} = (\mathbf{TP})^{\odot} = \begin{bmatrix} \mathbf{I} & -\mathbf{P}'^{\wedge} \\ \mathbf{0}^T & \mathbf{0}^T \end{bmatrix}. \tag{6.44}$$

In the definition of $\mathbf{P}'$, we took out the first 3 dimensions, so we get:

$$\frac{\partial \mathbf{P}'}{\partial \delta \boldsymbol{\xi}} = \left[\mathbf{I}, -\mathbf{P}'^{\wedge}\right]. \tag{6.45}$$

Multiply these two items together, we get the 2×6 Jacobian matrix:

$$\frac{\partial \mathbf{e}}{\partial \delta \boldsymbol{\xi}} = -\begin{bmatrix} \frac{f_x}{Z'} & 0 & -\frac{f_x X'}{Z'^2} & -\frac{f_x X' Y'}{Z'^2} & f_x + \frac{f_x X'^2}{Z'^2} & -\frac{f_x Y'}{Z'} \\ 0 & \frac{f_y}{Z'} & -\frac{f_y Y'}{Z'^2} & -f_y - \frac{f_y Y'^2}{Z'^2} & \frac{f_y X' Y'}{Z'^2} & \frac{f_y X'}{Z'} \end{bmatrix}. \tag{6.46}$$

This Jacobian matrix describes the first-order derivative of the reprojection error with respect to the left perturbation model. We keep the negative sign in front of it because the error is defined by the observed value minus the predicted value. It can also be reversed and defined in the form of the predicted value minus the observed value. In that case, just remove the negative sign in front. Besides, if the definition of $\mathfrak{se}(3)$ is a rotation followed by translation, just swap the first 3 columns and the last 3 columns of this Jacobian matrix.

On top of optimizing the pose, we want to optimize the spatial position of the feature points. Therefore, we also need to discuss the derivative of $\mathbf{e}$ with respect to the space point $\mathbf{P}$. Fortunately, this derivative matrix is relatively easy. Still using the chain rule, there are:

$$\frac{\partial \mathbf{e}}{\partial \mathbf{P}} = \frac{\partial \mathbf{e}}{\partial \mathbf{P}'} \frac{\partial \mathbf{P}'}{\partial \mathbf{P}}. \tag{6.47}$$

The first item has been deduced before, and the second item is defined as:

$$\mathbf{P}' = (\mathbf{TP})_{1:3} = \mathbf{RP} + \mathbf{t},$$

So only $\mathbf{R}$ is left in the derivative:

$$\frac{\partial \mathbf{e}}{\partial \mathbf{P}} = -\begin{bmatrix} \frac{f_x}{Z'} & 0 & -\frac{f_x X'}{Z'^2} \\ 0 & \frac{f_y}{Z'} & -\frac{f_y Y'}{Z'^2} \end{bmatrix} \mathbf{R}. \tag{6.48}$$

We have derived the two Jacobian matrices of the observation camera equation with respect to the camera pose and feature points. They are very important to provide gradient directions in the optimization and guide the iteration of optimization.

6.8 Practice: Solving PnP

6.8.1 Use EPnP to Solve the Pose

In the following, we will have a deeper understanding of the PnP process through practice. First, we demonstrate how to use OpenCV's EPnP to solve the PnP problem and then solve it again through nonlinear optimization. In the second edition of the book, we also add a handwriting optimization practice. Since PnP needs to use 3D points, to avoid the trouble of initialization, we use the depth map (1_depth.png) in the RGB-D camera as the 3D position of the feature points. First look at the PnP function provided by OpenCV:

Listing 6.9 slambook2/ch7/pose_estimation_3d2d.cpp (part)

```
int main( int argc, char** argv ) {
    Mat r, t;
    solvePnP(pts_3d, pts_2d, K, Mat(), r, t, false); // Call OpenCV's PnP,
         you can choose from EPNP, DLS and other methods
    Mat R;
    cv::Rodrigues(r, R); // r is in the form of rotation vector, and
         converted to a rotation matrix by Rodrigues formula
    cout << "R=" << endl << R << endl;
    cout << "t=" << endl << t << endl;
}
```

In the example, after obtaining the matched feature points, we get their depth in the depth map of the first image and find their spatial position. Taking this spatial position as a 3D point and then taking the second image's pixel position as a 2D observation, we call EPnP to solve the PnP problem. The program output is as follows:

Listing 6.10 Terminal output:

```
% build/pose_estimation_3d2d 1.png 2.png d1.png d2.png
-- Max dist : 95.000000
-- Min dist : 4.000000
In total, we get 79 set of feature points
3d-2d pairs: 76
R=
[0.9978662025826269, -0.05167241613316376, 0.03991244360207524;
0.0505958915956335, 0.998339762771668, 0.02752769192381471;
-0.04126860182960625, -0.025449547736074, 0.998823919929363]
t=
[-0.1272259656955879;
-0.007507297652615337;
0.06138584177157709]
```

Readers can compare **R**, **t** solved in the previous 2D-2D case to see the difference. It can be seen that when 3D information is involved, the estimated **R** is almost the same, while the **t** is quite different. This is due to the inclusion of depth information. However, since the depth map collected by Kinect has some noise, the 3D points here are not accurate. In a larger-scale BA, we would like to optimize the pose and all 3D feature points at the same time.

6.8.2 Pose Estimation from Scratch

The following demonstrates how to use nonlinear optimization to calculate the camera pose. We first write a PnP of the Gauss-Newton method and then demonstrate how to solve it by *g2o*.

Listing 6.11 slambook2/ch7/pose_estimation_3d2d.cpp (part)

```
void bundleAdjustmentGaussNewton(
const VecVector3d &points_3d,
const VecVector2d &points_2d,
const Mat &K,
Sophus::SE3d &pose) {
  typedef Eigen::Matrix<double, 6, 1> Vector6d;
  const int iterations = 10;
  double cost = 0, lastCost = 0;
  double fx = K.at<double>(0, 0);
  double fy = K.at<double>(1, 1);
  double cx = K.at<double>(0, 2);
  double cy = K.at<double>(1, 2);

  for (int iter = 0; iter < iterations; iter++) {
    Eigen::Matrix<double, 6, 6> H = Eigen::Matrix<double, 6, 6>::Zero();
    Vector6d b = Vector6d::Zero();

    cost = 0;
    // compute cost
    for (int i = 0; i < points_3d.size(); i++) {
      Eigen::Vector3d pc = pose * points_3d[i];
      double inv_z = 1.0 / pc[2];
      double inv_z2 = inv_z * inv_z;
      Eigen::Vector2d proj(fx * pc[0] / pc[2] + cx, fy * pc[1] / pc[2] + cy);
      Eigen::Vector2d e = points_2d[i] - proj;
      cost += e.squaredNorm();
      Eigen::Matrix<double, 2, 6> J;
      J << -fx * inv_z,
      0,
      fx * pc[0] * inv_z2,
      fx * pc[0] * pc[1] * inv_z2,
      -fx - fx * pc[0] * pc[0] * inv_z2,
      fx * pc[1] * inv_z,
      0,
      -fy * inv_z,
      fy * pc[1] * inv_z,
      fy + fy * pc[1] * pc[1] * inv_z2,
      -fy * pc[0] * pc[1] * inv_z2,
      -fy * pc[0] * inv_z;

      H += J.transpose() * J;
      b += -J.transpose() * e;
    }

    Vector6d dx;
    dx = H.ldlt().solve(b);

    if (isnan(dx[0])) {
      cout << "result is nan!" << endl;
      break;
    }

    if (iter > 0 && cost >= lastCost) {
      // cost increase, update is not good
      cout << "cost: " << cost << ", last cost: " << lastCost << endl;
      break;
```

```
    }

    // update your estimation
    pose = Sophus::SE3d::exp(dx) * pose;
    lastCost = cost;

    cout << "iteration " << iter << " cost=" << cout.precision(12) << cost <<
          endl;
    if (dx.norm() < 1e-6) {
      // converge
      break;
    }
  }

  cout << "pose by g-n: \n" << pose.matrix() << endl;
}
```

In this function, we implement a simple Gauss-Newton iterated optimization based on the previous derivation. Then we will compare the efficiency of OpenCV, handwritten implementation, and the *g2o* implementation.

6.8.3 Optimization by g2o

After handwriting the optimization process, let's look at how to achieve the same functionality using the library *g2o* (it is similar to Ceres). The basic knowledge of *g2o* has been introduced in lecture 5. Before using *g2o*, we have to model the problem as a graph optimization problem, as shown in Fig. 6.15.

In this graph optimization, the nodes and edges are defined as follows:

1. **Node**: The pose of the second camera $\mathbf{T} \in \mathrm{SE}(3)$.

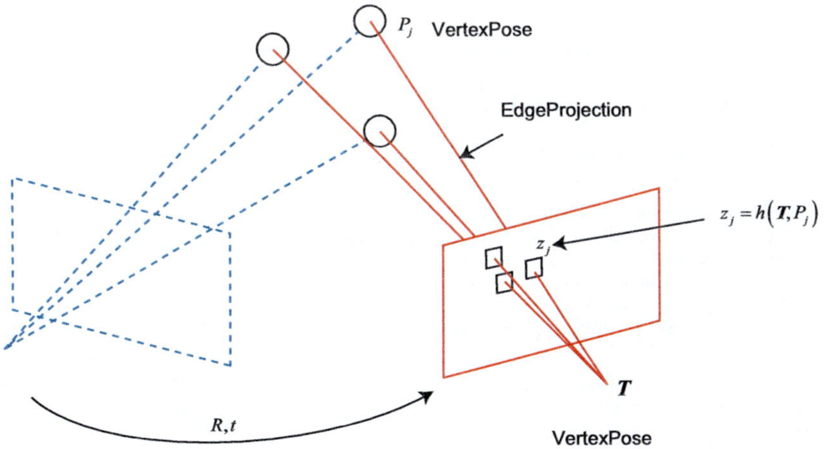

Fig. 6.15 PnP's graph structure

2. **Edge**: The projection of each 3D point in the second camera, described by the observation equation:

$$\mathbf{z}_j = h(\mathbf{T}, \mathbf{P}_j).$$

Since the pose of the first camera is fixed to identity, we excluded it from the optimization variables. But in normal cases, we will consider estimations of many cameras. Now we estimate the second camera pose based on a set of 3D points and the 2D projection in the second image. We drew the first camera as a dotted line to indicate that we don't want to consider it.

g2o provides many nodes and edges about BA. For example, "g2o/types/sba/types_six_dof_expmap.h" provides nodes and edges expressed by Lie algebra. In the second edition of the book, we implement a VertexPose vertex and EdgeProjection edge ourselves, as follows:

Listing 6.12 slambook2/ch7/pose_estimation_3d2d.cpp (part)

```
/// vertex and edges used in g2o ba
class VertexPose : public g2o::BaseVertex<6, Sophus::SE3d> {
  public:
  EIGEN_MAKE_ALIGNED_OPERATOR_NEW;

  virtual void setToOriginImpl() override {
    _estimate = Sophus::SE3d();
  }

  /// left multiplication on SE3
  virtual void oplusImpl(const double *update) override {
    Eigen::Matrix<double, 6, 1> update_eigen;
    update_eigen << update[0], update[1], update[2], update[3], update[4],
         update[5];
    _estimate = Sophus::SE3d::exp(update_eigen) * _estimate;
  }

  virtual bool read(istream &in) override {}

  virtual bool write(ostream &out) const override {}
};

class EdgeProjection : public g2o::BaseUnaryEdge<2, Eigen::Vector2d,
     VertexPose> {
  public:
  EIGEN_MAKE_ALIGNED_OPERATOR_NEW;

  EdgeProjection(const Eigen::Vector3d &pos, const Eigen::Matrix3d &K) :
       _pos3d(pos), _K(K) {}

  virtual void computeError() override {
    const VertexPose *v = static_cast<VertexPose *> (_vertices[0]);
    Sophus::SE3d T = v->estimate();
    Eigen::Vector3d pos_pixel = _K * (T * _pos3d);
    pos_pixel /= pos_pixel[2];
    _error = _measurement - pos_pixel.head<2>();
  }

  virtual void linearizeOplus() override {
    const VertexPose *v = static_cast<VertexPose *> (_vertices[0]);
    Sophus::SE3d T = v->estimate();
    Eigen::Vector3d pos_cam = T * _pos3d;
    double fx = _K(0, 0);
    double fy = _K(1, 1);
    double cx = _K(0, 2);
```

```
    double cy = _K(1, 2);
    double X = pos_cam[0];
    double Y = pos_cam[1];
    double Z = pos_cam[2];
    double Z2 = Z * Z;
    _jacobianOplusXi
    << -fx / Z, 0, fx * X / Z2, fx * X * Y / Z2, -fx - fx * X * X / Z2, fx * Y / Z,
    0, -fy / Z, fy * Y / (Z * Z), fy + fy * Y * Y / Z2, -fy * X * Y / Z2, -fy * X / Z;
  }

  virtual bool read(istream &in) override {}

  virtual bool write(ostream &out) const override {}

  private:
  Eigen::Vector3d _pos3d;
  Eigen::Matrix3d _K;
};
```

This implements vertex update and edge error calculation. The following is to combine them into a graph optimization problem:

Listing 6.13 slambook2/ch7/pose_estimation_3d2d.cpp (part)

```
void bundleAdjustmentG2O(
const VecVector3d &points_3d,
const VecVector2d &points_2d,
const Mat &K,
Sophus::SE3d &pose) {
  // Build graph optimization, first let's define g2o
  typedef g2o::BlockSolver<g2o::BlockSolverTraits<6, 3>> BlockSolverType; // pose is 6, landmark is 3
  typedef g2o::LinearSolverDense<BlockSolverType::PoseMatrixType> LinearSolverType;
  // Gradient descent method, you can choose from GN, LM, DogLeg
  auto solver = new g2o::OptimizationAlgorithmsGaussNewton(
    g2o::make_unique<BlockSolverType>(g2o::make_unique<LinearSolverType>()));
  g2o::SparseOptimizer optimizer; // Graph model
  optimizer.setAlgorithm(solver); // Set up the solver
  optimizer.setVerbose(true); // Turn on verbose output for debugging

  // vertex
  VertexPose *vertex_pose = new VertexPose(); // camera vertex_pose
  vertex_pose->setId(0);
  vertex_pose->setEstimate(Sophus::SE3d());
  optimizer.addVertex(vertex_pose);

  // K
  Eigen::Matrix3d K_eigen;
  K_eigen <<
  K.at<double>(0, 0), K.at<double>(0, 1), K.at<double>(0, 2),
  K.at<double>(1, 0), K.at<double>(1, 1), K.at<double>(1, 2),
  K.at<double>(2, 0), K.at<double>(2, 1), K.at<double>(2, 2);

  // edges
  int index = 1;
  for (size_t i = 0; i < points_2d.size(); ++i) {
    auto p2d = points_2d[i];
    auto p3d = points_3d[i];
    EdgeProjection *edge = new EdgeProjection(p3d, K_eigen);
    edge->setId(index);
    edge->setVertex(0, vertex_pose);
    edge->setMeasurement(p2d);
    edge->setInformation(Eigen::Matrix2d::Identity());
```

```
    optimizer.addEdge(edge);
    index++;
  }

  chrono::steady_clock::time_point t1 = chrono::steady_clock::now();
  optimizer.setVerbose(true);
  optimizer.initializeOptimization();
  optimizer.optimize(10);
  chrono::steady_clock::time_point t2 = chrono::steady_clock::now();
  chrono::duration<double> time_used = chrono::duration_cast<chrono::duration
      <double>>(t2 - t1);
  cout << "optimization costs time: " << time_used.count() << " seconds." <<
      endl;
  cout << "pose estimated by g2o =\n" << vertex_pose->estimate().matrix() <<
      endl;
  pose = vertex_pose->estimate();
}
```

The program is similar to *g2o* in lecture 6. We first declare the *g2o* graph optimizer and configure the solver and gradient descent method. Then, based on the estimated feature points, we put the pose and spatial points into the graph. Finally, the optimization function is called. The partial output of the run is as follows:

Listing 6.14 Terminal output:

```
./build/pose_estimation_3d2d 1.png 2.png 1_depth.png 2_depth.png
-- Max dist : 95.000000
-- Min dist : 4.000000
In total, we get 79 set of feature points
3d-2d pairs: 76
solve pnp in opencv cost time: 0.000332991 seconds.
R=
[0.9978662025826269, -0.05167241613316376, 0.03991244360207524;
0.0505958915956335, 0.998339762771668, 0.02752769192381471;
-0.04126860182960625, -0.025449547736074, 0.998823919929363]
t=
[-0.1272259656955879;
-0.007507297652615337;
0.06138584177157709]
calling bundle adjustment by gauss newton
iteration 0 cost=645538.1857253
iteration 1 cost=12750.239874896
iteration 2 cost=12301.774589343
iteration 3 cost=12301.427574651
iteration 4 cost=12301.426806652
pose by g-n:
0.99786618832 -0.0516873580423 0.039893448423 -0.127218696289
0.0506143671126 0.998340854865 0.0274540224544 -0.00738695798083
-0.0412462852904 -0.0253762590968 0.998826706403 0.0617019263823
0        0        0        1
solve pnp by gauss newton cost time: 0.000159492 seconds.
calling bundle adjustment by g2o
iteration= 0 chi2= 413.390599 time= 2.7291e-05 cumTime= 2.7291e-05 edges= 76
     schur= 0 lambda= 79.000412 levenbergIter= 1
iteration= 1 chi2= 301.367030 time= 1.47e-05 cumTime= 4.1991e-05 edges= 76
     schur= 0 lambda= 26.333471 levenbergIter= 1
iteration= 2 chi2= 301.365779 time= 1.7794e-05 cumTime= 5.9785e-05 edges= 76
     schur= 0 lambda= 17.555647 levenbergIter= 1
iteration= 3 chi2= 301.365779 time= 1.4875e-05 cumTime= 7.466e-05 edges= 76
     schur= 0 lambda= 11.703765 levenbergIter= 1
iteration= 4 chi2= 301.365779 time= 1.3132e-05 cumTime= 8.7792e-05 edges= 76
     schur= 0 lambda= 7.802510 levenbergIter= 1
iteration= 5 chi2= 301.365779 time= 2.0379e-05 cumTime= 0.000108171 edges=
     76 schur= 0 lambda= 41.613386 levenbergIter= 3
```

```
iteration= 6 chi2= 301.365779 time= 3.4186e-05 cumTime= 0.000142357 edges=
    76 schur= 0 lambda= 2859650082279.672363 levenbergIter= 8
optimization costs time: 0.000763649 seconds.
pose estimated by g2o =
0.997866202583 -0.0516724161336 0.0399124436024 -0.127225965696
0.050595891596 0.998339762772 0.0275276919261 -0.00750729765631
-0.04126860183 -0.0254495477384 0.998823919929 0.0613858417711
0       0       0       1
solve pnp by g2o cost time: 0.000923095 seconds.
```

Those three results are basically the same. In terms of efficiency, the Gauss-Newton method implemented by ourselves ranked first with 0.15 milliseconds, followed by OpenCV's PnP, and finally, the implementation of *g2o*. Nonetheless, the time usages are all within 1 millisecond, which shows that the pose estimation algorithm does not really consume computational effort.

Bundle Adjustment is a common method rather than a special task. It may not be limited to two images. We can put the poses, and spatial points matched by multiple images for iterative optimization and even put the entire SLAM process in. We will deal with the large-scale problem again in lecture 8. We usually consider a small bundle adjustment problem regarding local camera poses and feature points at the frontend, aiming to solve and optimize it in real-time.

6.9 3D–3D Iterative Closest Point (ICP)

In the end, we will introduce the 3D-3D pose estimation problem. Suppose we have a set of matched 3D points (for example, we matched two RGB-D images):

$$\mathbf{P} = \{\mathbf{p}_1, \cdots, \mathbf{p}_n\}, \quad \mathbf{P}' = \{\mathbf{p}'_1, \cdots, \mathbf{p}'_n\},$$

Now, we want to find an Euclidean transformation $\mathbf{R}, \mathbf{t}$, which is[8]:

$$\forall i, \mathbf{p}_i = \mathbf{R}\mathbf{p}'_i + \mathbf{t}.$$

This problem can be solved by the iterative closest point (ICP). The camera model does not appear in the 3D–3D pose estimation, meaning that when only the transformation between two sets of 3D points is considered, it has nothing to do with the camera. Therefore, ICP is also feasible in lidar SLAM. But since lidar features are not rich enough, it is hard to know the matching relationship between the two pointsets. We can only assume the closest points are the same. So this method is called the iterative closest point. The feature points provide us with a better matching relationship in the computer vision, so the whole problem becomes easier to solve. In RGB-D SLAM, the camera pose can also be estimated in this way.

[8] The notation in this section is slightly different from the symbols in the previous two sections. You can consider $\mathbf{p}_i$ as the data in the second image, and $\mathbf{p}'_i$ as the data in the first image, and consistently geting $\mathbf{R}, \mathbf{t}$.

In the following, we use ICP to refer to the motion estimation problem between the two sets of **matched** points.

Similar to PnP, the solution to ICP can be divided into two ways: using linear algebra (mainly SVD) and using nonlinear optimization (similar to Bundle Adjustment). They will be introduced separately below.

6.9.1 *Using Linear Algebra (SVD)*

First, let's look at the SVD on behalf of the algebraic method. According to the ICP problem described above, we first define the error term for the point i as:

$$\mathbf{e}_i = \mathbf{p}_i - (\mathbf{R}\mathbf{p}'_i + \mathbf{t}). \tag{6.49}$$

Then, construct a least-square problem to find the $\mathbf{R}, \mathbf{t}$ by minimization of sum of the squared errors:

$$\min_{\mathbf{R},\mathbf{t}} \frac{1}{2}\sum_{i=1}^{n} \| (\mathbf{p}_i - (\mathbf{R}\mathbf{p}_i{}' + \mathbf{t})) \|_2^2. \tag{6.50}$$

To solve this problem, we define the centroids of the two sets of points as:

$$\mathbf{p} = \frac{1}{n}\sum_{i=1}^{n}(\mathbf{p}_i), \quad \mathbf{p}' = \frac{1}{n}\sum_{i=1}^{n}(\mathbf{p}'_i). \tag{6.51}$$

The centroids are not subscripted. Then, in the error function:

$$\begin{aligned}
\tfrac{1}{2}\sum_{i=1}^{n} \left\| \mathbf{p}_i - (\mathbf{R}\mathbf{p}_i{}' + \mathbf{t}) \right\|^2 &= \tfrac{1}{2}\sum_{i=1}^{n} \left\| \mathbf{p}_i - \mathbf{R}\mathbf{p}_i{}' - \mathbf{t} - \mathbf{p} + \mathbf{R}\mathbf{p}' + \mathbf{p} - \mathbf{R}\mathbf{p}' \right\|^2 \\
&= \tfrac{1}{2}\sum_{i=1}^{n} \left\| (\mathbf{p}_i - \mathbf{p} - \mathbf{R}(\mathbf{p}_i{}' - \mathbf{p}')) + (\mathbf{p} - \mathbf{R}\mathbf{p}' - \mathbf{t}) \right\|^2 \\
&= \tfrac{1}{2}\sum_{i=1}^{n} (\left\| \mathbf{p}_i - \mathbf{p} - \mathbf{R}(\mathbf{p}_i{}' - \mathbf{p}') \right\|^2 + \left\| \mathbf{p} - \mathbf{R}\mathbf{p}' - \mathbf{t} \right\|^2 + \\
&\quad 2(\mathbf{p}_i - \mathbf{p} - \mathbf{R}(\mathbf{p}_i{}' - \mathbf{p}'))^T (\mathbf{p} - \mathbf{R}\mathbf{p}' - \mathbf{t})).
\end{aligned}$$

Since $(\mathbf{p}_i - \mathbf{p} - \mathbf{R}(\mathbf{p}_i{}' - \mathbf{p}'))$ is zero after the summation, so the optimization objective function can be simplified to

$$\min_{\mathbf{R},\mathbf{t}} J = \frac{1}{2}\sum_{i=1}^{n} \left\| \mathbf{p}_i - \mathbf{p} - \mathbf{R}(\mathbf{p}_i{}' - \mathbf{p}') \right\|^2 + \left\| \mathbf{p} - \mathbf{R}\mathbf{p}' - \mathbf{t} \right\|^2. \tag{6.52}$$

Carefully observe those two terms, we find that the first term is only related to the rotation matrix **R**, while the second has both **R** and **t**, but only related to the centroid. As long as we get **R**, we can get **t** by making the second term zero. Therefore, ICP can be solved in the following three steps:

1. Calculate the centroids of the two groups of points $\mathbf{p}$, $\mathbf{p}'$, and then calculate the **de-centroid coordinates** of each point:
$$\mathbf{q}_i = \mathbf{p}_i - \mathbf{p}, \quad \mathbf{q}'_i = \mathbf{p}'_i - \mathbf{p}'.$$
2. The rotation matrix is calculated according to the following optimization problem:
$$\mathbf{R}^* = \arg\min_{\mathbf{R}} \frac{1}{2}\sum_{i=1}^{n} \left\| \mathbf{q}_i - \mathbf{R}\mathbf{q}'_i \right\|^2. \tag{6.53}$$
3. Calculate **t** according to **R** in step 2:
$$\mathbf{t}^* = \mathbf{p} - \mathbf{R}\mathbf{p}'. \tag{6.54}$$

We find that once the rotation between the two sets of points is found, and the translation is easy to obtain. So we focus on the calculation of **R**. Expand the error term about **R**, we get:

$$\frac{1}{2}\sum_{i=1}^{n} \left\| \mathbf{q}_i - \mathbf{R}\mathbf{q}'_i \right\|^2 = \frac{1}{2}\sum_{i=1}^{n} \underbrace{\mathbf{q}_i^T\mathbf{q}_i}_{\text{not relavant}} + \mathbf{q}_i'^T \underbrace{\mathbf{R}^T\mathbf{R}}_{=\mathbf{I}} \mathbf{q}'_i - 2\mathbf{q}_i^T\mathbf{R}\mathbf{q}'_i. \tag{6.55}$$

The first item is not relevant to **R**. The second item can also be ignored since $\mathbf{R}^T\mathbf{R} = \mathbf{I}$. Therefore, the actual optimization objective function becomes the form of:

$$\sum_{i=1}^{n} -\mathbf{q}_i^T\mathbf{R}\mathbf{q}'_i = \sum_{i=1}^{n} -\text{tr}\left(\mathbf{R}\mathbf{q}'_i\mathbf{q}_i^T\right) = -\text{tr}\left(\mathbf{R}\sum_{i=1}^{n}\mathbf{q}'_i\mathbf{q}_i^T\right). \tag{6.56}$$

Next, we introduce how to solve the optimal **R** in the above problem through SVD. The proof of optimality is more complicated, and interested readers can refer to the literature [62, 63]. To find **R**, first define the matrix:

$$\mathbf{W} = \sum_{i=1}^{n} \mathbf{q}_i\mathbf{q}_i'^{\mathrm{T}}. \tag{6.57}$$

W is a 3×3 matrix. Performing SVD decomposition on **W**, we get:

$$\mathbf{W} = \mathbf{U}\mathbf{\Sigma}\mathbf{V}^T. \tag{6.58}$$

$\mathbf{\Sigma}$ is a diagonal matrix composed of singular values, the diagonal elements are arranged from large to small, and $\mathbf{U}$, $\mathbf{V}$ are diagonal matrices. When $\mathbf{W}$ is of full rank, $\mathbf{R}$ is:

$$\mathbf{R} = \mathbf{U}\mathbf{V}^T. \tag{6.59}$$

Once solving $\mathbf{R}$, we can use the equation (6.54) to solve $\mathbf{t}$. If the determinant of $\mathbf{R}$ is negative, then $-\mathbf{R}$ is taken as the optimal value.

6.9.2 Using Non-linear Optimization

Another way to solve ICP is to use nonlinear optimization to find the optimal value iteratively. This method is similar to the PnP we described earlier. When expressing the poses in Lie algebra, the objective function can be written as

$$\min_{\xi} = \frac{1}{2}\sum_{i=1}^{n} \| (\mathbf{p}_i - \exp(\xi^\wedge)\, \mathbf{p}_i') \|_2^2. \tag{6.60}$$

The derivative of a single error term with respect to the pose has been derived above, using the Lie algebra perturbation model:

$$\frac{\partial \mathbf{e}}{\partial \delta \xi} = -(\exp(\xi^\wedge)\, \mathbf{p}_i{}')^{\odot}. \tag{6.61}$$

Therefore, in nonlinear optimization, it only needs to iterate continuously to find the minimum value. Moreover, it can be proved that [5] the ICP problem has a unique solution or an infinite number of solutions. In the case of a unique solution, as long as the minimum value solution can be found, this minimum value is the global minimum. This also means that the initial value of the ICP solution can be arbitrarily selected. This is a big advantage of solving ICP when the points have been matched already.

It should be noted that the ICP we are talking about here refers to the problem of pose estimation when the matching is given by the image features. In the case of known matching, this least-squares problem actually has an analytical solution [64–66], so iterative optimization is not necessary. ICP researchers tend to be more concerned about the unknown matching situation. So, why do we introduce optimization-based ICP? Because in some cases, such as in RGB-D SLAM, the depth of a pixel may or may not be measured, so we can combine PnP and ICP optimization. For feature points with known depth, model their 3D-3D errors; for feature points with unknown depths, model 3D-2D reprojection errors. Therefore, all errors can be considered in the same problem, making the solution more convenient.

6.10 Practice: Solving ICP

6.10.1 Using SVD

Let's demonstrate how to use SVD and nonlinear optimization to solve ICP. In this section, we use two RGB-D images to obtain two sets of 3D points through feature matching, and finally, we use ICP to calculate their pose transformation. Since OpenCV does not currently have a method to calculate two sets of ICPs with matching points, and its principle is not complicated, we will implement an ICP by ourselves.

Listing 6.15 slambook2/ch7/pose_estimation_3d3d.cpp (part)

```
void pose_estimation_3d3d(
const vector<Point3f> &pts1,
const vector<Point3f> &pts2,
Mat &R, Mat &t) {
  Point3f p1, p2; // center of mass
  int N = pts1.size();
  for (int i = 0; i < N; i++) {
    p1 += pts1[i];
    p2 += pts2[i];
  }
  p1 = Point3f(Vec3f(p1) / N);
  p2 = Point3f(Vec3f(p2) / N);
  vector<Point3f> q1(N), q2(N); // remove the center
  for (int i = 0; i < N; i++) {
    q1[i] = pts1[i] - p1;
    q2[i] = pts2[i] - p2;
  }

  // compute q1*q2^T
  Eigen::Matrix3d W = Eigen::Matrix3d::Zero();
  for (int i = 0; i < N; i++) {
    W += Eigen::Vector3d(q1[i].x, q1[i].y, q1[i].z) * Eigen::Vector3d(q2[i].x
        , q2[i].y, q2[i].z).transpose();
  }
  cout << "W=" << W << endl;

  // SVD on W
  Eigen::JacobiSVD<Eigen::Matrix3d> svd(W, Eigen::ComputeFullU | Eigen::
      ComputeFullV);
  Eigen::Matrix3d U = svd.matrixU();
  Eigen::Matrix3d V = svd.matrixV();

  cout << "U=" << U << endl;
  cout << "V=" << V << endl;

  Eigen::Matrix3d R_ = U * (V.transpose());
  if (R_.determinant() < 0) {
    R_ = -R_;
  }
  Eigen::Vector3d t_ = Eigen::Vector3d(p1.x, p1.y, p1.z) - R_ * Eigen::
      Vector3d(p2.x, p2.y, p2.z);

  // convert to cv::Mat
  R = (Mat_<double>(3, 3) <<
    R_(0, 0), R_(0, 1), R_(0, 2),
    R_(1, 0), R_(1, 1), R_(1, 2),
```

```
    R_(2, 0), R_(2, 1), R_(2, 2)
  );
  t = (Mat_<double>(3, 1) << t_(0, 0), t_(1, 0), t_(2, 0));
}
```

The implementation of ICP is consistent with the previous theoretical part. We call *Eigen* for SVD and then calculate the $\mathbf{R}, \mathbf{t}$ matrix. We output the matched result, but please note that since the previous derivation is based on $\mathbf{p}_i = \mathbf{R}\mathbf{p}_i' + \mathbf{t}$, here is $\mathbf{R}, \mathbf{t}$ is the transformation from the second frame to the first frame, which is the opposite of the previous theoritical part. So in the output result, we also printed the inverse transform:

Listing 6.16 Terminal output:

```
./build/pose_estimation_3d3d 1.png 2.png 1_depth.png 2_depth.png
-- Max dist : 95.000000
-- Min dist : 4.000000
In total, we get 79 set of feature points
3d-3d pairs: 74
W= 11.9404 -0.567258 1.64182
-1.79283 4.31299 -6.57615
3.12791 -6.55815 10.8576
U= 0.474144 -0.880373 -0.0114952
-0.460275 -0.258979 0.849163
0.750556 0.397334 0.528006
V= 0.535211 -0.844064 -0.0332488
-0.434767 -0.309001 0.84587
0.724242 0.438263 0.532352
ICP via SVD results:
R = [0.9972395977366739, 0.05617039856770099, -0.04855997354553433;
-0.05598345194682017, 0.9984181427731508, 0.005202431117423125;
0.0487753812298326, -0.002469515369266572, 0.9988067198811421]
t = [0.1417248739257469;
-0.05551033302525193;
-0.03119093188273858]
R_inv = [0.9972395977366739, -0.05598345194682017, 0.0487753812298326;
0.05617039856770099, 0.9984181427731508, -0.002469515369266572;
-0.04855997354553433, 0.005202431117423125, 0.9988067198811421]
t_inv = [-0.1429199667309695;
0.04738475446275858;
0.03832465717628181]
```

Readers can compare the difference between ICP and PnP and the motion estimation results of epipolar geometry. We are using more and more information in this process (no depth → the depth of one image → the depth of two images). Therefore, when the depth is accurate, the estimates will be more and more accurate. However, due to the noise in Kinect's depth map and the possibility of data loss, we have to discard some feature points without depth data. This may cause the estimation of ICP to be inaccurate, and if too many feature points are discarded, it may cause a situation where motion estimation cannot be performed due to too few feature points.

6.10.2 Using Non-linear Optimization

Now consider using nonlinear optimization to calculate ICP. We still use Lie algebra to optimize the camera pose. The RGB-D camera can observe the 3D position of the landmarks every time, thereby generating 3D observation data. We use the VertexPose in the previous practice and then define the unary edges of 3D-3D:

Listing 6.17 slambook2/ch7/pose_estimation_3d3d.cpp (part)

```
/// g2o edge
class EdgeProjectXYZRGBDPoseOnly : public g2o::BaseUnaryEdge<3, Eigen::
    Vector3d, VertexPose> {
public:
  EIGEN_MAKE_ALIGNED_OPERATOR_NEW;

  EdgeProjectXYZRGBDPoseOnly(const Eigen::Vector3d &point) : _point(point) {}

  virtual void computeError() override {
    const VertexPose *pose = static_cast<const VertexPose *> ( _vertices[0] )
        ;
    _error = _measurement - pose->estimate() * _point;
  }

  virtual void linearizeOplus() override {
    VertexPose *pose = static_cast<VertexPose *>(_vertices[0]);
    Sophus::SE3d T = pose->estimate();
    Eigen::Vector3d xyz_trans = T * _point;
    _jacobianOplusXi.block<3, 3>(0, 0) = -Eigen::Matrix3d::Identity();
    _jacobianOplusXi.block<3, 3>(0, 3) = Sophus::SO3d::hat(xyz_trans);
  }

  bool read(istream &in) {}

  bool write(ostream &out) const {}

protected:
  Eigen::Vector3d _point;
};
```

They are unary edges, similar to the g2o::EdgeSE3ProjectXYZ mentioned earlier, but the observation has changed from 2 to 3 dimensions. There is no camera model involved, and only one vertex is related. Please pay attention to the form of the Jacobian matrix here. It must be consistent with our previous derivation. The Jacobian matrix gives the derivative of the camera pose and is 3×6.

The code for using *g2o* for optimization is similar. We just set the nodes and edges for graph optimization. Readers are suggested to check the source file for this part of the code, which is not listed here. Now, let's take a look at the results of the optimization:

Listing 6.18 Terminal output:

```
iteration= 0 chi2= 1.811539 time= 1.7046e-05 cumTime= 1.7046e-05 edges= 74
    schur= 0
iteration= 1 chi2= 1.811051 time= 1.0422e-05 cumTime= 2.7468e-05 edges= 74
    schur= 0
iteration= 2 chi2= 1.811050 time= 9.589e-06 cumTime= 3.7057e-05 edges= 74
    schur= 0
...
iteration= 9 chi2= 1.811050 time= 9.113e-06 cumTime= 0.000100604 edges= 74
    schur= 0
optimization costs time: 0.000559208 seconds.

after optimization:
T=
0.99724 0.0561704 -0.04856 0.141725
```

```
-0.0559834 0.998418 0.00520242 -0.0555103
0.0487754 -0.0024695 0.998807 -0.0311913
0 0 0 1
```

We found that the overall error has become stable after only one iteration, indicating that the algorithm has converged after only one iteration. From the result of the pose, it can be seen that it is almost the same as the pose calculated by the previous SVD, which shows that SVD has already given an analytical solution to the optimization problem. Therefore, in this practice, it can be considered that the result given by SVD is the optimal value of the camera pose.

It should be noted that in the practice of ICP, we used feature points that have depth readings in both images. However, as long as the depth of one of the images is determined, we can use errors similar to PnP to add them to the optimization. In addition to the camera pose, considering the spatial points as optimization variables is also a way to solve it. We should be clear that the actual solution is very flexible and does not need to be bound to a certain fixed form. If you consider points and cameras simultaneously, the whole problem becomes more flexible, and you may get other solutions. For example, you can make the camera rotate less and move the point more. This reflects that we would like to have as many constraints as possible in bundle adjustment because multiple observations will bring more information and enable us to estimate each variable more accurately.

6.11 Summary

This chapter introduces several important issues in visual odometry based on feature points, including:

1. How the feature points are extracted and matched.
2. How to estimate camera motion through 2D-2D feature points.
3. How to estimate the spatial position of a point from a 2D–2D match.
4. 3D–2D PnP problem and its linear solution and bundle adjustment solution.
5. 3D—3D ICP problem and its linear solution and bundle adjustment solution.

This chapter is complicated in content and combines the basic knowledge of the previous lectures. If readers find it difficult to understand, please go back to review the previous chapters. It is best to do the practice yourself to entirely understand the content of the motion estimation.

What needs to be mentioned here is we have omitted a lot of discussion about some special situations. For example, what happens if the given feature points are coplanar during the epipolar geometry solution (this is mentioned in the homography matrix **H**)? What happens to collinear? If we solve such a solution in PnP and ICP, what will happen? Can the algorithms recognize these exceptional cases and report that the resulting solution may be unreliable? Can you give the estimated uncertainty of **T**? Although they are all worthy of research and exploration, their discussion is more suitable in specific papers. The goal of this book is to cover basic knowledge of visual SLAM. We will not expand on these issues for now. At the same time, these situations rarely occur in engineering. If you care about these rare situations, you can read books such as [2].

Exercises

1. In addition to the ORB feature points introduced in this book, what other feature points do you know? Please elaborate on the principles of SIFT or SURF, and compare their advantages and disadvantages with ORB.
2. Write a program to call other types of feature points in OpenCV. Compare their time spent on your machine when extracting 1000 feature points.
3.* We found that the ORB feature points provided by OpenCV are not evenly distributed in the image. Can you find or propose a way to make the distribution of feature points more evenly?
4. Investigate why FLANN can quickly handle matching problems. In addition to FLANN, what other ways to accelerate matching?
5. Substitute the EPnP used in the demo program with other PnP methods and investigate their working principles.
6. In PnP optimization, taking the first camera's observation into consideration, how should the problem/program be formed? How will the final result change?
7. In the ICP program, if the spatial points are also considered optimization variables, how should the program be written? How will the final result change?
8.* In the feature point matching, mismatches will inevitably be encountered. What happens if we put the wrong match into PnP or ICP? What methods can you think of to avoid mismatches?
9.* Use the SE3 class in Sophus to write the nodes and edges of *g2o* by yourself and implement the optimization of PnP and ICP.
10.* Implement the optimization of PnP and ICP in Ceres.

Chapter 7
Visual Odometry: Part II

> **Goal of Study**
> 1. Study the principle of optical flow.
> 2. Learn how to use the direct method to estimate the camera pose.
> 3. Use *g2o* to implement the direct method.

Different from the feature method, the direct method is also another important approach in VO. Despite that it has not been the mainstream of VO, the direct method can compete with the feature method after years of development. In this chapter, we will introduce the principle of the direct method and implement its core algorithm.

7.1 The Motivation of the Direct Method

In the last chapter, we introduced using features to estimate camera motion. Although the feature point method plays a crucial role in visual odometry, researchers still believe that it has at least the following shortcomings:

1. The extraction of key points and the calculation of the descriptor are very time-consuming. In practice, SIFT currently cannot be calculated in real-time on the CPU, and ORB also requires nearly 20ms of calculation. If the entire SLAM runs at a speed of 30 Hz, more than half of the time will be spent on feature calculation.
2. In the feature method, not all the information is used. An image has hundreds of thousands of pixels but only a few hundred feature points. Using only feature points discards most of the possibly useful image information.
3. The camera sometimes moves to featureless places, where there is often no obvious texture information. For example, sometimes we will face a white wall or an empty corridor. The number of feature points in these scenes will be significantly reduced, and we may not find enough matching points to calculate camera motion.

X. Gao and T. Zhang, *Introduction to Visual SLAM*,
https://doi.org/10.1007/978-981-16-4939-4_7

Now we see that there are indeed some problems with using feature points. Is there any way to overcome these shortcomings? We have the following ideas:

- Keep feature points, but discard their descriptors. At the same time, use the optical flow to track the motion of the key points. This can avoid the time taken by the descriptors. The time spent on calculating optical flow itself is less than the descriptor calculation and matching.
- Only calculate key points, not descriptors. Simultaneously, use the direct method to estimate the camera motion as well as the tracked pixels. This can also save the time spent on the calculation of the descriptor and the optical flow.

The first approach still uses feature points but substitutes the descriptor matching with optical flow tracking. It still uses epipolar geometry, PnP, or ICP algorithms to estimate camera motion. This requires that the extracted key points are distinguishable, which means the system still relies on corner points. In the direct method, we will estimate the camera motion and the projection of the points simultaneously according to the pixel's grayscale. The corner points are no longer necessary. As you will see later, they can even be used for randomly selected points.

In the feature method, we regard feature points as fixed points in three-dimensional space. The camera motion is optimized by minimizing the reprojection error. In this process, we need to know exactly the spatial point's projected position (through feature matching), which is sometimes hard to find. Meanwhile, computing and matching features require a lot of computation. In contrast, in the direct method, we do not need to know the correspondence between points in advance but find it by minimizing the *photometric error*.

We will focus on the direct method in this chapter. Its (original) motivation is to overcome the shortcomings of the feature point method listed above. The direct method estimates the camera motion based on the pixels' brightness information and can completely eliminate the calculation of key points and descriptors. Therefore, it saves the calculation time of features and solves the problems caused by lacking features. As long as there are brightness changes in the scene (it can be an edge, not a corner, even gradual change without forming a local image gradient), the direct method will work. According to the number of pixels used, the direct method can be categorized into sparse, semi-dense, and dense. Compared with the feature point method that can only reconstruct sparse feature points (sparse map), the direct method also can restore semi-dense or dense structures.

Historically, there were also early uses of the direct method [67]. With the emergence of some open-source projects that use the direct method, such as SVO [68], LSD-SLAM [69], DSO [70], etc., the direct method became a more and more important part of the visual odometry.

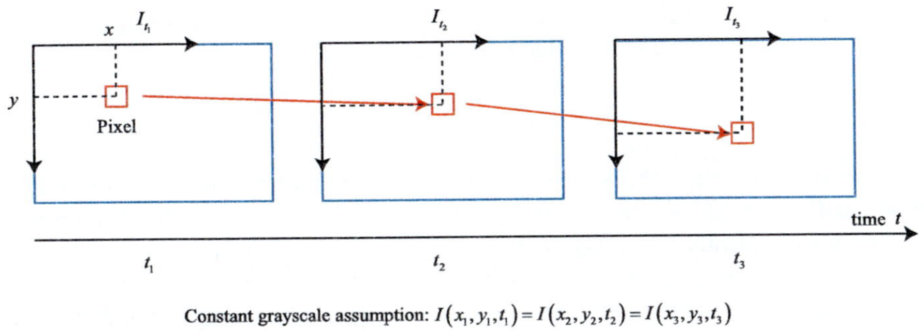

Fig. 7.1 Optical flow of a single pixel

7.2 2D Optical Flow

The direct method was inspired by the optical flow. It is similar to optical flow to some extent and uses the same assumptions. Optical flow describes the motion of pixels in the image, and the direct method is accompanied by a camera motion model. Before the direct method, we will introduce optical flow first.

Optical flow is a method of describing pixels' movement between images, as shown in Fig. 7.1. If a dynamic object moves in the camera, we will expect the pixels to also move in the image over time, and we want to track its movement. The calculation of a part of the pixel's motion is called sparse optical flow, and the calculation of all pixels in an image is called dense optical flow. A well-known sparse optical flow method is called Lucas-Kanade optical flow [71]. It can be used to track the position of feature points in SLAM. Dense optical flow is represented by Horn-Schunck optical flow [72]. This section mainly introduces Lucas-Kanade optical flow (LK flow) since it is more useful in SLAM.

7.2.1 Lucas-Kanade Optical Flow

In the LK optical flow, we assume the image changes over time. The image can be regarded as a function of time $\mathbf{I}(t)$. Then, for a pixel at (x, y) at time t, its grayscale can be written as:

$$\mathbf{I}(x, y, t).$$

In this way, the image is regarded as a function of position and time, and its value is the grayscale. Now consider a 2D pixel. Its coordinates at time t are x, y. Due to the movement of the camera, its image coordinates will change. We want to estimate the position of this pixel at other times. But how to estimate it? Here we will introduce the basic assumption of the optical flow method.

Constant grayscale assumption: The pixel's grayscale is constant in each image.

For the pixel at (x, y) at time t, suppose it moves to $(x + \mathrm{d}x, y + \mathrm{d}y)$ at time $t + \mathrm{d}t$. Since the grayscale is unchanged, we have:

$$\mathbf{I}(x + \mathrm{d}x, y + \mathrm{d}y, t + \mathrm{d}t) = \mathbf{I}(x, y, t). \tag{7.1}$$

Note that most of the time, the assumption of constant brightness is not true in practice. If you have learned some computer graphics knowledge, the final brightness of a pixel is determined by lots of parameters. Due to the objects' various materials, the pixels will have highlights and shadows; sometimes, the camera will automatically adjust its exposure parameters to make the overall image brighter or darker. At these times, the assumption of constant brightness is invalid, so the result of optical flow is not necessarily reliable. However, on the other hand, all algorithms work under certain assumptions. If we do not make any assumptions, we cannot design practical algorithms. So, let us temporarily accept this assumption and see how to calculate the motion of the pixels.

Carry out the Taylor expansion on the left side and only keep the first-order term. We have:

$$\mathbf{I}\left(x + \mathrm{d}x, y + \mathrm{d}y, t + \mathrm{d}t\right) \approx \mathbf{I}\left(x, y, t\right) + \frac{\partial \mathbf{I}}{\partial x}\mathrm{d}x + \frac{\partial \mathbf{I}}{\partial y}\mathrm{d}y + \frac{\partial \mathbf{I}}{\partial t}\mathrm{d}t. \tag{7.2}$$

Because we assume that the brightness does not change, so we have:

$$\frac{\partial \mathbf{I}}{\partial x}\mathrm{d}x + \frac{\partial \mathbf{I}}{\partial y}\mathrm{d}y + \frac{\partial \mathbf{I}}{\partial t}\mathrm{d}t = 0. \tag{7.3}$$

Divide both sides by $\mathrm{d}t$[1]:

$$\frac{\partial \mathbf{I}}{\partial x}\frac{\mathrm{d}x}{\mathrm{d}t} + \frac{\partial \mathbf{I}}{\partial y}\frac{\mathrm{d}y}{\mathrm{d}t} = -\frac{\partial \mathbf{I}}{\partial t}, \tag{7.4}$$

where $\mathrm{d}x/\mathrm{d}t$ is the speed of the pixel on the x axis, and $\mathrm{d}y/\mathrm{d}t$ is the speed on the y axis. We denote them as u, v. At the same time, $\partial \mathbf{I}/\partial x$ is the gradient of the image in the x direction at this point, and $\partial \mathbf{I}/\partial y$ is the gradient in the y direction, denoted as $\mathbf{I}_x, \mathbf{I}_y$. Denote the change of the image brightness with respect to time as $\mathbf{I}_t$. Then the above equation can be written in the matrix form:

$$\begin{bmatrix} \mathbf{I}_x & \mathbf{I}_y \end{bmatrix} \begin{bmatrix} u \\ v \end{bmatrix} = -\mathbf{I}_t. \tag{7.5}$$

What we want is to calculate the motion u, v of the pixel, but this formula is a linear equation with two variables, and we cannot find u, v by a single pixel. Therefore,

[1] Well, this is not strictly correct to say so. We can also start from the time derivative and get the same result.

additional constraints are needed to calculate u, v. In LK optical flow, we also assume the pixels in a certain window have the same motion.

Consider a window of size $w \times w$, which contains w^2 pixels. Since the pixels in this window are assumed to have the same motion, we have a total of w^2 equations:

$$\begin{bmatrix} \mathbf{I}_x & \mathbf{I}_y \end{bmatrix}_k \begin{bmatrix} u \\ v \end{bmatrix} = -\mathbf{I}_{tk}, \quad k = 1, \ldots, w^2. \tag{7.6}$$

Stacking them:

$$\mathbf{A} = \begin{bmatrix} [\mathbf{I}_x, \mathbf{I}_y]_1 \\ \vdots \\ [\mathbf{I}_x, \mathbf{I}_y]_k \end{bmatrix}, \mathbf{b} = \begin{bmatrix} \mathbf{I}_{t1} \\ \vdots \\ \mathbf{I}_{tk} \end{bmatrix}. \tag{7.7}$$

Then the whole equation is:

$$\mathbf{A} \begin{bmatrix} u \\ v \end{bmatrix} = -\mathbf{b}. \tag{7.8}$$

This is an overdetermined linear equation about u, v. And we can find its least-square solution.

$$\begin{bmatrix} u \\ v \end{bmatrix}^* = -\left(\mathbf{A}^T\mathbf{A}\right)^{-1}\mathbf{A}^T\mathbf{b}. \tag{7.9}$$

In this way, the speed u, v of pixels between images is obtained. When t takes discrete moments instead of continuous-time, we can estimate the position of a block of pixels in several images. Since the pixel gradient is only valid locally, if one iteration does not produce a good result, we will iterate this calculation several times. In SLAM, LK optical flow is often used to track the motion of corner points. We can have a deeper understanding of it through the program.

7.3 Practice: LK Optical Flow

7.3.1 LK Flow in OpenCV

We will use several sample images to track their feature points with OpenCV's optical flow in practice. At the same time, we will also manually implement an LK optical flow for a comprehensive understanding. We use two sample images from the Euroc dataset, extract the corner points in the first image, and then use optical flow to track their second image position. First, let's use the LK optical flow in OpenCV:

Listing 7.1 slambook2/ch8/optical_flow.cpp (part)

```
// use opencv's flow for validation
vector<Point2f> pt1, pt2;
for (auto &kp: kp1) pt1.push_back(kp.pt);
vector<uchar> status;
vector<float> error;
cv::calcOpticalFlowPyrLK(img1, img2, pt1, pt2, status, error);
```

The optical flow in OpenCV is very easy to use. We only need to call the cv::calcOpticalFlowPyrLK function, provide two images and the corresponding feature points, get the tracked points, the status, and each point's error. We can determine whether the corresponding point is tracked correctly according to whether the status variable is 1. This function also has some optional parameters, but we only use the default parameters in the demonstration. We omit other codes that mention features and draw results here, which have been shown in the previous code part.

7.3.2 *Optical Flow with Gauss-Newton Method*

7.3.2.1 Single-Layer Optical Flow

Optical flow can also be seen as an optimization problem: by minimizing the grayscale error, the optimal pixel shift is estimated. Therefore, similar to those previously implemented Gauss-Newton methods, we now also implement an optical flow based on the Gauss-Newton method.[2]

Listing 7.2 slambook2/ch8/optical_flow.cpp (part))

```
class OpticalFlowTracker {
public:
  OpticalFlowTracker(
    const Mat &img1_,
    const Mat &img2_,
    const vector<KeyPoint> &kp1_,
    vector<KeyPoint> &kp2_,
    vector<bool> &success_,
    bool inverse_ = true, bool has_initial_ = false) :
    img1(img1_), img2(img2_), kp1(kp1_), kp2(kp2_), success(success_),
        inverse(inverse_),
    has_initial(has_initial_) {}

  void calculateOpticalFlow(const Range &range);

private:
  const Mat &img1;
  const Mat &img2;
  const vector<KeyPoint> &kp1;
  vector<KeyPoint> &kp2;
  vector<bool> &success;
  bool inverse = true;
  bool has_initial = false;
};

```

[2] This example requires OpenCV 4.0 or higher because the interface of OpenCV has changed.

```
void OpticalFlowSingleLevel(
  const Mat &img1,
  const Mat &img2,
  const vector<KeyPoint> &kp1,
  vector<KeyPoint> &kp2,
  vector<bool> &success,
  bool inverse, bool has_initial) {
  kp2.resize(kp1.size());
  success.resize(kp1.size());
  OpticalFlowTracker tracker(img1, img2, kp1, kp2, success, inverse, has_initial);
  parallel_for_(Range(0, kp1.size()),
    std::bind(&OpticalFlowTracker::calculateOpticalFlow, &tracker, placeholders::_1));
}

void OpticalFlowTracker::calculateOpticalFlow(const Range &range) {
  // parameters
  int half_patch_size = 4;
  int iterations = 10;
  for (size_t i = range.start; i < range.end; i++) {
    auto kp = kp1[i];
    double dx = 0, dy = 0; // dx,dy need to be estimated
    if (has_initial) {
      dx = kp2[i].pt.x - kp.pt.x;
      dy = kp2[i].pt.y - kp.pt.y;
    }

    double cost = 0, lastCost = 0;
    bool succ = true; // indicate if this point succeeded

    // Gauss-Newton iterations
    Eigen::Matrix2d H = Eigen::Matrix2d::Zero(); // hessian
    Eigen::Vector2d b = Eigen::Vector2d::Zero(); // bias
    Eigen::Vector2d J; // jacobian
    for (int iter = 0; iter < iterations; iter++) {
      if (inverse == false) {
        H = Eigen::Matrix2d::Zero();
        b = Eigen::Vector2d::Zero();
      } else {
        // only reset b
        b = Eigen::Vector2d::Zero();
      }

      cost = 0;

      // compute cost and jacobian
      for (int x = -half_patch_size; x < half_patch_size; x++)
      for (int y = -half_patch_size; y < half_patch_size; y++) {
        double error = GetPixelValue(img1, kp.pt.x + x, kp.pt.y + y) -
          GetPixelValue(img2, kp.pt.x + x + dx, kp.pt.y + y + dy);; // Jacobian
        if (inverse == false) {
          J = -1.0 * Eigen::Vector2d(
            0.5 * (GetPixelValue(img2, kp.pt.x + dx + x + 1, kp.pt.y + dy + y) -
              GetPixelValue(img2, kp.pt.x + dx + x - 1, kp.pt.y + dy + y)),
            0.5 * (GetPixelValue(img2, kp.pt.x + dx + x, kp.pt.y + dy + y + 1) -
              GetPixelValue(img2, kp.pt.x + dx + x, kp.pt.y + dy + y - 1))
          );
        } else if (iter == 0) {
          // in inverse mode, J keeps same for all iterations
          // NOTE this J does not change when dx, dy is updated, so we can store it and only compute error
          J = -1.0 * Eigen::Vector2d(
            0.5 * (GetPixelValue(img1, kp.pt.x + x + 1, kp.pt.y + y) -
```

```
                GetPixelValue(img1, kp.pt.x + x - 1, kp.pt.y + y)),
              0.5 * (GetPixelValue(img1, kp.pt.x + x, kp.pt.y + y + 1) -
                GetPixelValue(img1, kp.pt.x + x, kp.pt.y + y - 1))
            );
          }
          // compute H, b and set cost;
          b += -error * J;
          cost += error * error;
          if (inverse == false || iter == 0) {
            // also update H
            H += J * J.transpose();
          }
        }

        // compute update
        Eigen::Vector2d update = H.ldlt().solve(b);

        if (std::isnan(update[0])) {
          // sometimes occurred when we have a black or white patch and H is irreversible
          cout << "update is nan" << endl;
          succ = false;
          break;
        }

        if (iter > 0 && cost > lastCost) {
          break;
        }

        // update dx, dy
        dx += update[0];
        dy += update[1];
        lastCost = cost;
        succ = true;

        if (update.norm() < 1e-2) {
          // converge
          break;
        }
      }

      success[i] = succ;

      // set kp2
      kp2[i].pt = kp.pt + Point2f(dx, dy);
    }
}
```

We have implemented a single-layer optical flow function in the OpticalFlowSingleLevel function, in which cv::parallel_for_ is called to parallelize the OpticalFlowTracker::calculateOpticalFlow function. It calculates the optical flow of feature points within a specified range. This parallel for loop is internally implemented by the Intel tbb library. We only need to define the function body according to its interface and then pass the function to it as a std::function object.

In the implementation of calculateOpticalFlow, we solve such a problem:

$$\min_{\Delta x, \Delta y} \left\| \mathbf{I}_1(x, y) - \mathbf{I}_2(x + \Delta x, y + \Delta y) \right\|_2^2. \tag{7.10}$$

Therefore, the residual is the part inside the brackets, and the corresponding Jacobian is the gradient of the second image at $x + \Delta x, y + \Delta y$. In addition, according to [73], the gradient can also be replaced by the gradient $\mathbf{I}_1(x, y)$ of the first image. This is called the inverse optical flow method. In inverse optical flow, the gradient of $\mathbf{I}_1(x, y)$ remains unchanged, so we can use the result calculated in the first iteration in the subsequent iterations. When the Jacobian remains unchanged, the $\mathbf{H}$ matrix is unchanged, and only the residual is calculated for each iteration, which can save a lot of calculation.

7.3.2.2 Multi-layer Optical Flow

Since we write optical flow as an optimization problem, we must assume that the initial value of optimization is close to the optimal value to ensure the algorithm's convergence. Therefore, if the camera moves faster and the difference between the two images is obvious, the single-layer image optical flow method can be easily stuck at a local minimum. But it can be resolved to some extent by image pyramids.

Image pyramid refers to scaling the image to get samples in different resolutions, as shown in Fig. 7.2. The original image is used as the bottom layer of the pyramid. Every time we go one layer up, the lower layer image is scaled to a certain magnification, and then a pyramid is obtained. When calculating the optical flow, we start from the top layer image and go down one by one using the previous layer's tracking result as the initial value of the next layer's optical flow. Since the upper layer image is relatively rough, this process is also called coarse-to-fine optical flow, which is also the usual optical flow process in practice.

The advantage of going from coarse to fine is that when the pixel motion of the original image is large, the motion is still within a small range from the image at the top of the pyramid. For example, suppose the original image's feature points move by 20 pixels. In that case, it is easy for the optimization to be trapped in the local minimum due to the image's non-convexity. But now, suppose there is a pyramid with a zoom magnification of 0.5 times. In the upper two layers of images, the pixel

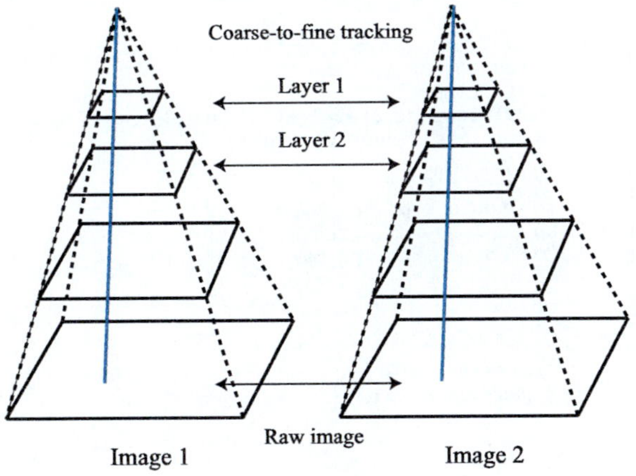

Fig. 7.2 Image pyramid and coarse-to-fine process

movement is only 5 pixels, and the result is obviously better than directly optimizing the original image.

We have implemented multi-layer optical flow in the program:

Listing 7.3 slambook2/ch8/optical_flow.cpp (part)

```
void OpticalFlowMultiLevel(
  const Mat &img1,
  const Mat &img2,
  const vector<KeyPoint> &kp1,
  vector<KeyPoint> &kp2,
  vector<bool> &success,
  bool inverse) {

  // parameters
  int pyramids = 4;
  double pyramid_scale = 0.5;
  double scales[] = {1.0, 0.5, 0.25, 0.125};

  // create pyramids
  vector<Mat> pyr1, pyr2; // image pyramids
  for (int i = 0; i < pyramids; i++) {
    if (i == 0) {
      pyr1.push_back(img1);
      pyr2.push_back(img2);
    } else {
      Mat img1_pyr, img2_pyr;
      cv::resize(pyr1[i - 1], img1_pyr,
      cv::Size(pyr1[i - 1].cols * pyramid_scale, pyr1[i - 1].rows *
          pyramid_scale));
      cv::resize(pyr2[i - 1], img2_pyr,
      cv::Size(pyr2[i - 1].cols * pyramid_scale, pyr2[i - 1].rows *
          pyramid_scale));
      pyr1.push_back(img1_pyr);
      pyr2.push_back(img2_pyr);
    }
  }

  // coarse-to-fine LK tracking in pyramids
  vector<KeyPoint> kp1_pyr, kp2_pyr;
  for (auto &kp:kp1) {
    auto kp_top = kp;
    kp_top.pt *= scales[pyramids - 1];
    kp1_pyr.push_back(kp_top);
    kp2_pyr.push_back(kp_top);
  }

  for (int level = pyramids - 1; level >= 0; level--) {
    // from coarse to fine
    success.clear();
    OpticalFlowSingleLevel(pyr1[level], pyr2[level], kp1_pyr, kp2_pyr,
        success, inverse, true);

    if (level > 0) {
      for (auto &kp: kp1_pyr)
      kp.pt /= pyramid_scale;
      for (auto &kp: kp2_pyr)
      kp.pt /= pyramid_scale;
    }
  }

  for (auto &kp: kp2_pyr)
    kp2.push_back(kp);
}
```

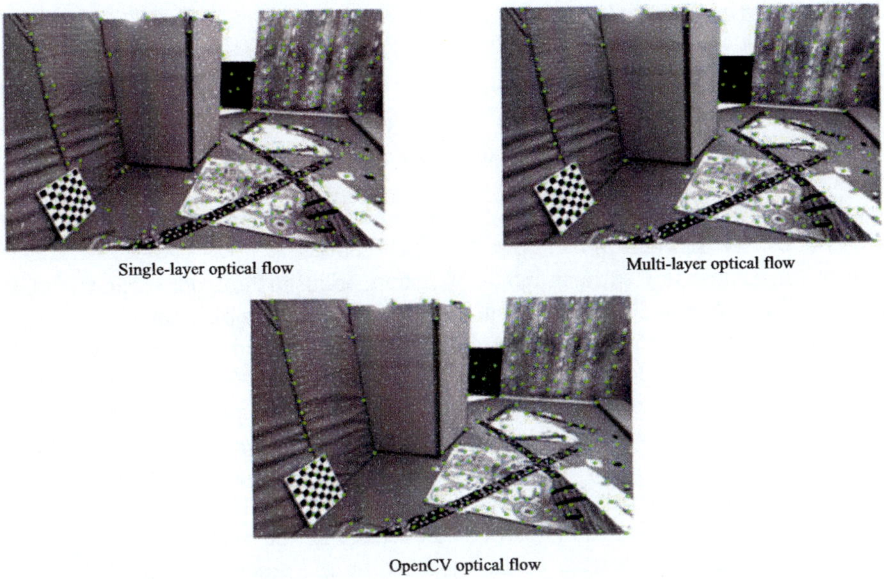

Fig. 7.3 Comparison of the results of various optical flows

This code constructs a four-layer pyramid with a scaling rate of 0.5 and calls the single-layer optical flow function to achieve the multi-layer optical flow. In the main function, we tested the performance of OpenCV's optical flow, single-layer optical flow, and multi-layer optical flow on two images and recorded their runtime:

Listing 7.4 Terminal input:

```
./build/optical_flow
build pyramid time: 0.000150349
track pyr 3 cost time: 0.000304633
track pyr 2 cost time: 0.000392889
track pyr 1 cost time: 0.000382347
track pyr 0 cost time: 0.000375099
optical flow by gauss-newton: 0.00189268
optical flow by opencv: 0.00220134
```

In terms of runtime, the multi-layer optical flow method takes roughly the same time as OpenCV. Since the parallelized program's performance varies from run to run, these numbers will not be exactly the same on the reader's machine. For the result of optical flow, see Fig. 7.3. It can be seen that the multi-layer optical flow has the same effect as OpenCV, and the single-layer optical flow performs obviously worse than the multi-layer optical flow.

7.3.3 Summary of the Optical Flow Practice

We see that LK optical flow can directly obtain the corresponding relationship of feature points. This correspondence is just like matching descriptors, except that optical flow requires higher image continuity and light stability. We can use PnP, ICP, or epipolar geometry to estimate the camera motion through the feature points tracked by optical flow. These methods were introduced in the previous lecture and will not be discussed here.

In terms of runtime, it extracts about 230 feature points in the experiment. OpenCV and multi-layer optical flow need about 2 milliseconds to complete the tracking (the CPU I use is Intel I7-8550U), which is quite fast. If we use key points like FAST, then the entire optical flow calculation can be done in about 5 milliseconds, which is very fast compared to feature matching. However, if the position of the corner point is not good, the optical flow is also easy to be lost or give wrong results, which requires the subsequent algorithm to have a certain outlier removal mechanism. We leave the relevant discussion to the later chapter.

In summary, the optical flow method can accelerate the visual odometry calculation method based on feature points by avoiding the process of calculating and matching descriptors but requires smoother camera movement (or higher collection frequency).

7.4 Direct Method

Next, let's discuss the direct method, which is somehow similar to the optical flow method. We first introduce the principle of the direct method and then implement it.

7.4.1 Derivation of the Direct Method

In the optical flow, we will first track the location of features and then determine the camera's movement based on these locations. Such a two-step plan is difficult to guarantee overall optimality. We can ask, can we adjust the previous results in the latter step? For example, if the camera has turned 15 degrees to the right, can the optical flow use this 15-degree motion as the initial value to adjust the optical flow calculation? This idea is reflected in the direct method.

In Fig. 7.4, consider a spatial point P and camera at two timestamps. The world coordinates of P are $[X, Y, Z]$, and the pixel coordinates of its imaging on two cameras are $\mathbf{p}_1, \mathbf{p}_2$.

Our goal is to find the relative pose transformation from the first camera to the second camera. We take the first camera as the frame of reference, and set the rotation and translation of the second camera as $\mathbf{R}, \mathbf{t}$ (corresponding to the $\mathbf{T}$ in SE(3)). At

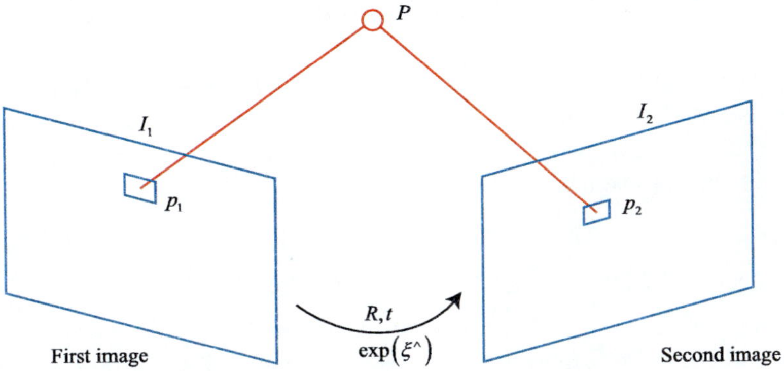

Fig. 7.4 The direct method

the same time, the intrinsics of the two cameras are the same, denoted as **K**. Let's write down the complete projection equation:

$$\mathbf{p}_1 = \begin{bmatrix} u \\ v \\ 1 \end{bmatrix}_1 = \frac{1}{Z_1}\mathbf{KP},$$

$$\mathbf{p}_2 = \begin{bmatrix} u \\ v \\ 1 \end{bmatrix}_2 = \frac{1}{Z_2}\mathbf{K}\left(\mathbf{RP}+\mathbf{t}\right) = \frac{1}{Z_2}\mathbf{K}\left(\mathbf{TP}\right)_{1:3},$$

where Z_1 is the depth of P, and Z_2 is the depth of P in the second camera frame, which is the third coordinate of $\mathbf{RP}+\mathbf{t}$. Since $\mathbf{T}$ can only be multiplied with homogeneous coordinates, we need to take out the first 3 elements after multiplying. This is consistent with the content of Chap. 4.

Recall that in the feature point method, since we know the pixel positions of $\mathbf{p}_1, \mathbf{p}_2$ through matching descriptors, we can calculate the reprojection position. But in the direct method, since there is no feature matching, we have no way of knowing which $\mathbf{p}_2$ and $\mathbf{p}_1$ correspond to the same point. The direct method's idea is to find the position of $\mathbf{p}_2$ according to the current camera pose estimation. But if the camera pose is not good enough, the appearance of $\mathbf{p}_2$ and $\mathbf{p}_1$ will be significantly different. Therefore, to reduce this difference, we optimize the camera's pose to find $\mathbf{p}_2$ that is more similar to $\mathbf{p}_1$. This can also be done by solving an optimization problem, but at this time, it is not to minimize the reprojection error but to minimize the *photometric error*, which is the brightness error of the two pixels of P:

$$e = \mathbf{I}_1\left(\mathbf{p}_1\right) - \mathbf{I}_2\left(\mathbf{p}_2\right). \tag{7.11}$$

Note that e is a scalar here. Similarly, the optimization is with respect to the $\mathcal{L}_2$ norm of the error, taking the unweighted form for now, as:

$$\min_{\mathbf{T}} J(\mathbf{T}) = \|e\|^2. \tag{7.12}$$

The optimization is still based on the **constant brightness assumption**. We assume that the grayscale of a spatial point imaged at various viewing points is constant. If we have many (for example, N) space points P_i, then the whole camera pose estimation problem becomes:

$$\min_{\mathbf{T}} J(\mathbf{T}) = \sum_{i=1}^{N} e_i^T e_i, \quad e_i = \mathbf{I}_1(\mathbf{p}_{1,i}) - \mathbf{I}_2(\mathbf{p}_{2,i}). \tag{7.13}$$

The variable to be optimized here is the camera pose $\mathbf{T}$, instead of the motion of each feature point in the optical flow. To solve this optimization problem, we are concerned about how the error e changes with the camera pose $\mathbf{T}$, and we need to analyze their derivative relationship. First, define two intermediate variables:

$$\mathbf{q} = \mathbf{TP},$$
$$\mathbf{u} = \frac{1}{Z_2}\mathbf{Kq}.$$

Here, $\mathbf{q}$ is the coordinates of P in the second camera coordinate system, and $\mathbf{u}$ is its pixel coordinates. Obviously $\mathbf{q}$ is a function of $\mathbf{T}$, and $\mathbf{u}$ is a function of $\mathbf{q}$, and thus is also a function of $\mathbf{T}$. Consider the left perturbation model of Lie algebra, using the first-order Taylor expansion:

$$e(\mathbf{T}) = \mathbf{I}_1(\mathbf{p}_1) - \mathbf{I}_2(\mathbf{u}), \tag{7.14}$$

Then we get:

$$\frac{\partial e}{\partial \mathbf{T}} = -\frac{\partial \mathbf{I}_2}{\partial \mathbf{u}} \frac{\partial \mathbf{u}}{\partial \mathbf{q}} \frac{\partial \mathbf{q}}{\partial \delta \boldsymbol{\xi}}, \tag{7.15}$$

where $\delta\boldsymbol{\xi}$ is the left disturbance of $\mathbf{T}$. We see that the first derivative is divided into three terms due to the chain rule, and these three terms are easy to obtain:

1. $\partial \mathbf{I}_2 / \partial \mathbf{u}$ is the grayscale gradient at pixel $\mathbf{u}$.
2. $\partial \mathbf{u} / \partial \mathbf{q}$ is the derivative of the projection equation with respect to the three-dimensional point in the camera frame. Let $\mathbf{q} = [X, Y, Z]^T$, according to Chap. 6, the derivative is:

$$\frac{\partial \mathbf{u}}{\partial \mathbf{q}} = \begin{bmatrix} \frac{\partial u}{\partial X} & \frac{\partial u}{\partial Y} & \frac{\partial u}{\partial Z} \\ \frac{\partial v}{\partial X} & \frac{\partial v}{\partial Y} & \frac{\partial v}{\partial Z} \end{bmatrix} = \begin{bmatrix} \frac{f_x}{Z} & 0 & -\frac{f_x X}{Z^2} \\ 0 & \frac{f_y}{Z} & -\frac{f_y Y}{Z^2} \end{bmatrix}. \tag{7.16}$$

3. $\partial \mathbf{q} / \partial \delta \boldsymbol{\xi}$ is the derivative of the transformed three-dimensional point with respect to the transformation, which was introduced in chapter 3:

$$\frac{\partial \mathbf{q}}{\partial \delta \boldsymbol{\xi}} = \left[\mathbf{I}, -\mathbf{q}^{\wedge}\right]. \tag{7.17}$$

In practice, the last two items are only related to the three-dimensional point $\mathbf{q}$, which is irrelevant to the image. We often combine them together:

$$\frac{\partial \mathbf{u}}{\partial \delta \boldsymbol{\xi}} = \begin{bmatrix} \frac{f_x}{Z} & 0 & -\frac{f_x X}{Z^2} & -\frac{f_x XY}{Z^2} & f_x + \frac{f_x X^2}{Z^2} & -\frac{f_x Y}{Z} \\ 0 & \frac{f_y}{Z} & -\frac{f_y Y}{Z^2} & -f_y - \frac{f_y Y^2}{Z^2} & \frac{f_y XY}{Z^2} & \frac{f_y X}{Z} \end{bmatrix}. \tag{7.18}$$

This 2×6 matrix also appeared in the last chapter. Therefore, we derive the Jacobian of residual with respect to Lie algebra:

$$\mathbf{J} = -\frac{\partial \mathbf{I}_2}{\partial \mathbf{u}} \frac{\partial \mathbf{u}}{\partial \delta \boldsymbol{\xi}}. \tag{7.19}$$

For the problem of N points, we can use this method to calculate the Jacobian of the optimization problem and then use the Gauss-Newton method or Levenberg-Marquardt method to calculate the increments and iteratively solve it. So far, we have introduced the entire process of the direct method to estimate the camera pose. Let's implement the direct method in a program.

7.4.2 Discussion of Direct Method

In the above derivation, P is a spatial point with a known location. Where does it come from? We can reproject any pixel into the three-dimensional space under the RGB-D camera and then project it into the next image. If it is binocular, the pixel depth can also be calculated based on the parallax. If in a monocular camera, this matter is more complicated because we must also consider the uncertainty caused by the depth of P. Depth estimation will be elaborated on in Chap. 11. Now let's consider the simple case first, i.e., when the depth of P is known.

According to the source of P, we can classify the direct method as:

1. P comes from the sparse key point, which we call the sparse direct method. Usually, we use hundreds to thousands of key points, and like L-K optical flow, it is assumed that the surrounding pixels are also unchanged. This sparse direct method does not need to calculate descriptors and only uses hundreds of pixels, so it is the fastest, but it can only calculate sparse reconstruction.
2. P comes from some pixels not necessarily being features. In the formula (7.19), if the pixel gradient is zero, the entire Jacobian is zero, which will not contribute to the problem. Therefore, we can consider only using pixels with high gradients and discarding areas where the pixel gradients are not obvious. This is called a semi-dense direct method, which can reconstruct a semi-dense structure.

3. P is all pixels, which is called the dense direct method. Dense reconstruction needs to calculate all pixels (generally hundreds of thousands to several million), so most of them cannot be calculated in real-time on the existing CPU and require GPU acceleration. However, as discussed above, the points with inconspicuous pixel gradients will not contribute much to motion estimation, and it will be difficult to estimate the position during reconstruction.

It can be seen that the reconstruction from sparse to dense can be calculated by the direct method. Their computational complexity is gradually increasing. The sparse method can quickly solve the camera pose, while the dense method can build a complete map. Which method to use depends on the objective of the application. In particular, on simple computing platforms, the sparse direct method can achieve very fast results and is suitable for occasions with high real-time performance and limited computing resources [70].

7.5 Practice: Direct method

7.5.1 Single-Layer Direct Method

Now, let's demonstrate how to use the sparse direct method. Since this book does not involve GPU programming, the dense direct method is omitted. Meanwhile, to keep the program simple, we use depth data instead of monocular data to omit the monocular depth recovery part. The depth recovery based on feature points (i.e., triangulation) has been introduced in the previous chapter, and the depth recovery based on block matching will be introduced later. In this section, we will consider the sparse direct method of binocular cameras.

Since solving the direct method is finally equivalent to solving an optimization problem, we can use optimization libraries such as *g2o* or Ceres to help us or implement the Gauss-Newton method by ourselves. Like the optical flow, the direct method can also be divided into a single-layer direct method and a pyramid-like multi-layer direct method. We also first implement the single-layer direct method and then extend it to multiple layers.

In the single-layer direct method, similar to the parallel optical flow, we can also calculate each pixel's error and Jacobian in parallel. For this reason, we define a class for calculating Jacobian:

Listing 7.5 slambook2/ch8/direct_method.cpp (part)

```
/// class for accumulator jacobians in parallel
class JacobianAccumulator {
public:
  JacobianAccumulator(
    const cv::Mat &img1_,
    const cv::Mat &img2_,
    const VecVector2d &px_ref_,
    const vector<double> depth_ref_,
    Sophus::SE3d &T21_) :
```

```
  img1(img1_), img2(img2_), px_ref(px_ref_), depth_ref(depth_ref_), T21(T21_)
      {
    projection = VecVector2d(px_ref.size(), Eigen::Vector2d(0, 0));
  }

  /// accumulate jacobians in a range
  void accumulate_jacobian(const cv::Range &range);

  /// get hessian matrix
  Matrix6d hessian() const { return H; }

  /// get bias
  Vector6d bias() const { return b; }

  /// get total cost
  double cost_func() const { return cost; }

  /// get projected points
  VecVector2d projected_points() const { return projection; }

  /// reset h, b, cost to zero
  void reset() {
    H = Matrix6d::Zero();
    b = Vector6d::Zero();
    cost = 0;
  }

private:
  const cv::Mat &img1;
  const cv::Mat &img2;
  const VecVector2d &px_ref;
  const vector<double> depth_ref;
  Sophus::SE3d &T21;
  VecVector2d projection; // projected points

  std::mutex hessian_mutex;
  Matrix6d H = Matrix6d::Zero();
  Vector6d b = Vector6d::Zero();
  double cost = 0;
};

void JacobianAccumulator::accumulate_jacobian(const cv::Range &range) {

  // parameters
  const int half_patch_size = 1;
  int cnt_good = 0;
  Matrix6d hessian = Matrix6d::Zero();
  Vector6d bias = Vector6d::Zero();
  double cost_tmp = 0;

  for (size_t i = range.start; i < range.end; i++) {
    // compute the projection in the second image
    Eigen::Vector3d point_ref =
    depth_ref[i] * Eigen::Vector3d((px_ref[i][0] - cx) / fx, (px_ref[i][1] -
        cy) / fy, 1);
    Eigen::Vector3d point_cur = T21 * point_ref;
    if (point_cur[2] < 0) // depth invalid
      continue;

    float u = fx * point_cur[0] / point_cur[2] + cx, v = fy * point_cur[1] /
        point_cur[2] + cy;
    if (u < half_patch_size || u > img2.cols - half_patch_size || v <
        half_patch_size ||
    v > img2.rows - half_patch_size)
      continue;

    projection[i] = Eigen::Vector2d(u, v);
```

```
      double X = point_cur[0], Y = point_cur[1], Z = point_cur[2],
      Z2 = Z * Z, Z_inv = 1.0 / Z, Z2_inv = Z_inv * Z_inv;
      cnt_good++;

      // and compute error and jacobian
      for (int x = -half_patch_size; x <= half_patch_size; x++)
      for (int y = -half_patch_size; y <= half_patch_size; y++) {
        double error = GetPixelValue(img1, px_ref[i][0] + x, px_ref[i][1] + y)
            -
          GetPixelValue(img2, u + x, v + y);
        Matrix26d J_pixel_xi;
        Eigen::Vector2d J_img_pixel;

        J_pixel_xi(0, 0) = fx * Z_inv;
        J_pixel_xi(0, 1) = 0;
        J_pixel_xi(0, 2) = -fx * X * Z2_inv;
        J_pixel_xi(0, 3) = -fx * X * Y * Z2_inv;
        J_pixel_xi(0, 4) = fx + fx * X * X * Z2_inv;
        J_pixel_xi(0, 5) = -fx * Y * Z_inv;

        J_pixel_xi(1, 0) = 0;
        J_pixel_xi(1, 1) = fy * Z_inv;
        J_pixel_xi(1, 2) = -fy * Y * Z2_inv;
        J_pixel_xi(1, 3) = -fy - fy * Y * Y * Z2_inv;
        J_pixel_xi(1, 4) = fy * X * Y * Z2_inv;
        J_pixel_xi(1, 5) = fy * X * Z_inv;

        J_img_pixel = Eigen::Vector2d(
          0.5 * (GetPixelValue(img2, u + 1 + x, v + y) - GetPixelValue(img2, u
              - 1 + x, v + y)),
          0.5 * (GetPixelValue(img2, u + x, v + 1 + y) - GetPixelValue(img2, u
              + x, v - 1 + y))
        );

        // total jacobian
        Vector6d J = -1.0 * (J_img_pixel.transpose() * J_pixel_xi).transpose();
        hessian += J * J.transpose();
        bias += -error * J;
        cost_tmp += error * error;
      }
    }

    if (cnt_good) {
      // set hessian, bias and cost
      unique_lock<mutex> lck(hessian_mutex);
      H += hessian;
      b += bias;
      cost += cost_tmp / cnt_good;
    }
}
```

In the accumulate_jacobian function of this class, we calculate the pixel residual and Jacobian according to the previous derivation for the pixels in the specified range and finally add it to the overall **H** matrix. Then, define a function to iterate this process:

Listing 7.6 slambook2/ch8/direct_method.cpp (part)

```
void DirectPoseEstimationSingleLayer(
  const cv::Mat &img1,
  const cv::Mat &img2,
  const VecVector2d &px_ref,
  const vector<double> depth_ref,
  Sophus::SE3d &T21) {
  const int iterations = 10;
```

```
  double cost = 0, lastCost = 0;
  JacobianAccumulator jaco_accu(img1, img2, px_ref, depth_ref, T21);

  for (int iter = 0; iter < iterations; iter++) {
    jaco_accu.reset();
    cv::parallel_for_(cv::Range(0, px_ref.size()),
      std::bind(&JacobianAccumulator::accumulate_jacobian, &jaco_accu, std::
           placeholders::_1));
    Matrix6d H = jaco_accu.hessian();
    Vector6d b = jaco_accu.bias();

    // solve update and put it into estimation
    Vector6d update = H.ldlt().solve(b);;
    T21 = Sophus::SE3d::exp(update) * T21;
    cost = jaco_accu.cost_func();

    if (std::isnan(update[0])) {
      // sometimes occurred when we have a black or white patch and H is
           irreversible
      cout << "update is nan" << endl;
      break;
    }
    if (iter > 0 && cost > lastCost) {
      cout << "cost increased: " << cost << ", " << lastCost << endl;
      break;
    }
    if (update.norm() < 1e-3) {
      // converge
      break;
    }

    lastCost = cost;
    cout << "iteration: " << iter << ", cost: " << cost << endl;
  }
}
```

This function calculates the corresponding pose updates according to the calculated **H** and **b** and then updates it to the current estimated value. We have introduced the details clearly in the theoretical part. This part of the code does not seem difficult.

7.5.2 Multi-layer Direct Method

Then, similar to optical flow, we extend the direct method to the pyramid and use the coarse-to-fine process to calculate relative transformation. This part of the code is also similar to optical flow:

Listing 7.7 slambook2/ch8/direct_method.cpp (part)

```
void DirectPoseEstimationMultiLayer(
  const cv::Mat &img1,
  const cv::Mat &img2,
  const VecVector2d &px_ref,
  const vector<double> depth_ref,
  Sophus::SE3d &T21) {
  // parameters
  int pyramids = 4;
  double pyramid_scale = 0.5;
  double scales[] = {1.0, 0.5, 0.25, 0.125};

```

```
  // create pyramids
  vector<cv::Mat> pyr1, pyr2; // image pyramids
  for (int i = 0; i < pyramids; i++) {
    if (i == 0) {
      pyr1.push_back(img1);
      pyr2.push_back(img2);
    } else {
      cv::Mat img1_pyr, img2_pyr;
      cv::resize(pyr1[i - 1], img1_pyr,
        cv::Size(pyr1[i - 1].cols * pyramid_scale, pyr1[i - 1].rows *
            pyramid_scale));
      cv::resize(pyr2[i - 1], img2_pyr,
        cv::Size(pyr2[i - 1].cols * pyramid_scale, pyr2[i - 1].rows *
            pyramid_scale));
      pyr1.push_back(img1_pyr);
      pyr2.push_back(img2_pyr);
    }
  }

  double fxG = fx, fyG = fy, cxG = cx, cyG = cy; // backup the old values
  for (int level = pyramids - 1; level >= 0; level--) {
    VecVector2d px_ref_pyr; // set the key points in this pyramid level
    for (auto &px: px_ref) {
      px_ref_pyr.push_back(scales[level] * px);
    }

    // scale fx, fy, cx, cy in different pyramid levels
    fx = fxG * scales[level];
    fy = fyG * scales[level];
    cx = cxG * scales[level];
    cy = cyG * scales[level];
    DirectPoseEstimationSingleLayer(pyr1[level], pyr2[level], px_ref_pyr,
        depth_ref, T21);
  }
}
```

It should be noted that, because the direct method of Jacobian takes the camera's intrinsic parameters, and when the pyramid scales the image, the corresponding internal parameters also need to be multiplied by the corresponding ratio.

7.5.3 *Discussion*

Finally, we use some sample pictures to test the results of the direct method. We use several images of the Kitti [74] autonomous driving dataset. First, we read the first image left.png, in the corresponding disparity map disparity.png, calculate the depth corresponding to each pixel, and then use the direct method to calculate the camera poses for the five images "000001.png"-"000005.png". In order to show the insensitivity of the direct method to the feature points, we randomly select some points in the first image without using any corner points or feature point extraction algorithms.

Listing 7.8 slambook2/ch8/direct_method.cpp (part)

```
int main(int argc, char **argv) {
  cv::Mat left_img = cv::imread(left_file, 0);
  cv::Mat disparity_img = cv::imread(disparity_file, 0);

```

```
  // let's randomly pick pixels in the first image and generate some 3d points in the first image's frame
  cv::RNG rng;
  int nPoints = 2000;
  int boarder = 20;
  VecVector2d pixels_ref;
  vector<double> depth_ref;

  // generate pixels in ref and load depth data
  for (int i = 0; i < nPoints; i++) {
    int x = rng.uniform(boarder, left_img.cols - boarder); // don't pick pixels close to boarder
    int y = rng.uniform(boarder, left_img.rows - boarder); // don't pick pixels close to boarder
    int disparity = disparity_img.at<uchar>(y, x);
    double depth = fx * baseline / disparity; // you know this is disparity to depth
    depth_ref.push_back(depth);
    pixels_ref.push_back(Eigen::Vector2d(x, y));
  }

  // estimates 01~05.png's pose using this information
  Sophus::SE3d T_cur_ref;

  for (int i = 1; i < 6; i++) { // 1~10
    cv::Mat img = cv::imread((fmt_others % i).str(), 0);
    DirectPoseEstimationMultiLayer(left_img, img, pixels_ref, depth_ref, T_cur_ref);
  }
  return 0;
}
```

Readers can run this program on your machine. It will output the tracking points on each level of each image's pyramid and print the running time. The result of the multi-layer direct method is shown in Fig. 7.5. According to the program's output, we can see that the fifth image is about when the camera moves 3.8 m forward. It can be seen that even if we randomly select the points, the direct method can correctly track most of the pixels and estimate the camera motion. It does not include any feature extraction, matching, or optical flow. At 2000 points, it takes 1-2 milliseconds for each layer of the direct method to iterate in terms of running time, so the four-layer pyramid takes about 8 milliseconds. In contrast, the optical flow of 2000 points takes about ten milliseconds, excluding the subsequent pose estimation. Therefore, the direct method is usually faster than the traditional feature points and optical flow.

Below we briefly explain the iterative process of the direct method. Compared with the feature point method, the direct method completely relies on the optimization solver. It can be seen from the formula (7.19) that the pixel gradient guides the direction of optimization. If you want to get the correct optimization results, you must ensure that most pixel gradients can guide the optimization in the right direction.

What does it mean? Assume that for the reference image, we measured a pixel with a gray value of 229. And, since we know its depth, we can infer the position of the space point P (Fig. 7.6 shown as the grayscale measured in I_1).

Now, we have got a new image and need to estimate its camera pose. This pose is obtained by continuous optimization iterations of an initial value. Assuming that our initial value is relatively poor. Under this initial value, the gray value after the projection of the space point P is 126. Therefore, the error of this pixel is $229 - 126 =$

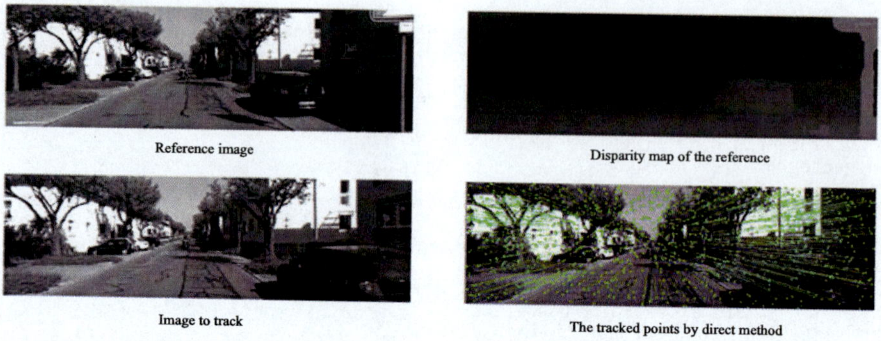

Fig. 7.5 Experimental results of the direct method. upper left: original image; upper right: disparity map corresponding to the original image; lower left: fifth tracking image; lower right: tracking result

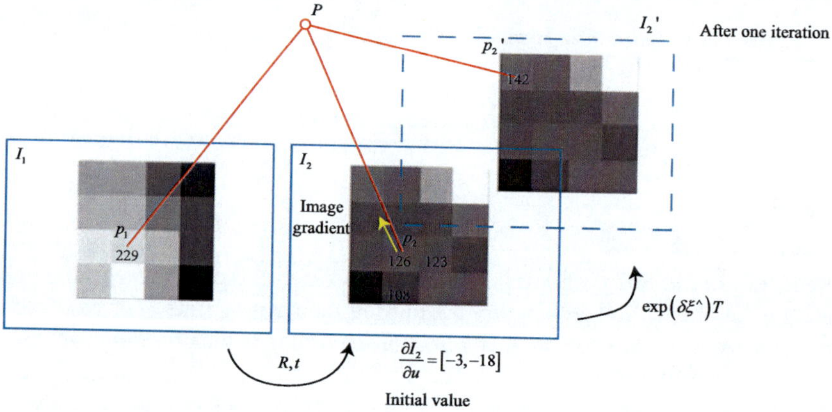

Fig. 7.6 One iteration in the direct method

103. To reduce this error, we hope to fine-tune the camera's pose to make the pixels brighter.

How do I know where to fine-tune the pixels to make them brighter? This requires the use of local pixel gradients. We found in the image that if we take a step forward along the u axis, the gray value at that point becomes 123. That is, 3 is subtracted. Similarly, if you take a step forward along the v axis, the gray value is reduced by 18 and becomes 108. Around this pixel, the gradient is $[-3, -18]$. In order to increase the brightness, we will suggest optimizing the algorithm to fine-tune in a direction so that the image of P moves to the top left. In this process, we use the local gradient of the pixel to approximate the grayscale distribution near it, but please note that the real image is not smooth, so this gradient is not valid at a distance.

However, the optimization can't just follow just one pixel's behavior and need to get track of other pixels. After considering many pixels, the optimization algorithm chose a place not far from the direction we suggested and calculated an updated

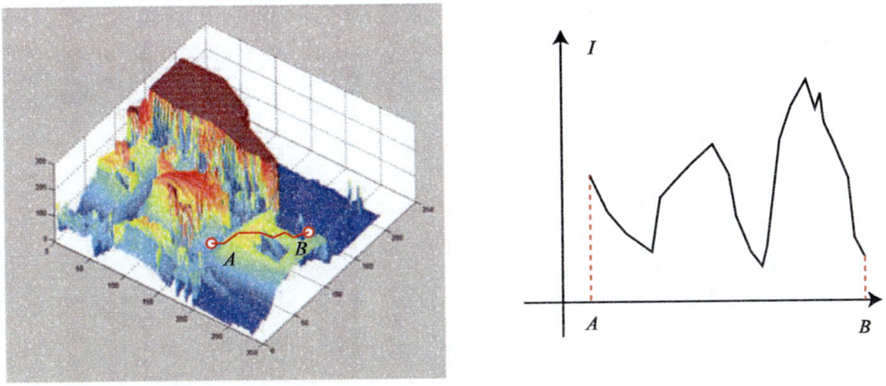

Fig. 7.7 Three-dimensional visualization of an image. The path from one point in the image to another point is not necessarily a "straight downhill road", but needs to be "climbing over the mountains" frequently. This reflects the non-convexity of the image function

amount $\exp(\boldsymbol{\xi}^\wedge)$. After adding the updated amount, the image has moved from I_2 to I_2', and the projection position of the pixel has also changed to a brighter place. We see that with this update, an error has become smaller. Under ideal circumstances, we expect the error to continue to decrease and eventually converge.

But is this actually the case? Do we really only need to walk along the gradient direction to reach an optimal value? Note that the gradient of the direct method is directly determined by the image gradient, so we must ensure that the photometric error will continue to decrease when walking along the image gradient. However, the image is usually a very strong non-convex function, as shown in Fig. 7.7. In practice, if we move along the image gradient, it is easy to fall into a local minimum due to the non-convexity (or noise) of the image itself, and we cannot continue to optimize. The direct method can only be established when the camera movement is very small, and the gradient in the image will not have strong non-convexity.

In the example, we only calculated the difference of a single pixel, and this difference is obtained by directly subtracting the grayscale. However, a single pixel is not distinguishable, and there are probably many pixels around with similar brightness. Therefore, we sometimes use small patches and use more complex difference measures, such as the normalized cross-correlation (NCC). For the sake of simplicity, the example uses the sum of squares of errors to maintain consistency with the derivation.

7.5.4 Advantages and Disadvantages of the Direct Method

Finally, we summarize the advantages and disadvantages of the direct method. In general, its advantages are as follows:

- It can save time in calculating feature points and descriptors.
- Only pixel gradients are required. No feature points are required. Therefore, the direct method can be used in the absence of features. An extreme example is an image with only edge gradients. It may not be able to extract corner features, but its motion can be estimated by a direct method. In the demonstration experiment, we see that the direct method can also work normally for randomly selected points. This is very important in practice because practical scenes may not have many corner points to use.
- It is possible to construct semi-dense or even dense maps, which cannot be achieved by the feature point method.

On the other hand, its shortcomings are also obvious:

- **Non-convexity**. The direct method completely relies on gradient search and reduces the objective function to calculate the camera pose. The objective function needs to take the pixel's gray value, and the image is a strongly non-convex function. This makes the optimization algorithm easy to be stuck at a local minimum, and the direct method can only succeed when the movement is small. Against this, the pyramids can reduce the impact of non-convexity to a certain extent.
- **A single pixel is not separable**. Many pixels just look alike. So we either calculate on image patches or the complex correlations. Since each pixel may have inconsistent "opinions" about the camera's adjustment, we need many pixels to make a fair judgment. But how many pixels are enough? This is hard to answer. The performance of the direct method decreases when there are fewer selected points. We usually recommend using more than 500 points, but this is just an empirical choice.
- **Constant brightness is a strong assumption**. Constant brightness is a strong assumption. If the camera is automatically exposed, it will make the overall image brighter or darker when it adjusts the exposure parameters. This situation also occurs when the light changes. The feature point method has a certain tolerance to illumination. In contrast, the direct method calculates the difference of brightness, and the overall brightness change will destroy the brightness constant assumption and make the algorithm fail. In response to this, the direct method will also estimate the camera's exposure parameters [70] to still work when the exposure time changes.

Exercises

1. In addition to LK optical flow, do you know other optical flow methods? What are their characteristics?
2. In the program to calculate the image gradient, we simply calculate the difference between the brightness of $u + 1$ and $u - 1$ divided by 2 as the gradient in the direction of u. What are the disadvantages of this approach? Hint: For features closer together, the changes should be faster; while for features farther

away it changes more slowly in the image, can this information be used when calculating the gradient?

3. Can the direct method be implemented in an "inverse" way like optical flow? That is, use the gradient of the original image instead of the gradient of the target image?
4.* Use Ceres or *g2o* to implement sparse direct method and semi-dense direct method.
5. Compared with the direct method of RGB-D, the monocular direct method is often more complicated. In addition to the unknown matching, the distance of the pixel also needs to be estimated. We need to use pixel depth as an optimization variable during optimization. Refer to the literature [69, 75], can you implement its calculation?

Chapter 8
Filters and Optimization Approaches: Part I

> **Goal of Study**
> 1. Learn how to formulate the backend problem into a filter or least-square optimization problem.
> 2. Learn how to use the sparse structure in the bundle adjustment problem.
> 3. Solve a BA problem with *g2o* and *Ceres*.

From this lecture, we turn to another important module: backend optimization. We see that the frontend visual odometry can give a short-term trajectory and map. Still, due to the inevitable accumulation of errors, this map is inaccurate for a long time. Therefore, based on visual odometry, we also hope to construct a larger-scale optimization problem to consider the optimal trajectory and map over a long time. However, considering the balance of accuracy and performance, there are many different approaches in practice.

8.1 Introduction

8.1.1 State Estimation from Probabilistic Perspective

As mentioned in the lecture 1, the visual odometry only has a short memory, but we hope that the system can maintain the entire motion trajectory in an optimal state for a long time. We may use the latest knowledge to update an old state. At that time, it seems that future information tells you, "where you should be now." Therefore, in the backend optimization, we usually consider the problem of state estimation for a longer time (or all-time) and not only use the past information to update our current state but also use future information to update ourselves. Such a method might be called "batch." Otherwise, if the current state is only determined by the past, or even only by the last moment, it might also be called "incremental."

X. Gao and T. Zhang, *Introduction to Visual SLAM*,
https://doi.org/10.1007/978-981-16-4939-4_8

We already know that the SLAM process can be described by the motion and observation equations. Suppose in the time from $t = 0$ to $t = N$, we have the poses from $\mathbf{x}_0$ to $\mathbf{x}_N$ and observation $\mathbf{y}_1, \ldots, \mathbf{y}_M$. According to the equations in Chap. 1, we write them as:

$$\begin{cases} \mathbf{x}_k = f(\mathbf{x}_{k-1}, \mathbf{u}_k) + \mathbf{w}_k \\ \mathbf{z}_{k,j} = h(\mathbf{y}_j, \mathbf{x}_k) + \mathbf{v}_{k,j} \end{cases} \quad k = 1, \ldots, N, \; j = 1, \ldots, M. \tag{8.1}$$

Note that in the SLAM problem, we have the following characteristics:

1. In the observation equation, only when $\mathbf{x}_k$ sees $\mathbf{y}_j$ will we have a real observation equation. In fact, we can usually see only a small part of the landmarks in one location. Moreover, due to a large number of visual SLAM feature points, the number of observation equations in practice will be much larger than that of motion equations.
2. We may not have a device to measure the motion, so there may not be a motion equation. In this case, there are several ways to deal with it:

   - Assume that there is really no motion equation.
   - Assume that the camera does not move.
   - Assume that the camera is moving at a constant speed.

   These several methods are all feasible. In the absence of motion equations, the entire optimization problem consists of only observation equations. This is very similar to the SfM (Structure from Motion) problem, which is equivalent to restoring the motion and structure through a set of images. The difference with SfM is that the SLAM images have a chronological order, while SfM allows the use of completely unrelated images.

We know that every measurement is affected by noise, so the poses $\mathbf{x}$ and landmarks $\mathbf{y}$ here are regarded as *random variables that obey a certain probability distribution* instead of a single number. Therefore, the question becomes: when I have some motion data $\mathbf{u}$ and observation data $\mathbf{z}$, how to determine the state $\mathbf{x}$ and landmarks $\mathbf{y}$'s distribution? Furthermore, if new data is obtained, how to update our estimation? In more common and reasonable cases, we assume that the state quantity and noise terms obey Gaussian distribution-which means that only their mean and covariance matrix need to be stored in the program. The mean can be regarded as an estimate of the state variable's optimal value, and the covariance matrix measures its uncertainty. Then, the question becomes: when there are some motion and observation data, how do we estimate the Gaussian distribution of the states?

We still put ourselves in the role of a robot. When there is only the equation of motion, it is equivalent to walking blindfolded in an unknown place. Although we know how far we have taken each step, we will become more and more uncertain about where we are as time grows. This reflects that when the input data is affected by noise, the error is gradually accumulated, and our estimate of the position variance will become larger and larger. However, when we open our eyes, we will become more confident because we can continuously observe the external scene, making

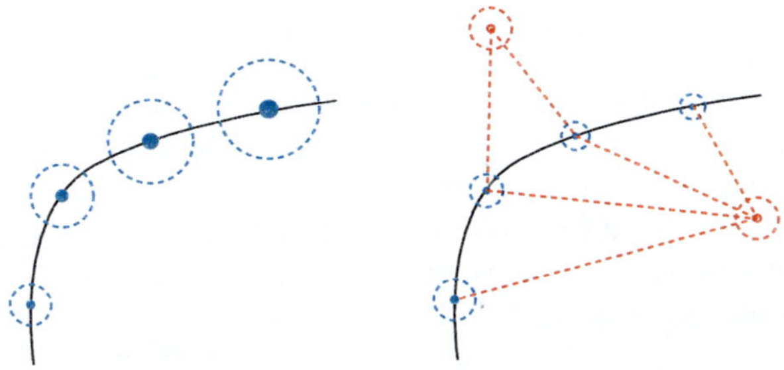

Fig. 8.1 An intuitive description of uncertainty. Left: When there is only the motion equation, the pose at the next moment adds noise to the previous moment, so the uncertainty is getting bigger and bigger. Right: When there are road signs, the uncertainty will be significantly reduced. Please note that this is only an intuitive diagram, not real data

the uncertainty of position estimation smaller. If we use an ellipse to express the covariance matrix intuitively, this process is a bit like walking in a mobile phone map software. Taking Fig. 8.1 as an example, readers can imagine that when there is no observation data, the circle will become larger and larger with the movement, and if there are correct observations, the circle will shrink to a certain size and keep stable.

The above statements explain the problem of state estimation in a metaphorical form. Below we will look at it in a quantitative way. In lecture 5, we introduced the maximum likelihood estimation where we say that the problem of *batch state estimation* can be transformed into a *maximum likelihood estimation problem* and solved by the *least-square method*. In this section, we will explore how to apply this conclusion to progressive problems and get some classic conclusions. At the same time, we will investigate the special structure of the least-square method in visual SLAM.

First of all, since the poses and landmarks are all optimize variables, we change the notation a bit and let $\mathbf{x}_k$ be all the unknowns at the moment of k. It contains the current camera pose and m landmarks. With a little abuse of the notation, we rewrite the equations into:

$$\mathbf{x}_k \triangleq \{\mathbf{x}_k, \mathbf{y}_1, \ldots, \mathbf{y}_m\}. \tag{8.2}$$

At the same time, all observations at time k are denoted as $\mathbf{z}_k$. Therefore, we can write the motion and observation equations more concisely. There will be no $\mathbf{y}$ here, but we should remember that the previous $\mathbf{y}$ is already included in the $\mathbf{x}$:

$$\begin{cases} \mathbf{x}_k = f\left(\mathbf{x}_{k-1}, \mathbf{u}_k\right) + \mathbf{w}_k \\ \mathbf{z}_k = h\left(\mathbf{x}_k\right) + \mathbf{v}_k \end{cases} \quad k = 1, \ldots, N. \tag{8.3}$$

Now consider the situation at time k. We hope to use the data from 0 to k to estimate the current state distribution:

$$P(\mathbf{x}_k|\mathbf{x}_0, \mathbf{u}_{1:k}, \mathbf{z}_{1:k}). \tag{8.4}$$

The subscript $1:k$ represents all the data from time 0 to time k. Please note that $\mathbf{z}_k$ represents all the observation data at time k. It may be more than one, but we use a single notation here. At the same time, also note that $\mathbf{x}_k$ is related to the previous states like $\mathbf{x}_{k-1}, \mathbf{x}_{k-2}$, but this formula does not explicitly write them out.

Let's see how to do the state estimation. According to Bayes' rule, swap the positions of $\mathbf{z}_k$ and $\mathbf{x}_k$, we have:

$$P(\mathbf{x}_k|\mathbf{x}_0, \mathbf{u}_{1:k}, \mathbf{z}_{1:k}) \propto \underbrace{P(\mathbf{z}_k|\mathbf{x}_k)}_{\text{likelihood}} \underbrace{P(\mathbf{x}_k|\mathbf{x}_0, \mathbf{u}_{1:k}, \mathbf{z}_{1:k-1})}_{\text{prior}}. \tag{8.5}$$

You may think this is familiar with what we have discussed in the previous sections. Here the first term is called *likelihood*, and the second term is called *prior*. The observation equation gives the likelihood. For the prior part, we should know that the current state $\mathbf{x}_k$ is estimated based on all past states. Let's first consider the effect of $\mathbf{x}_{k-1}$. It is expanded according to the conditional probability of $\mathbf{x}_{k-1}$ moment:

$$P(\mathbf{x}_k|\mathbf{x}_0, \mathbf{u}_{1:k}, \mathbf{z}_{1:k-1}) = \int P(\mathbf{x}_k|\mathbf{x}_{k-1}, \mathbf{x}_0, \mathbf{u}_{1:k}, \mathbf{z}_{1:k-1}) P(\mathbf{x}_{k-1}|\mathbf{x}_0, \mathbf{u}_{1:k}, \mathbf{z}_{1:k-1}) \mathrm{d}\mathbf{x}_{k-1}. \tag{8.6}$$

There are generally some methodology differences for the next steps: (i) We assume the *Markov property*. The simple first-order Markov property assumes that the state at time k is only related to the state at time $k-1$ and is not related to the earlier states. If such an assumption is made, we will get a filter method represented by *Extended Kalman Filter* (EKF). In the filtering method, we estimate from the state at a specific moment and derive it to the next moment. (ii) Another technique is to keep the relationship between the state at k and *all* the previous states (and also, the future states). At this time, we will obtain an optimization framework with *nonlinear optimization* as the main body. The basic knowledge of nonlinear optimization has been introduced in the previous sections. At present, the mainstream of visual SLAM is to use the nonlinear optimization method. However, in order to make this book more comprehensive, let's first introduce the principles of the Kalman filter and EKF.

8.1.2 Linear Systems and the Kalman Filter

Let's look at the filter model first. When the Markov property is assumed, what changes will happen from a mathematical perspective? The current state is only related to the previous moment. The first part on the right side of the formula (8.6) can be further simplified as:

$$P(\mathbf{x}_k|\mathbf{x}_{k-1}, \mathbf{x}_0, \mathbf{u}_{1:k}, \mathbf{z}_{1:k-1}) = P(\mathbf{x}_k|\mathbf{x}_{k-1}, \mathbf{u}_k), \tag{8.7}$$

where we omit the states earlier than $k-1$ since they are not related to the kth state.

The second part can be simplified as:

$$P\left(\mathbf{x}_{k-1}|\mathbf{x}_0, \mathbf{u}_{1:k}, \mathbf{z}_{1:k-1}\right) = P\left(\mathbf{x}_{k-1}|\mathbf{x}_0, \mathbf{u}_{1:k-1}, \mathbf{z}_{1:k-1}\right), \tag{8.8}$$

where we drop the $\mathbf{u}_k$ since it has nothing to do with $\mathbf{x}_{k-1}$.

It can be seen that this item is actually the state distribution at time $k-1$. Therefore, this series of equations shows that what we are actually doing is "how to derive the state distribution from time $k-1$ to time k". In other words, we only need to maintain the current state estimation and update it incrementally. Furthermore, if it is assumed that the state quantity obeys a Gaussian distribution, we only need to consider the state variable's mean and covariance. You can imagine that the robot's localization module has been outputting two pieces of information: the estimated pose (mean) and the uncertainty (covariance), which is exactly the common case in many practical localization libraries.

We start from the simplest linear Gaussian system, and finally, we will reach the famous Kalman filter. A linear Gaussian system is such a system where the motion and observation equations are all linear so that we can write them in a matrix form:

$$\begin{cases} \mathbf{x}_k = \mathbf{A}_k \mathbf{x}_{k-1} + \mathbf{u}_k + \mathbf{w}_k \\ \mathbf{z}_k = \mathbf{C}_k \mathbf{x}_k + \mathbf{v}_k \end{cases} \quad k = 1, \ldots, N, \tag{8.9}$$

and we assume the states and noises are all Gaussian, so that:

$$\mathbf{w}_k \sim N(\mathbf{0}, \mathbf{R}). \quad \mathbf{v}_k \sim N(\mathbf{0}, \mathbf{Q}), \tag{8.10}$$

where we omit the subscripts of $\mathbf{R}$ and $\mathbf{Q}$ for brevity. Now, using the Markov property, suppose we know the posterior state estimation at time $k-1$ $\hat{\mathbf{x}}_{k-1}$ and its covariance $\hat{\mathbf{P}}_{k-1}$, now we want to determine the posterior distribution of $\mathbf{x}_k$ based on the input and the observation data at time k. In order to distinguish the prior and the posterior variables, we make a difference in the notation: we use the up hat $\hat{\mathbf{x}}_k$ represents the posterior, and use the down hat $\check{\mathbf{x}}_k$ to represent the prior distribution.

The Kalman filter's first step is to determine the distribution of $\mathbf{x}_k$ through the equation of motion, which is called the prior at time k. It is very easy to transform the Gaussians in the linear system:

$$P\left(\mathbf{x}_k|\mathbf{x}_0, \mathbf{u}_{1:k}, \mathbf{z}_{1:k-1}\right) = N\left(\mathbf{A}_k \hat{\mathbf{x}}_{k-1} + \mathbf{u}_k, \mathbf{A}_k \hat{\mathbf{P}}_{k-1} \mathbf{A}_k^T + \mathbf{R}\right). \tag{8.11}$$

This step is called the *prediction*. Please refer to the Appendix A.3 if you are not familiar with the linear transform here. It shows how to infer the current state distribution based on the noisy input from the previous state. Let's note it as:

$$\check{\mathbf{x}}_k = \mathbf{A}_k \hat{\mathbf{x}}_{k-1} + \mathbf{u}_k, \quad \check{\mathbf{P}}_k = \mathbf{A}_k \hat{\mathbf{P}}_{k-1} \mathbf{A}_k^T + \mathbf{R}, \tag{8.12}$$

which is very natural. Obviously, the state's uncertainty in this step will become larger because the motion noise is added to our system. On the other hand, from the observation equation, we can calculate what kind of observation data should be generated in a certain state:

$$P\left(\mathbf{z}_k|\mathbf{x}_k\right) = N\left(\mathbf{C}_k\mathbf{x}_k, \mathbf{Q}\right), \tag{8.13}$$

which is exactly the likelihood we talked about before. Remember what we want to get is the posterior $P\left(\mathbf{x}_k|\mathbf{z}_k\right)$. In order to do that, we need to calculate the product of the prior and the likelihood according to the Bayesian formula (8.5). However, although we know that we will finally get a Gaussian distribution of $\mathbf{x}_k$ in the end, it is a little bit troublesome in the calculation. Let's set the result as $\mathbf{x}_k \sim N(\hat{\mathbf{x}}_k, \hat{\mathbf{P}}_k)$, then:

$$N(\hat{\mathbf{x}}_k, \hat{\mathbf{P}}_k) = \eta N\left(\mathbf{C}_k\mathbf{x}_k, \mathbf{Q}\right) \cdot N(\check{\mathbf{x}}_k, \check{\mathbf{P}}_k), \tag{8.14}$$

where η is a normalization number to make the integral of the distribution equal to one. This is a little tricky method. Please keep in mind that what we really want to know is the relationship between $\hat{\mathbf{x}}_k, \hat{\mathbf{P}}_k$ and $\check{\mathbf{x}}_k, \check{\mathbf{P}}_k$.

Since we know that both sides of the equation are Gaussian distributions, we only need to compare the exponential part and ignore the factor part in front because they can be merged to η. The exponential part is very similar to a quadratic form, and let's derive it. First, we expand the exponential part as[1]:

$$\left(\mathbf{x}_k - \hat{\mathbf{x}}_k\right)^T \hat{\mathbf{P}}_k^{-1}\left(\mathbf{x}_k - \hat{\mathbf{x}}_k\right) = \left(\mathbf{z}_k - \mathbf{C}_k\mathbf{x}_k\right)^T \mathbf{Q}^{-1}\left(\mathbf{z}_k - \mathbf{C}_k\mathbf{x}_k\right) + \left(\mathbf{x}_k - \check{\mathbf{x}}_k\right)^T \check{\mathbf{P}}_k^{-1}\left(\mathbf{x}_k - \check{\mathbf{x}}_k\right). \tag{8.15}$$

In order to compute the $\hat{\mathbf{x}}_k$ with $\hat{\mathbf{P}}_k$ on the left side, we expand the quadratics and compare their first-order and second-order coefficients of $\mathbf{x}_k$. For the second-order coefficients, we have:

$$\hat{\mathbf{P}}_k^{-1} = \mathbf{C}_k^T\mathbf{Q}^{-1}\mathbf{C}_k + \check{\mathbf{P}}_k^{-1}, \tag{8.16}$$

which gives the relationship of the covariance matrix.

We define an intermediate variable for convenience in the following derivation:

$$\mathbf{K} = \hat{\mathbf{P}}_k\mathbf{C}_k^T\mathbf{Q}^{-1}. \tag{8.17}$$

According to this definition, we multiply $\hat{\mathbf{P}}_k$ on both sides of the Eq. (8.16):

$$\mathbf{I} = \hat{\mathbf{P}}_k\mathbf{C}_k^T\mathbf{Q}^{-1}\mathbf{C}_k + \hat{\mathbf{P}}_k\check{\mathbf{P}}_k^{-1} = \mathbf{K}\mathbf{C}_k + \hat{\mathbf{P}}_k\check{\mathbf{P}}_k^{-1}. \tag{8.18}$$

Then we have[2]:

---

[1] The the equal sign is not strict here since we can put the variables that are not related to $\mathbf{x}_k$ into η.

[2] It seems to have a little circular definition here. We define the $\mathbf{K}$ by $\hat{\mathbf{P}}_k$, and then write $\hat{\mathbf{P}}_k$ as the expression of $\mathbf{K}$. However, $\mathbf{K}$ can also be calculated without relying on $\hat{\mathbf{P}}_k$, but this requires the introduction of the Sherman-Morrison-Woodbury identity [76], see the exercises in this lecture. In practice, we can also simply calculate the $\hat{\mathbf{P}}_k$ first and then use the result to calculate $\mathbf{K}$.

$$\hat{\mathbf{P}}_k = (\mathbf{I} - \mathbf{K}\mathbf{C}_k)\check{\mathbf{P}}_k. \tag{8.19}$$

Then we compare the first-order coefficients and get:

$$-2\hat{\mathbf{x}}_k^T \hat{\mathbf{P}}_k^{-1} \mathbf{x}_k = -2\mathbf{z}_k^T \mathbf{Q}^{-1} \mathbf{C}_k \mathbf{x}_k - 2\check{\mathbf{x}}_k^T \check{\mathbf{P}}_k^{-1} \mathbf{x}_k. \tag{8.20}$$

Reorganize it (take the coefficients and transpose them):

$$\hat{\mathbf{P}}_k^{-1} \hat{\mathbf{x}}_k = \mathbf{C}_k^T \mathbf{Q}^{-1} \mathbf{z}_k + \check{\mathbf{P}}_k^{-1} \check{\mathbf{x}}_k. \tag{8.21}$$

Multiply $\hat{\mathbf{P}}_k$ on both sides and take (8.17) into it:

$$\hat{\mathbf{x}}_k = \hat{\mathbf{P}}_k \mathbf{C}_k^T \mathbf{Q}^{-1} \mathbf{z}_k + \hat{\mathbf{P}}_k \check{\mathbf{P}}_k^{-1} \check{\mathbf{x}}_k \tag{8.22}$$

$$= \mathbf{K}\mathbf{z}_k + (\mathbf{I} - \mathbf{K}\mathbf{C}_k)\check{\mathbf{x}}_k = \check{\mathbf{x}}_k + \mathbf{K}\left(\mathbf{z}_k - \mathbf{C}_k \check{\mathbf{x}}_k\right). \tag{8.23}$$

So we get the expression of the posterior mean. In summary, the above two steps can be written into two steps: the "predict" and the "update":

1. *Predict*:

$$\check{\mathbf{x}}_k = \mathbf{A}_k \hat{\mathbf{x}}_{k-1} + \mathbf{u}_k, \quad \check{\mathbf{P}}_k = \mathbf{A}_k \hat{\mathbf{P}}_{k-1} \mathbf{A}_k^T + \mathbf{R}. \tag{8.24}$$

2. *Update*: Calculate $\mathbf{K}$, which is the Kalman gain:

$$\mathbf{K} = \check{\mathbf{P}}_k \mathbf{C}_k^T \left(\mathbf{C}_k \check{\mathbf{P}}_k \mathbf{C}_k^T + \mathbf{Q}_k\right)^{-1}, \tag{8.25}$$

and the posterior:

$$\begin{aligned} \hat{\mathbf{x}}_k &= \check{\mathbf{x}}_k + \mathbf{K}\left(\mathbf{z}_k - \mathbf{C}_k \check{\mathbf{x}}_k\right) \\ \hat{\mathbf{P}}_k &= (\mathbf{I} - \mathbf{K}\mathbf{C}_k)\check{\mathbf{P}}_k. \end{aligned} \tag{8.26}$$

So far, we have derived the classic Kalman filter (to my knowledge, in the simplest way). The Kalman filter has several derivation methods, and we use the form of maximum posterior probability estimation from a probability perspective. We see that in a linear Gaussian system, the Kalman filter constitutes the maximum posterior probability estimate. Moreover, since the Gaussian distribution still obeys the Gaussian distribution after linear transformation, we did not perform any approximation in the whole process. The Kalman filter constitutes the optimal unbiased estimate of the linear system. If readers are interested in other derivation methods, please refer to the state estimation books like [5].

8.1.3 Nonlinear Systems and the EKF

After introducing the Kalman filter, we must clarify one point: the motion equation and observation equation in SLAM are usually nonlinear functions, especially the camera model in visual SLAM, which requires the camera projection model and the pose represented by SE(3). A Gaussian distribution, after a nonlinear transformation, is often no longer Gaussian. Therefore, in a nonlinear system, we must take a certain approximation to transform the non-Gaussian distributions into Gaussians.

If we hope to extend the Kalman filter results to a nonlinear system, we will get an Extended Kalman Filter (EKF). The usual approach is to consider the first-order Taylor expansion of the motion equation and the observation equation near a certain point (working point), and only retain the first-order term, that is, the linear part, and then derive it according to the linear system.

Let's do the derivation just like for the Kalman filter. Let the mean and covariance matrix at time $k-1$ be $\hat{\mathbf{x}}_{k-1}, \hat{\mathbf{P}}_{k-1}$. At the moment k, we put the motion equation and the observation equation at $\hat{\mathbf{x}}_{k-1}, \hat{\mathbf{P}}_{k-1}$, and do the *linearization* (equivalent to first-order Taylor expansion):

$$\mathbf{x}_k \approx f\left(\hat{\mathbf{x}}_{k-1}, \mathbf{u}_k\right) + \left.\frac{\partial f}{\partial \mathbf{x}_{k-1}}\right|_{\hat{\mathbf{x}}_{k-1}} \left(\mathbf{x}_{k-1} - \hat{\mathbf{x}}_{k-1}\right) + \mathbf{w}_k. \tag{8.27}$$

Note the partial derivative here as:

$$\mathbf{F} = \left.\frac{\partial f}{\partial \mathbf{x}_{k-1}}\right|_{\hat{\mathbf{x}}_{k-1}}. \tag{8.28}$$

Similar for the observation model:

$$\mathbf{z}_k \approx h\left(\check{\mathbf{x}}_k\right) + \left.\frac{\partial h}{\partial \mathbf{x}_k}\right|_{\check{\mathbf{x}}_k} \left(\mathbf{x}_k - \check{\mathbf{x}}_k\right) + \mathbf{v}_k. \tag{8.29}$$

And note the partial derivative as:

$$\mathbf{H} = \left.\frac{\partial h}{\partial \mathbf{x}_k}\right|_{\check{\mathbf{x}}_k}. \tag{8.30}$$

Then the *prediction* part becomes:

$$P\left(\mathbf{x}_k | \mathbf{x}_0, \mathbf{u}_{1:k}, \mathbf{z}_{0:k-1}\right) = N(f\left(\hat{\mathbf{x}}_{k-1}, \mathbf{u}_k\right), \mathbf{F}\hat{\mathbf{P}}_{k-1}\mathbf{F}^T + \mathbf{R}_k), \tag{8.31}$$

which is almost same as the Kalman filter. We note the prior mean and covariance as:

$$\check{\mathbf{x}}_k = f\left(\hat{\mathbf{x}}_{k-1}, \mathbf{u}_k\right), \quad \check{\mathbf{P}}_k = \mathbf{F}\hat{\mathbf{P}}_{k-1}\mathbf{F}^T + \mathbf{R}_k. \tag{8.32}$$

Then, for the observation part, we have:

$$P\left(\mathbf{z}_k | \mathbf{x}_k\right) = N(h\left(\check{\mathbf{x}}_k\right) + \mathbf{H}\left(\mathbf{x}_k - \check{\mathbf{x}}_k\right), \mathbf{Q}_k). \tag{8.33}$$

Finally, according to the Bayesian formula, the posterior probability form of $\mathbf{x}_k$ can be derived. We omit the intermediate derivation process and only give the results. Readers can imitate the Kalman filter to derive the EKF prediction and update equations. In short, we will first define a *Kalman gain* $\mathbf{K}_k$:

$$\mathbf{K}_k = \check{\mathbf{P}}_k \mathbf{H}^T \left(\mathbf{H} \check{\mathbf{P}}_k \mathbf{H}^T + \mathbf{Q}_k \right)^{-1}. \tag{8.34}$$

The posterior can be written out based on the $\mathbf{K}$:

$$\hat{\mathbf{x}}_k = \check{\mathbf{x}}_k + \mathbf{K}_k \left(\mathbf{z}_k - h\left(\check{\mathbf{x}}_k \right) \right), \hat{\mathbf{P}}_k = \left(\mathbf{I} - \mathbf{K}_k \mathbf{H} \right) \check{\mathbf{P}}_k. \tag{8.35}$$

The EKF shows the change of the state variable distribution after linearization. In the linear system and Gaussian noise case, the Kalman filter gives an unbiased optimal estimate. In nonlinear systems like SLAM, it gives the maximum a posteriori estimate (MAP) under a single linearization step.

8.1.4 Discussion About KF and EKF

EKF is known for its simple form and wide application. It may seem to be a bit complicated at first glance, but you find it is really very easy to implement (less than 20 lines in Python or MATLAB). When we want to estimate a certain amount of uncertainty within a certain period of time, the first thing we think of is the EKF. In the early SLAM researches, EKF occupied a dominant position for a long time. Researchers implemented a lot of various filters in SLAM, such as IF (information filter) [77] , IKF [78] (Iterated KF), UKF [79] (Unscented KF) and particle filter [80–82], SWF (Sliding Window Filter) [83], etc. [16],[3] or use divide and conquer to improve the efficiency of EKF [84, 85]. To this day, although we realize that nonlinear optimization has obvious advantages over filters, EKF is still an effective way when the computing resources are limited, or the state dimension is relatively small.

So, is EKF enough for modern SLAM systems? Why are we using optimization approaches rather than the filters? The issues come from two aspects: the theoretical and practical.

1. First of all, the filter method always assumes a certain extent Markov property: the kth state is only related to $k-1$ (or finite moments before). It is not related to the state and observations before $k-1$. This is like considering only the relationship between two adjacent frames in visual odometry. If the current frame is indeed related to data long ago (for example, the robot returns to the origin after a long time), the filter method will be difficult to process in such cases.
   The optimization methods tend to use longer or all historical data. It not limited feature points and trajectories at adjacent moments but also considers the long

[3] The principle of particle filter is quite different from Kalman filter.

ago states, which is called the Full-SLAM. In this sense, nonlinear optimization methods use more information and, of course, require more calculations.

2. Compared with the optimization method introduced in Lecture 6, the EKF filter is only linearized *once* at $\hat{\mathbf{x}}_{k-1}$. Then we calculated the posterior directly based on the linearization result at this time. This is equivalent to saying that we believe that the linear approximation is still valid at the posterior point. In fact, when we are far away from the operating point, the first-order Taylor expansion may not be able to approximate the entire function, which depends on the nonlinearity of the motion model and the observation model. If they have a strong nonlinearity, then the linear approximation is only valid in a small range, and we cannot consider that the linear approximation can still be used at a long distance. This is known as the *nonlinear error* of EKF and is also a major problem.
   In the optimization problem, although we also make the first-order (fastest descent) or second-order (Gauss-Newton method or Levenberg-Marquardt method) approximation, after each iteration, we will re-do the Taylor expansion for the new estimated point, instead of only doing it once on a fixed point like EKF. This makes the optimization method applicable to a wider range, which can also be effective when the state changes greatly. So, in general, you can roughly think that *EKF is just one iteration of optimization.*[4]
3. From the point of view of implementation, EKF needs to store the state variable's mean and variance in the memory and update them iteratively. If the landmarks are also put into the states, where the number of landmarks is significantly larger than that of the poses in visual SLAM, this storage amount is considerable, and it grows squarely with the state amount (because the covariance matrix is also stored). Therefore, it is generally believed that EKF SLAM is not suitable for large-scale scenarios.
4. Finally, filter methods such as EKF have no outlier detection mechanism, which causes the system to diverge when there are outliers. Outliers are very common in visual SLAM: regardless of feature matching or optical flow method, it is easy to track or match to the wrong point. Lack of an outlier detection mechanism will make the system very unstable in practice.

Due to these obvious shortcomings of EKF, we usually think that with the same amount of calculation, nonlinear optimization can achieve better results than the filters in terms of accuracy and robustness [13]. Let's discuss the backend based on nonlinear optimization in the following sections. We will mainly introduce the graph optimization and demonstrate the backend optimization with *g2o* and *Ceres*.

---

[4] Technically, it is better than one iteration because the linearization of the update step is based on prediction. If the motion and observation model is linearized simultaneously at the time of prediction, it is the same as an iteration in the optimization.

8.2 Bundle Adjustment and Graph Optimization

If you have been working on visual 3D reconstruction, you should be familiar with this concept. The so-called Bundle Adjustment refers to optimizing both camera parameters (intrinsic and extrinsic) and 3D landmarks with images. Consider the bundles of light rays emitted from 3D points. They are projected into the image planes of several cameras and then detected as feature points. The purpose of optimization can be explained as to *adjust* the camera poses and the 3D points, to ensure the projected 2D features (bundles) match the detected results [34].

We have briefly introduced the principles of BA in lectures 4 and 6. This section aims to introduce the characteristics of its corresponding graph model structure and then introduce some general quick solutions.

8.2.1 The Projection Model and Cost Function

First, let's review the entire projection process. Starting from a point $\mathbf{p}$ in the world coordinate system, taking into account the internal and external parameters and distortion of the camera, and finally projecting into pixel coordinates, we get the following steps.

1. First, transform the world coordinates into the camera frame using extrinsics $(\mathbf{R}, \mathbf{t})$:
$$\mathbf{P}' = \mathbf{R}\mathbf{p} + \mathbf{t} = [X', Y', Z']^T. \tag{8.36}$$

2. Then, project $\mathbf{P}'$ into the normalized plane and get the normalized coordinates:
$$\mathbf{P}_c = [u_c, v_c, 1]^T = [X'/Z', Y'/Z', 1]^T. \tag{8.37}$$

3. Apply the distortion model. We only consider the radical distortion here:
$$\begin{cases} u_c' = u_c\left(1 + k_1 r_c^2 + k_2 r_c^4\right) \\ v_c' = v_c\left(1 + k_1 r_c^2 + k_2 r_c^4\right) \end{cases}. \tag{8.38}$$

4. And compute the pixel coordinates using intrinsics:
$$\begin{cases} u_s = f_x u_c' + c_x \\ v_s = f_y v_c' + c_y \end{cases}. \tag{8.39}$$

This series of calculation procedures seems a bit long for beginners. We use the process Fig. 8.2 to visualize the entire process to help readers understand. Readers should be able to understand that this process is exactly the *observation equation* mentioned earlier, and we note it as:

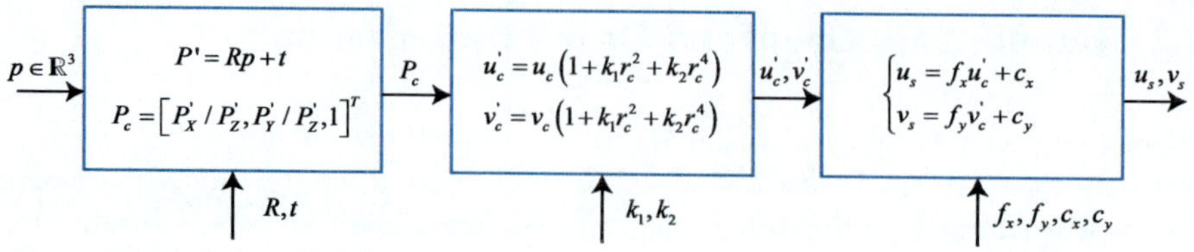

Fig. 8.2 Schematic diagram of the calculation process. The **p** on the left is the 3D point in the world frame, and the u_s, v_s on the right are the final pixel coordinates of the point on the image plane. $r_c^2 = u_c^2 + v_c^2$ in the intermediate distortion model

$$\mathbf{z} = h(\mathbf{x}, \mathbf{y}). \tag{8.40}$$

Now, we give its detailed parameterization process. Specifically, the **x** here refers to the pose of the camera. We may write it as **R**, **t**, or **T** in the corresponding Lie group, or the Lie algebra ξ. The landmark **y** is the three-dimensional point **p** here, and the observation data is the pixel coordinate $\mathbf{z} \triangleq [u_s, v_s]^T$. Consider from the perspective of least-squares, and we can write out the error of this observation:

$$\mathbf{e} = \mathbf{z} - h(\mathbf{T}, \mathbf{p}). \tag{8.41}$$

Then, let's consider the observations at other moments by adding a subscript to the error. Suppose $\mathbf{z}_{ij}$ is the data generated by observing landmark $\mathbf{p}_j$ at the pose $\mathbf{T}_i$, then the overall *cost function* is:

$$\frac{1}{2}\sum_{i=1}^{m}\sum_{j=1}^{n}\|\mathbf{e}_{ij}\|^2 = \frac{1}{2}\sum_{i=1}^{m}\sum_{j=1}^{n}\|\mathbf{z}_{ij} - h(\mathbf{T}_i, \mathbf{p}_j)\|^2. \tag{8.42}$$

Solving this least-squares is equivalent to adjusting the pose and road signs at the same time, which is the so-called BA. Next, we will gradually discuss the solution of the model based on the objective function and the nonlinear optimization content introduced in the lecture 5.

8.2.2 Solving Bundle Adjustment

Looking into the observation model $h(\mathbf{T}, \mathbf{p})$ in the previous section, it is easy to tell that the function is not a linear function. So we hope to use some nonlinear optimization methods introduced in Sect. 5.2 to optimize it. According to the principle of nonlinear optimization, we should start from a certain initial value and continuously look for the descending direction $\Delta\mathbf{x}$ to find the optimal solution of the objective function, that is, continuously solve the incremental Eq. (5.33) for the increment $\Delta\mathbf{x}$. Although a single error term is only for one pose and one landmark, the overall BA is about optimizing all variables together:

$$\mathbf{x} = [\mathbf{T}_1, \ldots, \mathbf{T}_m, \mathbf{p}_1, \ldots, \mathbf{p}_n]^T. \tag{8.43}$$

Correspondingly, the $\Delta\mathbf{x}$ in the increment equation is the increment of the overall variable. In this sense, when we give an increment to the optimization variable, the objective function becomes

$$\frac{1}{2}\left\|f(\mathbf{x}+\Delta\mathbf{x})\right\|^2 \approx \frac{1}{2}\sum_{i=1}^{m}\sum_{j=1}^{n}\left\|\mathbf{e}_{ij} + \mathbf{F}_{ij}\Delta\xi_i + \mathbf{E}_{ij}\Delta\mathbf{p}_j\right\|^2, \tag{8.44}$$

where $\mathbf{F}_{ij}$ is the partial derivative of the entire cost function to the ith pose, and $\mathbf{E}_{ij}$ is the partial derivative of the function to the jth landmark. We have introduced their specific forms in the 6.7.3 section, so we will not expand the derivation here. Now, put the camera pose variables together:

$$\mathbf{x}_c = [\xi_1, \xi_2, \ldots, \xi_m]^T \in \mathbb{R}^{6m}, \tag{8.45}$$

and also the landmarks together:

$$\mathbf{x}_p = [\mathbf{p}_1, \mathbf{p}_2, \ldots, \mathbf{p}_n]^T \in \mathbb{R}^{3n}, \tag{8.46}$$

then the Eq. (8.44) can be simplified as:

$$\frac{1}{2}\left\|f(\mathbf{x}+\Delta\mathbf{x})\right\|^2 = \frac{1}{2}\left\|\mathbf{e} + \mathbf{F}\Delta\mathbf{x}_c + \mathbf{E}\Delta\mathbf{x}_p\right\|^2. \tag{8.47}$$

It should be noted that this formula has changed from a sum of many small quadratic terms to a more overall form (sometimes mentioned as the "lifted form"). The Jacobian matrix $\mathbf{E}$ and $\mathbf{F}$ in this equation are the derivative of the overall objective function to the overall variable. It will be a large matrix that are patched by many small blocks like $\mathbf{F}_{ij}$ and $\mathbf{E}_{ij}$ from individual error terms. Then, whether we use the Gauss Newton method or the Levenberg-Marquardt method, we will finally face the incremental linear equation:

$$\mathbf{H}\Delta\mathbf{x} = \mathbf{g}. \tag{8.48}$$

According to the knowledge of 5, we know that the main difference between the Gauss Newton method and the Levenberg-Marquardt method is that the $\mathbf{H}$ here is taken from $\mathbf{J}^T\mathbf{J}$ or $\mathbf{J}^T\mathbf{J} + \lambda\mathbf{I}$. Since we have categorized the variables into poses and landmarks, the Jacobian matrix can be divided into two parts:

$$\mathbf{J} = [\mathbf{F}\ \mathbf{E}]. \tag{8.49}$$

So the Hessian matrix will be (take Gauss-Newton as an example):

$$\mathbf{H} = \mathbf{J}^T\mathbf{J} = \begin{bmatrix} \mathbf{F}^T\mathbf{F} & \mathbf{F}^T\mathbf{E} \\ \mathbf{E}^T\mathbf{F} & \mathbf{E}^T\mathbf{E} \end{bmatrix}. \tag{8.50}$$

Of course, in the Levenberg-Marquardt method we also need to calculate this matrix. It is not difficult to find that the dimension of this linear equation will be very large because all optimization variables are considered, including all camera poses and landmarks. Especially in visual SLAM, a single image will contain hundreds of feature points, which greatly increases the dimension of this linear equation. If we directly invert $\mathbf{H}$ to solve the incremental equation, such a matrix inversion is an operation [86] with $O(n^3)$ complexity, which is very expensive in computation. Fortunately, the $\mathbf{H}$ matrix here has a certain special structure. Using this special structure, we can speed up the solution process.

8.2.3 *Sparsity*

An important development of visual SLAM in the 21st century is to recognize the sparse structure of the matrix $\mathbf{H}$, and find that this structure can be naturally and explicitly represent by a graph [36, 87]. This section will discuss the matrix sparse structure in detail.

The sparsity of the $\mathbf{H}$ matrix is caused by the Jacobian matrix $\mathbf{J}(\mathbf{x})$. Consider one of the error terms $\mathbf{e}_{ij}$. Note that this error term only describes the residual about $\mathbf{p}_j$ in $\mathbf{T}_i$, and only involves the ith camera pose and the jth landmark. The derivatives of the remaining variables are all 0. Therefore, the Jacobian matrix corresponding to the error term has the following form:

$$\mathbf{J}_{ij}(\mathbf{x}) = \left(\mathbf{0}_{2\times6}, ...\mathbf{0}_{2\times6}, \frac{\partial \mathbf{e}_{ij}}{\partial \mathbf{T}_i}, \mathbf{0}_{2\times6}, ...\mathbf{0}_{2\times3}, ...\mathbf{0}_{2\times3}, \frac{\partial \mathbf{e}_{ij}}{\partial \mathbf{p}_j}, \mathbf{0}_{2\times3}, ...\mathbf{0}_{2\times3}\right), \tag{8.51}$$

where $\mathbf{0}_{2\times6}$ represents the $\mathbf{0}$ matrix with dimension 2×6, and the same for $\mathbf{0}_{2\times3}$. The partial derivative of the error term to the camera pose $\partial \mathbf{e}_{ij}/\partial \xi_i$ has a dimension of 2×6, and the partial derivative of the landmark $\partial \mathbf{e}_{ij}/\partial \mathbf{p}_j$ dimension is 2×3. The Jacobian matrix of this error term is zero except for the two non-zero blocks. This reflects the fact that the error term has nothing to do with other landmarks and poses except $\mathbf{T}_i$ and $\mathbf{p}_j$. From the perspective of graph optimization, this observation edge is only related to two vertices. So, how does it affect the incremental equation? Why does the $\mathbf{H}$ matrix have sparsity?

Let's look and the example of Fig. 8.3. Since $\mathbf{J}_{ij}$ only has non-zero blocks in column i, j, it will add four non-zero blocks into the overall Hessian matrix $\mathbf{H}$. The non-zero blocks are at (i, i), (i, j), (j, i), (j, j):

$$\mathbf{H} = \sum_{i,j} \mathbf{J}_{ij}^T \mathbf{J}_{ij}, \tag{8.52}$$

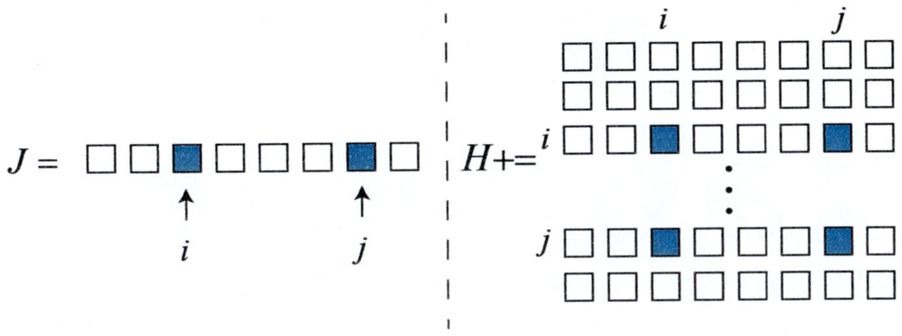

Fig. 8.3 When an error term has sparsity in its Jacobian **J**, it will also contribute to the sparsity of the Hessian **H**

Now let's divide **H** into blocks:

$$\mathbf{H} = \begin{bmatrix} \mathbf{H}_{11} & \mathbf{H}_{12} \\ \mathbf{H}_{21} & \mathbf{H}_{22} \end{bmatrix}. \tag{8.53}$$

where $\mathbf{H}_{11}$ is only related to camera poses and $\mathbf{H}_{22}$ only to landmarks. When we iterate over the i, j index, please note that the following properties are always hold:

1. No matter how i, j changes, $\mathbf{H}_{11}$ is always a block-diagonal matrix, with only non-zero blocks at $\mathbf{H}_{i,i}$.
2. Save for $\mathbf{H}_{22}$ since we only have non-zero blocks in $\mathbf{H}_{j,j}$.
3. For $\mathbf{H}_{12}$ and $\mathbf{H}_{21}$, they may be sparse or dense, depending on the specific observation data.

This shows the sparse structure of **H**. We will take advantage tof the sparsity in following section to solve the incremental equations.

8.2.4 Minimal Example of BA

Let us give an example to illustrate the situation intuitively. Suppose there are two camera poses (C_1, C_2) and six landmarks (P_1, P_2, P_3, P_4, P_5, P_6) in the scene. The variables corresponding to these cameras and point clouds are $\mathbf{T}_i, i = 1, 2$ and $\mathbf{p}_j, j = 1, \ldots, 6$. The camera C_1 observes the landmarks P_1, P_2, P_3, P_4, and the camera C_2 observes P_3, P_4, P_5, P_6. We draw this process as a schematic Fig. 8.4. Cameras and landmarks are represented by circular nodes. If the ith camera can observe the jth point, we will connect an edge at their corresponding node.

We can easily write out the overall cost function as:

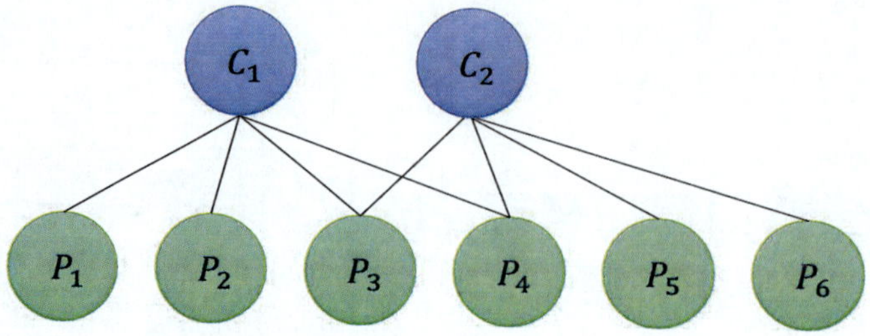

Fig. 8.4 A minimal example of bundle adjustment

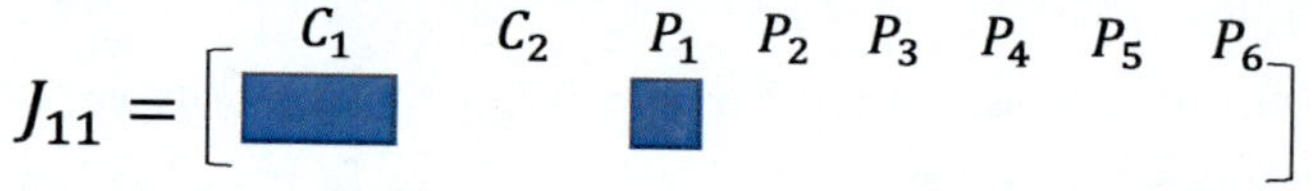

Fig. 8.5 The distribution of non-zero blocks of the $\mathbf{J}_{11}$ matrix. The upper mark indicates the variable corresponding to that column of the matrix. Since the pose dimension is larger than the landmark dimension, the matrix block corresponding to C_1 is wider than the matrix block corresponding to P_1

$$\frac{1}{2}\left(\|\mathbf{e}_{11}\|^2 + \|\mathbf{e}_{12}\|^2 + \|\mathbf{e}_{13}\|^2 + \|\mathbf{e}_{14}\|^2 + \|\mathbf{e}_{23}\|^2 + \|\mathbf{e}_{24}\|^2 + \|\mathbf{e}_{25}\|^2 + \|\mathbf{e}_{26}\|^2\right). \tag{8.54}$$

Here, the $\mathbf{e}_{ij}$ follows the previous definition of (8.42). Take $\mathbf{e}_{11}$ as an example. It describes observation of P_1 in C_1, which has nothing to do with other camera poses and landmarks. Let $\mathbf{J}_{11}$ be the Jacobian matrix corresponding to $\mathbf{e}_{11}$, and it is not difficult to see that the partial derivatives of ξ_2 and landmarks $\mathbf{p}_2, \ldots, \mathbf{p}_6$ are all zero. We rearrange the variables in the order of $\mathbf{x} = (\xi_1, \xi_2, \mathbf{p}_1, \ldots, \mathbf{p}_2)^T$, then we have:

$$\mathbf{J}_{11} = \frac{\partial \mathbf{e}_{11}}{\partial \mathbf{x}} = \left(\frac{\partial \mathbf{e}_{11}}{\partial \xi_1}, \mathbf{0}_{2\times 6}, \frac{\partial \mathbf{e}_{11}}{\partial \mathbf{p}_1}, \mathbf{0}_{2\times 3}, \mathbf{0}_{2\times 3}, \mathbf{0}_{2\times 3}, \mathbf{0}_{2\times 3}, \mathbf{0}_{2\times 3}\right). \tag{8.55}$$

In order to show the sparsity, we use a colored square to indicate that the matrix has a non-zero value in the square, and the remaining areas without color indicate that the matrix has a value of 0 at that place. Then the above $\mathbf{J}_{11}$ can be expressed as the pattern shown in Fig. 8.5. Similarly, other Jacobian matrices will have similar sparse patterns.

In order to obtain the Jacobian matrix corresponding to the objective function, we can list these $\mathbf{J}_{ij}$ as vectors in a certain order, then the overall Jacobian matrix and the corresponding $\mathbf{H}$ matrix The sparse situation is as shown in Fig. 8.6.

You may have noticed that the Hessian matrix in the Fig. 8.4 has the same structure with the corresponding *adjacency matrix* of the graph.[5] The above $\mathbf{H}$ matrix has a total of 8×8 matrix blocks. For the non-diagonal matrix block in the $\mathbf{H}$ matrix, if the matrix block is non-zero, then There will be an edge in the graph between

[5] The adjacency matrix is such a matrix that, its i, j element describes whether there is a connected edge at nodes i and j. If this edge exists, we set this element to 1, otherwise set to 0.

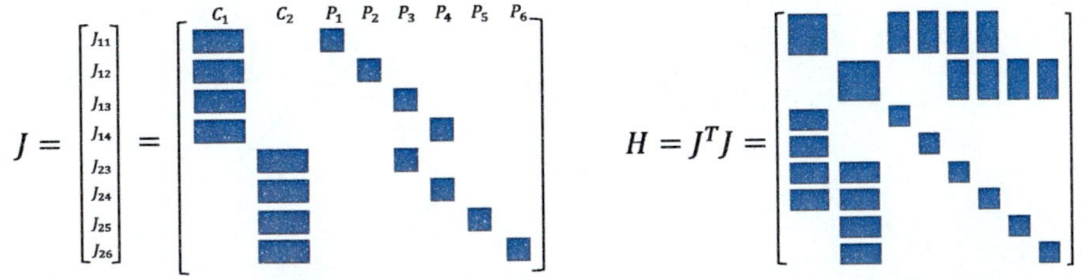

Fig. 8.6 The sparsity of the Jacobian matrix (left) and the sparsity of the **H** matrix (right). The colored squares indicate that the matrix has a non-zero value in the corresponding matrix block, and the rest of the uncolored parts indicate that the value of is always 0

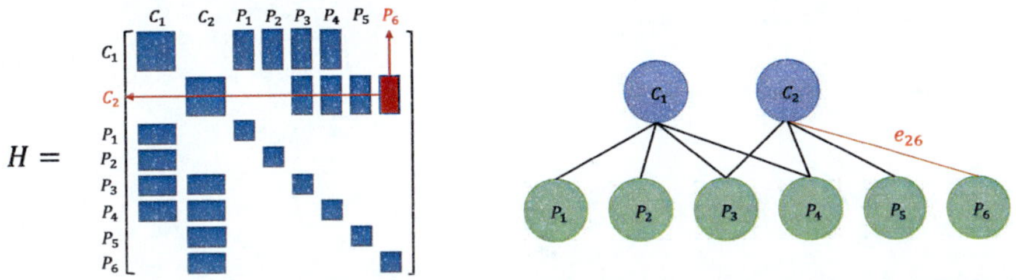

Fig. 8.7 The corresponding relationship between the non-zero matrix blocks in the **H** matrix and the edges in the graph. For example, the red matrix block on the right in the **H** matrix in the left picture indicates that there is an edge between the corresponding variables C_2 and P_6 in the right picture $\mathbf{e}_{26}$

the variables corresponding to their positions. We can clearly see this from Fig. 8.7. Therefore, the non-zero matrix block in the non-diagonal part of the **H** matrix can be understood as a connection between its corresponding two variables, or it can be called a constraint. Therefore, we found that the graph optimization structure is obviously related to the sparsity of the incremental equation.

Now consider the more general situation, suppose we have m camera poses and n landmarks. Since there are usually far more landmarks than cameras, we have $n \gg m$. From the above reasoning, the actual **H** matrix will be something like in Fig. 8.8. Its upper left corner block appears to be very small, while the lower right corner diagonal block takes up a lot of space. In addition, the non-diagonal part is distributed with scattered observation data. Because its shape is very similar to an arrow, it is also called an arrow-like matrix [5]. At the same time, it is also very similar to a pickaxe in Minecraft, so I also call it a pickaxe matrix.[6]

[6] This is just a joke, please do not write it like this in formal academic papers.

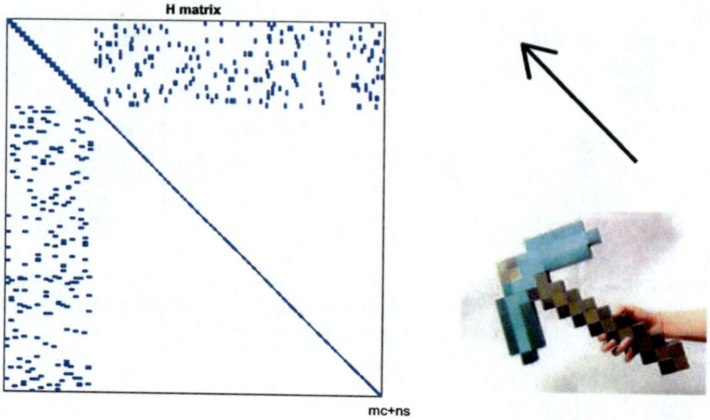

Fig. 8.8 **H** matrix in general cases

8.2.5 *Schur Trick*

For **H** with this sparse structure, what is the difference in the solution of the linear equation $\mathbf{H}\Delta\mathbf{x} = \mathbf{g}$? In fact, there are several ways to use the sparsity of **H** to accelerate the calculations. This section introduces one of the most commonly used methods in visual SLAM: Schur elimination, also known as *marginalization* in SLAM research.

Take a closer look at Fig. 8.8, we can easily find that this matrix can be divided into 4 blocks, which is consistent with (8.53). The upper left corner is a block-diagonal matrix, and the dimension of each block is the same as the dimension of the camera pose. The bottom-right corner is also a diagonal block matrix, and the dimension of each diagonal block is same as the dimension of the landmark. The structure of the off-diagonal block is related to the specific observation data. We first divide this matrix into regions as shown in Fig. 8.9. Readers can easily find that these 4 regions correspond to the 4 matrix blocks in the formula (8.50). For the convenience of subsequent analysis, we note these 4 blocks as $\mathbf{B}$, $\mathbf{E}$, $\mathbf{E}^T$, $\mathbf{C}$.

The linear equations $\mathbf{H}\Delta\mathbf{x} = \mathbf{g}$ can be rewritten as:

$$\begin{bmatrix} \mathbf{B} & \mathbf{E} \\ \mathbf{E}^{\mathrm{T}} & \mathbf{C} \end{bmatrix} \begin{bmatrix} \Delta\mathbf{x}_c \\ \Delta\mathbf{x}_p \end{bmatrix} = \begin{bmatrix} \mathbf{v} \\ \mathbf{w} \end{bmatrix}, \tag{8.56}$$

where **B** is a block-diagonal matrix. The dimension of each diagonal block is the same as the dimension of the camera poses. The number of diagonal blocks is the number of camera variables. Since the number of landmarks will be much larger than the number of camera variables, **C** is often much larger than **B**. Since we use 3D landmarks, the **C** matrix is a diagonal block matrix wich each block beging a 3×3 matrix. The difficulty of inverting a diagonal block matrix is much less than that of a general matrix, because we only need to invert those diagonal blocks separately.

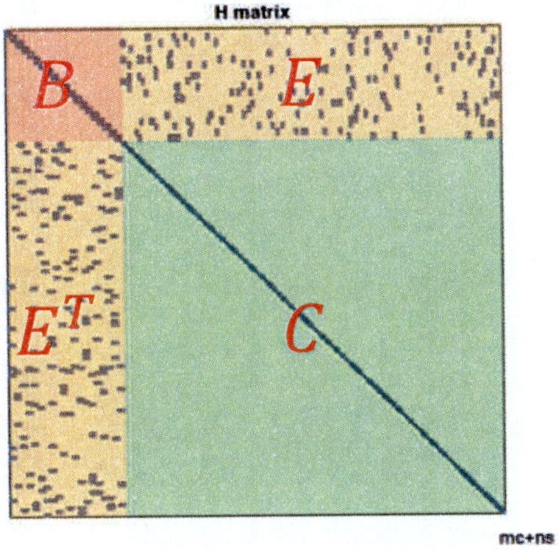

Fig. 8.9 Blocks in **H** matrix

Taking this feature into account, we perform Gaussian elimination on the linear equations. The goal is to eliminate the non-diagonal part **E** in the upper right corner, so we multiply a coefficient matrix on the left side:

$$\begin{bmatrix} \mathbf{I} & -\mathbf{E}\mathbf{C}^{-1} \\ \mathbf{0} & \mathbf{I} \end{bmatrix} \begin{bmatrix} \mathbf{B} & \mathbf{E} \\ \mathbf{E}^{\mathrm{T}} & \mathbf{C} \end{bmatrix} \begin{bmatrix} \Delta \mathbf{x}_c \\ \Delta \mathbf{x}_p \end{bmatrix} = \begin{bmatrix} \mathbf{I} & -\mathbf{E}\mathbf{C}^{-1} \\ \mathbf{0} & \mathbf{I} \end{bmatrix} \begin{bmatrix} \mathbf{v} \\ \mathbf{w} \end{bmatrix}. \tag{8.57}$$

Rearrange it:

$$\begin{bmatrix} \mathbf{B} - \mathbf{E}\mathbf{C}^{-1}\mathbf{E}^T & \mathbf{0} \\ \mathbf{E}^T & \mathbf{C} \end{bmatrix} \begin{bmatrix} \Delta \mathbf{x}_c \\ \Delta \mathbf{x}_p \end{bmatrix} = \begin{bmatrix} \mathbf{v} - \mathbf{E}\mathbf{C}^{-1}\mathbf{w} \\ \mathbf{w} \end{bmatrix}. \tag{8.58}$$

After the elimination, the first line of the equations becomes a term that has nothing to do with $\Delta \mathbf{x}_p$. Take it out separately and get the incremental equation about the pose part:

$$\left[\mathbf{B} - \mathbf{E}\mathbf{C}^{-1}\mathbf{E}^T\right] \Delta \mathbf{x}_c = \mathbf{v} - \mathbf{E}\mathbf{C}^{-1}\mathbf{w}. \tag{8.59}$$

The dimension of this linear equation is the same as the **B** matrix, and is much smaller than the overall equations. The Schur trick is to solve this equation first, then substitute the solved $\Delta \mathbf{x}_c$ into the original equation, and then solve $\Delta \mathbf{x}_p$. This process is called *marginalization* [83], or *Schur elimination* (Schur trick). Compared with the method of directly solving linear equations, it have several obvious advantages:

1. Since **C** is a block diagonal matrix, $\mathbf{C}^{-1}$ is easy to solve.
2. After solving $\Delta \mathbf{x}_c$, the incremental equation of the landmark part is given by $\Delta \mathbf{x}_p = \mathbf{C}^{-1}(\mathbf{w} - \mathbf{E}^T \Delta \mathbf{x}_c)$. This still uses the easy-to-solve feature of $\mathbf{C}^{-1}$.

Therefore, the main amount of calculation for marginalization is to solve the Eq. (8.59). There is not much we can say about this equation. It is just an ordinary linear equation, no special structure can be used. Let us denote the coefficient of this

Fig. 8.10 The sparse structure of the **S** matrix after Schur elimination. It looks like a QR code. Don't scan it

equation as **S**. How about its sparsity? Fig. 8.10 shows an instance of **S** after Schur elimination. It can be seen that its sparsity is somehow irregular.

As mentioned earlier, the non-zero elements at the non-diagonal blocks of the **H** matrix correspond to the association between the camera and the landmark. Then, do the sparsity of **S** have physical meaning after Schur elimination? The answer is yes. Here we say without proof that the non-zero matrix block on the off-diagonal line of the **S** matrix indicates that there is a co-observation between the two camera variables. It is called *co-visibility*. Conversely, if the block is zero, it means that the two cameras do not share any observation. For example, in the sparse matrix shown in Fig. 8.11, the first 4×4 matrix blocks in the upper left corner can indicate that there are common observations between the corresponding camera variables C_1, C_2, C_3, and C_4.

Therefore, the sparsity structure of the **S** matrix depends on the actual observation results, which we cannot predict in advance. In practice, for example, in the local mapping module of ORB-SLAM [88], we may deliberately select those frames with co-observations as keyframes, in this case the resulting **S** is a dense matrix. However, since this module is not executed in real time (normally runs in a separate thread in the backend), this approach is also acceptable. But in other methods, such as DSO [70], OKVIS [89], etc., they use the sliding window method. Such methods will use the previous **S** matrix as a prior constraint in the following BA steps, so they must use some techniques to maintain the sparsity of the **S** matrix by discarding some observations. Readers who want to go deeper into this area can refer to their papers. I won't talk about these detailed things here.

From a probabilistic persepctive, we call this step as *marginalization* because we actually converted the problem of seeking $(\Delta \mathbf{x}_c, \Delta \mathbf{x}_p)$ into fixing $\Delta \mathbf{x}_p$ first, and finding $\Delta \mathbf{x}_c$, and then finding $\Delta \mathbf{x}_p$. This step is equivalent to the conditional probability expansion:

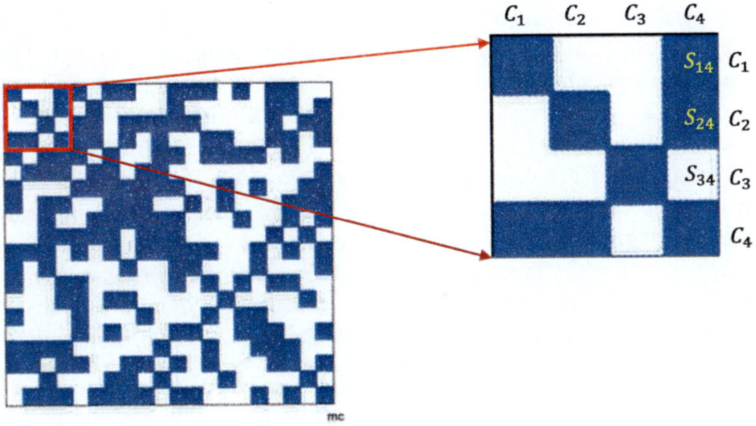

Fig. 8.11 Take the first 4 × 4 matrix blocks in the **S** matrix as an example, the matrix blocks in this area $\mathbf{S}_{14}, \mathbf{S}_{24}$ are not zero, meaning that there is at least one common observation between the camera C_4 and the cameras C_1 and C_2. Meanwhile, S_{34} is zero means that there is no common observation between C_3 and C_4

$$\underbrace{P(\mathbf{x}_c, \mathbf{x}_p)}_{\text{joint}} = \underbrace{P(\mathbf{x}_c|\mathbf{x}_p)}_{\text{conditional}} \underbrace{P(\mathbf{x}_p)}_{\text{marginal}}, \tag{8.60}$$

The result is to convert a joint distribution to a multiplication of a conditional distribution and marginal distribution. In the previous sparse BA process, we actually marginalized all the landmarks. But we can also choose which part to marginalization according to the actual situation. Please also note that Schur elimination is only one way to achieve marginalization, and Cholesky decomposition can also be used for marginalization.

Readers may continue to ask, after the Schur elimination, we still need to solve the linear equations (8.59). Are there any techniques for solving it? Unfortunately, this part belongs to the traditional numerical matrix solution, which is usually calculated by matrix decomposition. No matter which solution method is adopted, we recommend using the sparsity of **H** to perform Schur elimination first. This is not only because this can increase the speed, but also because the condition number of the **S** matrix after elimination is often smaller than the previous **H** matrix. Note that Schur elimination does not only mean to eliminate the landmarks. Eliminating camera variables is also a common method used in SLAM.

8.2.6 Robust Kernels

In the previous BA problem, we minimize the $\mathcal{L}_2$ norm of the error term in the objective function. Although this approach is intuitive, there is a serious problem:

what happens if the data given by a certain error term is wrong for reasons such as mismatch? We added an edge that shouldn't have been added to the graph. However, the optimization algorithm does not know that this is wrong data. It will treat all data as noisy observations. From the point of view of the algorithm, this is equivalent to that we find an observation almost impossible to happen. At this time, there will be an edge with large error in graph optimization, and its gradient is also large, which means that adjusting the variables related to it will make the objective function drop more. Therefore, the algorithm will try to adjust the estimated value of the nodes connected by this edge first to make them comply with the unreasonable requirements of this edge. Since the error of this edge is really large, it tends to eliminate out the influence of other correct edges, making the optimization algorithm focus on adjusting a wrong value. This is obviously not what we want to see.

The reason for this problem is that when the error is large, the $\mathcal{L}_2$ norm grows too fast. So people propose something called kernel functions. The kernel function guarantees that the error of each edge will not be too big to cover up the other edges. The specific method is to replace the original $\mathcal{L}_2$ norm metric of error with a function that does not grow that fast, while ensuring its own smoothness (otherwise we cannot compute derivatives). Because they make the overall optimization result more robust, they are also called robust kernel.

There are many kinds of robust kernel functions, such as the most commonly used Huber kernel:

$$H(e) = \begin{cases} \frac{1}{2}e^2 & \text{when} |e| \leqslant \delta, \\ \delta\left(|e| - \frac{1}{2}\delta\right) & \text{otherwise} \end{cases} \tag{8.61}$$

We see that when the error of e is greater than a certain threshold δ, the function growth changes from a quadratic form to a linear form, which is equivalent to limiting the maximum value of the gradient. At the same time, the Huber kernel function is smooth and can be easily derived. Figure 8.12 shows the comparison between Huber's kernel function and the quadratic function. It can be seen that the growth of Huber's kernel function is significantly lower than that of the quadratic function when the error is large.

In addition to Huber kernels, there are also Cauchy kernels, Tukey kernels, and so on. Please check which kernel functions are supported in *g2o* and *Ceres*.

8.2.7 Summary

In this section, we focus on the sparsity problem in BA. However, in practice, most libraries have implemented the detailed operations for us, and what we need to do is mainly to construct the Bundle Adjustment problem, set the marginalized part, and then call the dense or sparse matrix solver to optimize the variables. If readers want to have a deeper understanding of BA, please refer to the [34] or other related papers.

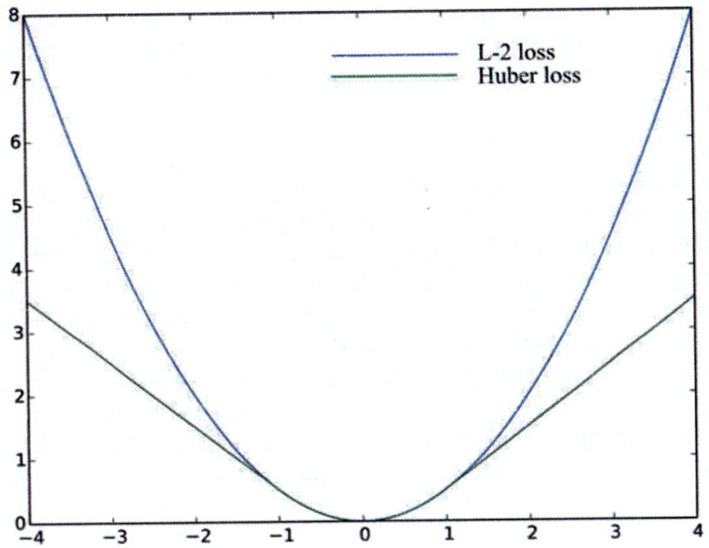

Fig. 8.12 The Huber kernel

In the following two sections, we will use the Ceres and *g2o* libraries to do Bundle Adjustment. In order to show the difference between them, we will use a public data set BAL [90], and use the common read and write code.

8.3 Practice: BA with Ceres

8.3.1 BAL Dataset

We use the BAL data set to demonstrate the BA experiments. The BAL data set provides several scenes. The camera and landmark information in each scene are given by a text file. We use the file problem-16-22106-pre.txt as an example. This file stores the BA problem information in a line-by-line manner. For the detailed format, see https://grail.cs.washington.edu/projects/bal. We use the BALProblem class defined in common.h to read in the content of the file, and then use Ceres and *g2o* to solve them.

It should be noted that the BAL data set has some special features:

1. The camera intrinsic model of BAL is given by the focal length f and the distortion parameters k_1, k_2, where f is same as the f_x and f_y mentioned before. Since the pixels of the photo are basically square, in many practical situations, f_x is very close to f_y, so it is possible to use the same value. In addition, there is no c_x, c_y in this model, because these two values have been removed from the stored data.

2. BAL data assumes that the projection plane is behind the optical center of the camera when projecting, so if we calculate according to the model we used before, we need to multiply -1 after projection. However, most data sets still use the projection plane in front of the optical center. We should read the format description carefully before using the data set.

After reading the data with the BALProblem class, we can call the Normalize function to normalize the original data, or add noise to the data through the Perturb function. Normalization means setting zero to the centers of all landmarks, and then scaling them to an appropriate scale. This will make the value in the optimization process more stable and prevent BA to get very large values in extreme cases.

Please read the other interfaces of the BALProblem class by yourself. Since these codes are only responsible for IO functions, in order to save space, we will not print them in the text. After solving the BA, we can also use this type of function to write the result into a ply file (a point cloud file format), and then use the meshlab software to view it. Meshlab can be installed via apt-get, and the installation method is not described here.

8.3.2 *Solving BA in Ceres*

In the bundle_adjustment_ceres.cpp file, we implemented the process of solving BA in Ceres. The key to using Ceres is to define the projection error model. This part of the code is given in SnavelyReprojectionError.h:

Listing 8.1 slambook2/ch9/SnavelyReprojectionError.cpp (part)

```
class SnavelyReprojectionError {
public:
  SnavelyReprojectionError(double observation_x, double observation_y) :
      observed_x(observation_x), observed_y(observation_y) {}

  template<typename T>
  bool operator()(const T *const camera,
    const T *const point,
    T *residuals) const {
    // camera[0,1,2] are the angle-axis rotation
    T predictions[2];
    CamProjectionWithDistortion(camera, point, predictions);
    residuals[0] = predictions[0] - T(observed_x);
    residuals[1] = predictions[1] - T(observed_y);

    return true;
  }

  // camera : 9 dims array
  // [0-2] : angle-axis rotation
  // [3-5] : translation
  // [6-8] : camera parameter, [6] focal length, [7-8] second and forth
      order radial distortion
  // point : 3D location.
  // predictions : 2D predictions with center of the image plane.
  template<typename T>
  static inline bool CamProjectionWithDistortion(const T *camera, const T *
      point, T *predictions) {
```

```
    // Rodrigues' formula
    T p[3];
    AngleAxisRotatePoint(camera, point, p);
    // camera[3,4,5] are the translation
    p[0] += camera[3];
    p[1] += camera[4];
    p[2] += camera[5];

    // Compute the center fo distortion
    T xp = -p[0] / p[2];
    T yp = -p[1] / p[2];

    // Apply second and fourth order radial distortion
    const T &l1 = camera[7];
    const T &l2 = camera[8];

    T r2 = xp * xp + yp * yp;
    T distortion = T(1.0) + r2 * (l1 + l2 * r2);

    const T &focal = camera[6];
    predictions[0] = focal * distortion * xp;
    predictions[1] = focal * distortion * yp;

    return true;
  }

  static ceres::CostFunction *Create(const double observed_x, const double
      observed_y) {
    return (new ceres::AutoDiffCostFunction<SnavelyReprojectionError, 2, 9,
        3>(
      new SnavelyReprojectionError(observed_x, observed_y)));
  }

private:
  double observed_x;
  double observed_y;
};
```

This overloaded bracket operator in this class implements the error calculation, whose main body is in the CamProjectionWithDistortion function. Note that in Ceres, we must store optimized variables in the form of a double array. Now each camera has a total of 6-dimensional pose, 1-dimensional focal length, and 2-dimensional distortion parameters, which are described by a total of 9-dimensional parameters. We must also store them in this order in actual storage. The stati cCreate function of this class acts as an external interface that directly returns a Ceres cost function that can be automatically derived. We only need to call the Create function and put the cost function into ceres::Problem.

Next we realize the part of building and solving BA:

Listing 8.2 slambook2/ch9/SnavelyReprojectionError.cpp (part)

```
void SolveBA(BALProblem &bal_problem) {
  const int point_block_size = bal_problem.point_block_size();
  const int camera_block_size = bal_problem.camera_block_size();
  double *points = bal_problem.mutable_points();
  double *cameras = bal_problem.mutable_cameras();

  // Observations is 2 * num_observations long array observations
  // [u_1, u_2, ... u_n], where each u_i is two dimensional, the x
  // and y position of the observation.
  const double *observations = bal_problem.observations();
  ceres::Problem problem;
```

```
  for (int i = 0; i < bal_problem.num_observations(); ++i) {
    ceres::CostFunction *cost_function;

    // Each Residual block takes a point and a camera as input
    // and outputs a 2 dimensional Residual
    cost_function =
    SnavelyReprojectionError::Create(observations[2 * i + 0], observations[2
        * i + 1]);

    // If enabled use Huber's loss function.
    ceres::LossFunction *loss_function = new ceres::HuberLoss(1.0);

    // Each observation corresponds to a pair of a camera and a point
    // which are identified by camera_index()[i] and point_index()[i]
    // respectively.
    double *camera = cameras + camera_block_size * bal_problem.camera_index
        ()[i];
    double *point = points + point_block_size * bal_problem.point_index()[i
        ];

    problem.AddResidualBlock(cost_function, loss_function, camera, point);
  }

  std::cout << "Solving ceres BA ... " << endl;
  ceres::Solver::Options options;
  options.linear_solver_type = ceres::LinearSolverType::SPARSE_SCHUR;
  options.minimizer_progress_to_stdout = true;
  ceres::Solver::Summary summary;
  ceres::Solve(options, &problem, &summary);
  std::cout << summary.FullReport() << "\n";
}
```

It can be seen that the problem building part is quite simple. If you want to add other cost functions, the entire process will not change much. Finally, in ceres::Solver::Options, we can set the solution method. Using SPARSE_SCHUR will make the actual solution process of Ceres same as we described earlier, that is, first perform Schur marginalization on the landmarks to solve this problem in an accelerated manner. However, in Ceres we cannot control which part of the variables are marginalized, which is automatically searched and calculated by the Ceres solver.

The output of Ceres BA is like:

Listing 8.3 Terminal output

```
./build/bundle_adjustment_ceres problem-16-22106-pre.txt
Header: 16 22106 83718bal problem file loaded...
bal problem have 16 cameras and 22106 points.
Forming 83718 observations.
Solving ceres BA ...
iter cost cost_change |gradient| |step| tr_ratio tr_radius ls_iter
    iter_time total_time
0 1.842900e+07 0.00e+00 2.04e+06 0.00e+00 0.00e+00 1.00e+04 0 6.10e-02 2.24
    e-01
1 1.449093e+06 1.70e+07 1.75e+06 2.16e+03 1.84e+00 3.00e+04 1 1.79e-01 4.03
    e-01
2 5.848543e+04 1.39e+06 1.30e+06 1.55e+03 1.87e+00 9.00e+04 1 1.56e-01 5.59
    e-01
3 1.581483e+04 4.27e+04 4.98e+05 4.98e+02 1.29e+00 2.70e+05 1 1.51e-01 7.10
    e-01
......
```

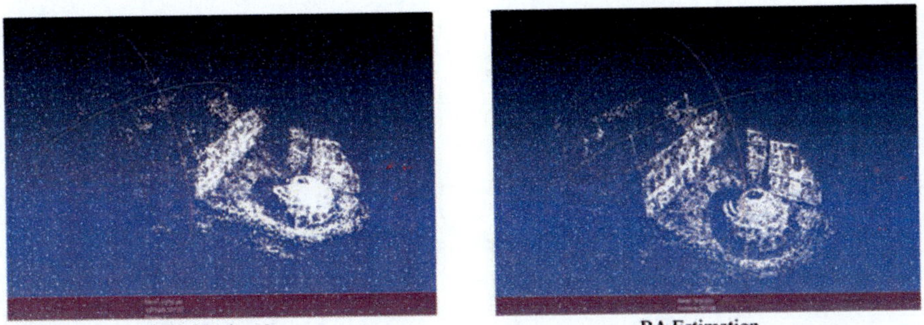

Initial Estimation BA Estimation

Fig. 8.13 Visualized point cloud before and after optimization. Left side: the initial value before optimization; Right side: the optimized value

The overall error should continue to decrease as the number of iterations increases. Finally, we output the point clouds before and after optimization as initial.ply and final.ply, which can be opened directly with meshlab. The result graph is shown in Fig. 8.13.

8.4 Practice: BA with *g2o*

Let's consider how to use *g2o* to solve this BA problem. As mentioned before, *g2o* uses a graph model to describe the problem's structure, so we need to use nodes to represent cameras and landmarks and then use edges to represent observations between them. We still use custom defined vertices and edges instead of built-in edges. For cameras and landmarks, we can define the following structure and use the override keyword to indicate the virtual functions:

Listing 8.4 slambook2/ch9/bundle_adjustment_g2o.cpp (part)

```
struct PoseAndIntrinsics {
  PoseAndIntrinsics() {}

  /// set from given data address
  explicit PoseAndIntrinsics(double *data_addr) {
    rotation = SO3d::exp(Vector3d(data_addr[0], data_addr[1], data_addr[2]))
    ;
    translation = Vector3d(data_addr[3], data_addr[4], data_addr[5]);
    focal = data_addr[6];
    k1 = data_addr[7];
    k2 = data_addr[8];
  }

  /// set to double array
  void set_to(double *data_addr) {
    auto r = rotation.log();
    for (int i = 0; i < 3; ++i) data_addr[i] = r[i];
    for (int i = 0; i < 3; ++i) data_addr[i + 3] = translation[i];
    data_addr[6] = focal;
    data_addr[7] = k1;
```

```
    data_addr[8] = k2;
  }

  SO3d rotation;
  Vector3d translation = Vector3d::Zero();
  double focal = 0;
  double k1 = 0, k2 = 0;
};

/// pose and f, k1, k2
class VertexPoseAndIntrinsics : public g2o::BaseVertex<9, PoseAndIntrinsics
    > {
  public:
  EIGEN_MAKE_ALIGNED_OPERATOR_NEW;

  VertexPoseAndIntrinsics() {}

  virtual void setToOriginImpl() override {
    _estimate = PoseAndIntrinsics();
  }

  virtual void oplusImpl(const double *update) override {
    _estimate.rotation = SO3d::exp(Vector3d(update[0], update[1], update[2])
        ) * _estimate.rotation;
    _estimate.translation += Vector3d(update[3], update[4], update[5]);
    _estimate.focal += update[6];
    _estimate.k1 += update[7];
    _estimate.k2 += update[8];
  }

  Vector2d project(const Vector3d &point) {
    Vector3d pc = _estimate.rotation * point + _estimate.translation;
    pc = -pc / pc[2];
    double r2 = pc.squaredNorm();
    double distortion = 1.0 + r2 * (_estimate.k1 + _estimate.k2 * r2);
    return Vector2d(_estimate.focal * distortion * pc[0],
    _estimate.focal * distortion * pc[1]);
  }

  virtual bool read(istream &in) {}
  virtual bool write(ostream &out) const {}
};

class VertexPoint : public g2o::BaseVertex<3, Vector3d> {
  public:
  EIGEN_MAKE_ALIGNED_OPERATOR_NEW;

  VertexPoint() {}

  virtual void setToOriginImpl() override {
    _estimate = Vector3d(0, 0, 0);
  }

  virtual void oplusImpl(const double *update) override {
    _estimate += Vector3d(update[0], update[1], update[2]);
  }

  virtual bool read(istream &in) {}

  virtual bool write(ostream &out) const {}
};

class EdgeProjection :
public g2o::BaseBinaryEdge<2, Vector2d, VertexPoseAndIntrinsics,
    VertexPoint> {
  public:
  EIGEN_MAKE_ALIGNED_OPERATOR_NEW;
```

```

  virtual void computeError() override {
    auto v0 = (VertexPoseAndIntrinsics *) _vertices[0];
    auto v1 = (VertexPoint *) _vertices[1];
    auto proj = v0->project(v1->estimate());
    _error = proj - _measurement;
  }

  // use numeric derivatives in g2o
  virtual bool read(istream &in) {}
  virtual bool write(ostream &out) const {}
};
```

We define the rotation, translation, focal length, and distortion parameters in the same camera vertex and then define the camera's observation edge and the landmark. Here we do not implement the Jacobian calculation function of the edge,[7] so *g2o* will automatically provide a numerical calculation of Jacobian. Finally, according to the data in BAL, we can set up the optimization problem of g2o:

Listing 8.5 slambook2/ch9/bundle_adjustment_g2o.cpp (part)

```
void SolveBA(BALProblem &bal_problem) {
  const int point_block_size = bal_problem.point_block_size();
  const int camera_block_size = bal_problem.camera_block_size();
  double *points = bal_problem.mutable_points();
  double *cameras = bal_problem.mutable_cameras();

  // pose dimension 9, landmark is 3
  typedef g2o::BlockSolver<g2o::BlockSolverTraits<9, 3>> BlockSolverType;
  typedef g2o::LinearSolverCSparse<BlockSolverType::PoseMatrixType>
       LinearSolverType;
  // use LM
  auto solver = new g2o::OptimizationAlgorithmLevenberg(
  g2o::make_unique<BlockSolverType>(g2o::make_unique<LinearSolverType>()));
  g2o::SparseOptimizer optimizer;
  optimizer.setAlgorithm(solver);
  optimizer.setVerbose(true);

  /// build g2o problem
  const double *observations = bal_problem.observations();
  // vertex
  vector<VertexPoseAndIntrinsics *> vertex_pose_intrinsics;
  vector<VertexPoint *> vertex_points;
  for (int i = 0; i < bal_problem.num_cameras(); ++i) {
    VertexPoseAndIntrinsics *v = new VertexPoseAndIntrinsics();
    double *camera = cameras + camera_block_size * i;
    v->setId(i);
    v->setEstimate(PoseAndIntrinsics(camera));
    optimizer.addVertex(v);
    vertex_pose_intrinsics.push_back(v);
  }
  for (int i = 0; i < bal_problem.num_points(); ++i) {
    VertexPoint *v = new VertexPoint();
    double *point = points + point_block_size * i;
    v->setId(i + bal_problem.num_cameras());
    v->setEstimate(Vector3d(point[0], point[1], point[2]));
    // in g2o we should manually set the marginalized part
    v->setMarginalized(true);
    optimizer.addVertex(v);
    vertex_points.push_back(v);
  }
```

[7] You can implement it by yourself if interested. Here we just want to make a fair comparison since we do not provide Jacobians in Ceres version.

```

  // edge
  for (int i = 0; i < bal_problem.num_observations(); ++i) {
    EdgeProjection *edge = new EdgeProjection;
    edge->setVertex(0, vertex_pose_intrinsics[bal_problem.camera_index()[i]]);
    edge->setVertex(1, vertex_points[bal_problem.point_index()[i]]);
    edge->setMeasurement(Vector2d(observations[2 * i + 0], observations[2 * i + 1]));
    edge->setInformation(Matrix2d::Identity());
    edge->setRobustKernel(new g2o::RobustKernelHuber());
    optimizer.addEdge(edge);
  }

  optimizer.initializeOptimization();
  optimizer.optimize(40);

  // set to bal problem
  for (int i = 0; i < bal_problem.num_cameras(); ++i) {
    double *camera = cameras + camera_block_size * i;
    auto vertex = vertex_pose_intrinsics[i];
    auto estimate = vertex->estimate();
    estimate.set_to(camera);
  }
  for (int i = 0; i < bal_problem.num_points(); ++i) {
    double *point = points + point_block_size * i;
    auto vertex = vertex_points[i];
    for (int k = 0; k < 3; ++k) point[k] = vertex->estimate()[k];
  }
}
```

The above defines the nodes and edges used in this problem. Next, we need to generate some nodes and edges based on the actual data in the BALProblem class and hand them to *g2o* for optimization. It is worth noting that to take full advantage of sparse BA, we must set the setMarginalized function in the landmarks here. The big difference between *g2o* and Ceres is that when using sparse optimization, *g2o* must manually set which vertices are marginalized; otherwise, it will report a runtime error (readers can try to comment out the v->setMarginalized(true) line). The rest is similar to the Ceres version, so that we won't introduce more. The g2o experiment will also output point clouds before and after optimization for comparison and review.

8.5 Summary

This lecture discusses the state estimation problem and the solution of graph optimization in depth. We see that SLAM can be regarded as a state estimation problem in the classic model. If we assume Markov property and only consider the current state, we get a filter model represented by EKF. If not, we can also choose to consider all motions and observations, which constitute a least-squares problem. In the case of only observation equations, this problem is called BA and can be solved by nonlinear optimization methods. We carefully discussed the sparsity problem in the solution process and pointed out the connection between the problem and graph optimization. Finally, we demonstrated how to use the *g2o* and Ceres libraries to solve the same optimization problem so that readers have an intuitive understanding of BA.

Exercises

1. Derive (8.25) using Sherman-Morrison-Woodbury identities [5, 76].
2. Compare the final cost of *g2o* and Ceres after optimization. Discuss why they are siminlar in Meshlab by have different numerical values.
3. Search for methods that can provide partial Schur elimination in Ceres, and show the difference.
4. Show that the **S** matrix is semi-definite.
5. Read [36] and discuss the difference between the kernel functions in *g2o* and the loss function in Ceres.
6. *In the practice part, we optimize the pose and intrinsics parametrized with f, k_1, k_2. Please use the full intrinsic and distortion model in lecture 5 and modify the programs to complete the optimization.

Chapter 9
Filters and Optimization Approaches: Part II

Goal of Study

1. Learn the principles of sliding window optimization;
2. Learn the basic knowledge about pose graph optimization.
3. Pose graph optimization with *g2o*.

In the last lecture, we focused on graph optimization based on BA. BA can accurately optimize the pose and feature point position of each camera. However, in a larger scene, the high dimensionality of landmarks will seriously reduce the calculation efficiency, resulting in an increasing amount of calculation that cannot be real-time. The first part of this lecture will introduce a simplified, yet widely used optimization approach: pose graph.

9.1 Sliding Window Filter and Optimization

9.1.1 Controlling the Structure of BA

The graph optimization with camera pose and spatial points is called BA, which can effectively solve large-scale positioning and mapping problems. This is very useful in SfM, but in the real-time SLAM process, we often need to control the problem's scale to maintain real-time calculations. If the computing power is unlimited, we might calculate the entire BA every moment-but that does not meet the reality. The reality is that we must limit the calculation time of the backend. For example, the BA scale cannot exceed 10,000 landmarks, the iterations cannot exceed 20 times, and the time used does not exceed 0.5 s, and so on. An algorithm that takes a week to reconstruct a city map like SfM is not necessarily effective in SLAM.

There are many ways to control the calculation scale, such as extracting keyframes from the continuous video [89], only constructing the BA between the keyframe and the landmarks. The non-keyframes are only used for localization and do not contribute to the mapping part. Even so, as time goes by, the number of keyframes will increase,

X. Gao and T. Zhang, *Introduction to Visual SLAM*,
https://doi.org/10.1007/978-981-16-4939-4_9

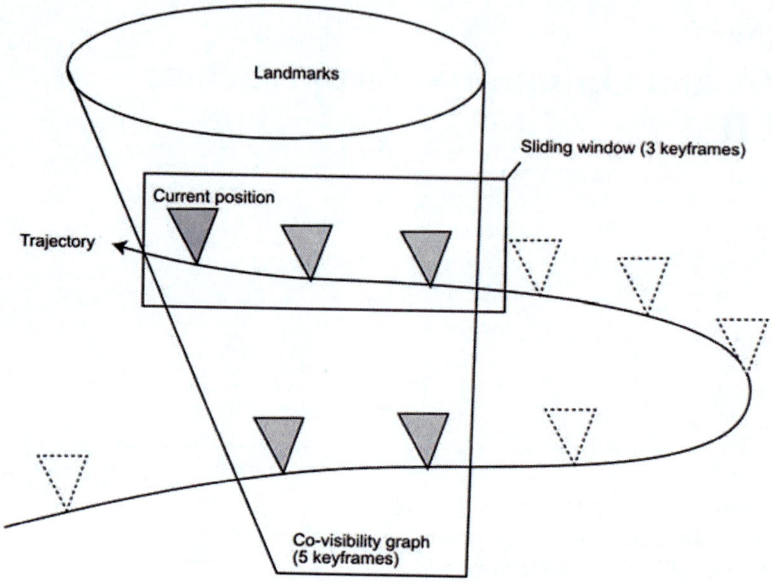

Fig. 9.1 Co-visibility graph in sliding window

and the scale of the map will continue to grow. For batch optimization methods like BA, the computational efficiency will (worryingly) continue to decline. To avoid this situation, we need to control the scale of the backend problem. These methods can be theoretical or engineering.

For example, the simplest way to control the BA scale is to keep only the *N* keyframes closest to the current moment and remove the earlier ones. Therefore, we will fix the BA within a time window, and those that leave this window will be discarded. This method is called the sliding window method [91]. Of course, we can change the criteria for selecting these *N* keyframes. For example, it is unnecessary to take the closest in time, but according to a certain principle, take the keyframes that are close in time and expandable in space. To ensure that even when the camera is stopped, the BA structure will not shrink into a single point (this can easily lead to some bad degradation). If we think more deeply about the structure of frames and frames, we can also define a structure called "covisibility graph" like ORB-SLAM2 [88] (see Fig. 9.1). The so-called co-visibility refers to those features that are observed together with the current keyframe). Therefore, in BA optimization, we take some keyframes and landmarks in the co-visibility graph to optimize. For example, we may pick about 20 co-visible keyframes with the current frame and leave the others unchanged. If we can construct the co-visibility relationship correctly, the optimization will remain optimal for a longer time.

No matter whether it is a sliding window or a co-visibility approach, in general, it is a kind of engineering trade-off between full SLAM and real-time computing. But in theory, they also introduce a new problem: when we talk about "discarding" variables outside the sliding window or "fixing" variables outside the co-visibility graph, what is the specific operation of "discarding" and "fixing"? "Fixed" seems to be easy to understand. We only need to keep the keyframe estimates unchanged during optimization. But "discarding" refers to completely abandoning it, which means that the variables outside the window totally do not affect the variables in the window? Or does the data outside the window should have some influence but is somehow ignored?

Next, we will talk about these issues. How should they be dealt with in theory, and whether can we simplify them in engineering?

9.1.2 Sliding Window

Now consider a sliding window. Assume there are N keyframes in this window, and their poses are denoted as

$$\boldsymbol{x}_1, \ldots, \boldsymbol{x}_N,$$

we assume that they are in the vector space, that is, using Lie Algebra expression, then, what can we talk about these keyframes?

Obviously, we care about the location of these keyframes and their uncertainty. This corresponds to their mean and covariance under the assumption of Gaussian distribution. If these keyframes also correspond to a local map, we can also ask for the entire local map's mean and covariance. Suppose there are M landmark points in this sliding window: $\boldsymbol{y}_1, \ldots, \boldsymbol{y}_N$, they form a local map together with the N keyframes. Obviously, we can use the bundle adjustment method introduced in the last lecture to deal with this sliding window, including building a graph optimization model, building an overall Hessian matrix, and then marginalizing all landmarks to speed up the solution. After the marginalization, we get the conditional distribution of the poses, namely

$$[\boldsymbol{x}_1, \ldots, \boldsymbol{x}_N | \boldsymbol{y}_1, \ldots, \boldsymbol{y}_M] \sim N([\boldsymbol{\mu}_1, \ldots, \boldsymbol{\mu}_N]^T, \boldsymbol{\Sigma}).$$

where $\boldsymbol{\mu}_k$ is mean of the kth keyframe, $\boldsymbol{\Sigma}$ is the covariance matrix of all keyframes. So obviously, the mean part refers to the optimal result after BA, and $\boldsymbol{\Sigma}$ is the result of the marginalization, i.e., the matrix $\boldsymbol{S}$ mentioned in the previous lecture. We think that readers are already familiar with this process.

In a sliding window, another question is to ask, when the window moves, how should these state variables change? This matter can be discussed in two parts:

1. We want to add a new keyframe into the window as well as its corresponding landmarks.
2. We need to delete an old keyframe in the window, and may also delete the landmarks it observes.

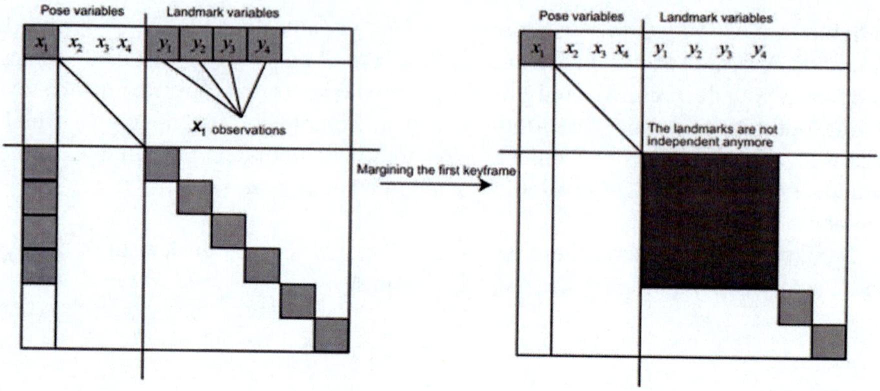

Fig. 9.2 Margining old keyframes will break the sparse structure of the Hessian

At this time, the difference between the sliding window method and the traditional BA is revealed. If processed as a traditional BA, then this only corresponds to two BAs with different structures, and there is no difference in the solution. But in the case of sliding windows, we have to discuss these specific details.

Adding New Keyframes
Considering that the sliding window has established N keyframes at the last moment, and we already know that they obey a certain Gaussian distribution, and their mean and variance are as described above. At this time, a new keyframe $\boldsymbol{x}_{N+1}$ has arrived, and the variables in the whole problem become a collection of $N + 1$ keyframes and more road signs. In fact, this is still ordinary. We only need to follow the normal BA process. When all points are marginalized, the Gaussian distribution of these $N + 1$ keyframe poses are obtained.

Removing Old Keyframes
When considering deleting old keyframes, a theoretical problem will emerge. For example, we want to delete the old keyframe $\boldsymbol{x}_1$. But $\boldsymbol{x}_1$ is not isolated. It will observe the same landmarks as other frames. After marginalizing $\boldsymbol{x}_1$, the whole problem will no longer be sparse. As in the previous lecture, let's take a schematic diagram, as shown in Fig. 9.2.

In this example, we assume that $\boldsymbol{x}_1$ sees the landmarks from $\boldsymbol{y}_1$ to $\boldsymbol{y}_4$, so before processing, the Hessian matrix of the BA problem should be like the left side of this figure. There are non-zero blocks in the columns $\boldsymbol{y}_1$ to $\boldsymbol{y}_4$ in the $\boldsymbol{x}_1$ row, which means $\boldsymbol{x}_1$ saw them. At this time, consider the marginalization $\boldsymbol{x}_1$. Please recall what we do in the Schur trick: We multiply the first row by the coefficient and adding it to the rows below the dividing line to eliminate the non-zero block in the first column; then use the first column to eliminate the non-zero block in the first row. But when we do this, the Hessian of the Landmark-Landmark part is filled with information by this operation, and it is no longer a diagonal block. This phenomenon is called (fill in) [91].

Intuitively, fill-in means that when you ask for $P(\boldsymbol{x}_2, \boldsymbol{x}_3, \boldsymbol{x}_4, \boldsymbol{y}_1, \ldots \boldsymbol{y}_4|\boldsymbol{x}_1)$, because the marginalized keyframe sees some landmarks, these landmarks have one more constraint saying "where they should be if $\boldsymbol{x}_1$ is set to the current value" (conditional distribution).Such a priori constraint describes the information about where these landmarks should be.

Recalling the marginalization mentioned in the previous lecture, when we marginalize the landmarks, the fill-in effect will appear in the pose block in the upper left corner. However, because BA does not require the pose block to be a diagonal block, the sparse BA solution is still feasible. However, when the keyframes are marginalized, it will destroy the diagonal block structure between the landmark points in the lower right corner. We cannot solve BA iteratively in the previous sparse method. This is obviously a terrible question. In fact, in the backend of the early EKF filter, people did maintain a dense Hessian matrix, which also made the backend of the EKF unable to handle higher dimension states.

However, if we make some modifications to the marginalization process, we can also maintain the sliding window BA's sparsity. For $\boldsymbol{y}_1$ to $\boldsymbol{y}_4$, they may fall into the three cases listed below:

1. The landmark is only observed in $\boldsymbol{x}_1$ and does not appear in the remaining keyframes. Then you can just throw away the landmark without any impact on the window. This landmark is isolated.
2. The landmark is seen in $\boldsymbol{x}_2$-$\boldsymbol{x}_4$, but will not be seen in the future. (This is just an assumption here, and it actually depends on the implementation of the frontend. Optical flow frontends like VINS [92] or DSO will not track the missing feature points, so we can just assume this is the case.) Then you can also marginalize this landmark. When road signs are marginalized, a priori of the pose-pose part is generated, so it becomes the prior information of the future pose estimation.
3. The landmark is seen in $\boldsymbol{x}_2$-$\boldsymbol{x}_4$ and may be seen in the future. Then this landmark should not be marginalized, because we will need to update its estimate later. In theory, we can only maintain the dense structure in the landmark-landmark part, but in engineering, we can pretend that the observation of this landmark by $\boldsymbol{x}_1$ can be simply discarded (equivalent to thinking that $\boldsymbol{x}_1$ did not see it). So we kept the diagonal structure of the Landmark part at a small cost. In this case you don't have to do anything.

Intuitive Explanation of the Marginalization in SWF

We know that the meaning of marginalization in probability refers to decomposing a joint distribution into a conditional and marginal distribution. So intuitively speaking, when we marginalize a keyframe, we mean "keep the current estimated value of this keyframe and find the conditional probability of other state variables conditioned on this keyframe." Therefore, when a keyframe is marginalized, the landmark points it observes will generate a priori information of "**where these landmarks should be**," which affects the estimated value of the rest. If these landmark points are then marginalized, their observers will get a priori information of "**where the keyframe to observe them should be**."

Mathematically, when we marginalize a certain keyframe, the description of the state variables in the entire window will change from a joint distribution to a conditional probability distribution. Taking the example above, it means:

$$p\left(\boldsymbol{x}_1, \ldots \boldsymbol{x}_4, \boldsymbol{y}_1, \ldots \boldsymbol{y}_6\right) = p\left(\boldsymbol{x}_2, \ldots, \boldsymbol{x}_4, \boldsymbol{y}_1, \ldots \boldsymbol{y}_6 | \boldsymbol{x}_1\right) \underbrace{p\left(\boldsymbol{x}_1\right)}_{\text{discarded}}. \tag{9.1}$$

Then discard the marginalized part of the information. After the variable is marginalized, we should no longer use it in engineering. Therefore, the sliding window method is more suitable for VO systems but not for large-scale mapping systems.

Since *g2o* and Ceres have not explicitly supported the marginal operation in the sliding window method,[1] we omit the corresponding experimental part in this section. I hope the theoretical part can help readers understand some SLAM systems based on sliding windows.

9.2 Pose Graph Optimization

9.2.1 *Definition of Pose Graph*

According to the previous discussion, we found that landmarks occupy most of the optimization time. In fact, after several observations, the spatial position of the landmarks will converge to a value and remain almost unchanged, while the divergent outliers are usually invisible. Optimizing the convergent landmarks seems a bit useless. Therefore, we are more inclined to fix the feature points after a few iterations and only regard them as constraints of pose estimation instead of actually optimizing their position.

Following this idea, we will think: Can we ignore the landmarks at all and only focus on the poses? We can construct a graph optimization with only pose variables. The edge between the pose vertices can be set with measurement by the ego-motion estimation obtained from feature matching. The difference is that once the initial estimation is completed, we no longer optimize the positions of those landmark points but only care about the connections between all camera poses. In this way, we save a lot of calculations for landmarks optimization and only keep the keyframe's trajectory, thus constructing the so-called pose graph (Pose Graph), shown in Fig. 9.3.

We know that the number of landmarks in BA is much greater than that of pose nodes. A keyframe is often associated with hundreds of key points. Even if the sparsity is used, the maximum calculation scale of real-time BA is generally around tens of thousands of points on the current mainstream CPU. This limits the application scenarios of SLAM. Therefore, when the robot moves in a larger range of time and space, we must consider some solutions: Either discard some historical data

[1] In the project, we can walk around *g2o* and Ceres's restrictions through some clever tricks, but this is often very troublesome and not suitable for demos.

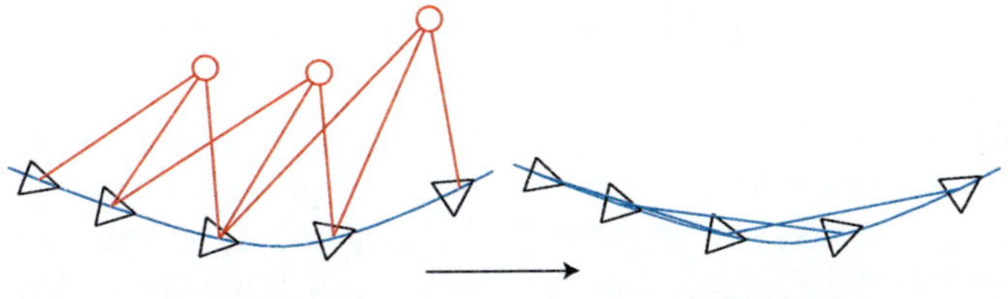

Fig. 9.3 Pose Graph diagram. When we no longer optimize the landmark in Bundle Adjustment, and only regard them as constraints on the pose nodes, we get a pose graph with a much reduced calculation scale

like the sliding window method [93]; or like the usage of pose graph, abandon the optimization of landmark, and only keep the edges between pose variables [94–96]. Besides, if we have additional sensors to measure pose (GPS, UWB), then the pose graph is also a common method of fusing pose measurements.

9.2.2 *Residuals and Jacobians*

So, how do we define the vertices and edges in a pose graph optimization? The node here represents the camera pose, expressed in $\boldsymbol{T}_1, \ldots, \boldsymbol{T}_n$. The edge is the estimation of the relative motion between the two pose nodes. The estimation may come from the feature point method or the direct method, or GPS or IMU integration, but no matter what, we estimate it, for example, a movement between $\boldsymbol{T}_i$ and $\boldsymbol{T}_j$ is $\Delta \boldsymbol{T}_{ij}$. We can express the movement in several ways. Let's take the a natural one:

$$\Delta \boldsymbol{\xi}_{ij} = \boldsymbol{\xi}_i^{-1} \circ \boldsymbol{\xi}_j = \ln \left(\boldsymbol{T}_i^{-1} \boldsymbol{T}_j \right)^{\vee}, \tag{9.2}$$

Or in SE(3)'s manner:

$$\boldsymbol{T}_{ij} = \boldsymbol{T}_i^{-1} \boldsymbol{T}_j. \tag{9.3}$$

According to the principle of graph optimization, the equation will not hold exactly in practice. Hence, we set up the least-square error and then, as before, discuss the derivative of the error with respect to the optimized variable. Here, we move the $\boldsymbol{T}_{ij}$ of the above equation to the right side of the equation to construct the error $\boldsymbol{e}_{ij}$:

$$\boldsymbol{e}_{ij} = \ln \left(\boldsymbol{T}_{ij}^{-1} \boldsymbol{T}_i^{-1} \boldsymbol{T}_j \right)^{\vee} \tag{9.4}$$

Note that there are two optimization variables: $\boldsymbol{\xi}_i$ and $\boldsymbol{\xi}_j$, so we find the derivative of $\boldsymbol{e}_{ij}$ about these two variables. According to the derivation method of Lie algebra, give $\boldsymbol{\xi}_i$ and $\boldsymbol{\xi}_j$ a left disturbance: $\delta \boldsymbol{\xi}_i$ and $\delta \boldsymbol{\xi}_j$. Then the error becomes:

$$\hat{e}_{ij} = \ln\left(T_{ij}^{-1} T_i^{-1} \exp((-\delta\xi_i)^\wedge) \exp(\delta\xi_j^\wedge) T_j\right)^\vee. \tag{9.5}$$

In this formula, the two disturbance terms are placed in the middle. We want to move the disturbance term to the equation's left or right side to use the BCH approximation. Recall the adjoint property in the fourth lecture, which is the formula (3.55). If you have not done this exercise before, use it as the correct conclusion for now:

$$\exp\left((\mathrm{Ad}(T)\xi)^\wedge\right) = T \exp(\xi^\wedge) T^{-1}. \tag{9.6}$$

Rearrange it:

$$\exp(\xi^\wedge) T = T \exp\left((\mathrm{Ad}(T^{-1})\xi)^\wedge\right). \tag{9.7}$$

This formula shows that by introducing an adjoint term, we can "exchange" the T on the left and right sides of the disturbance term. With this property, the disturbance terms can be moved to the right (of course, the left is also possible), and the Jacobian matrix in the form of right multiplication (left multiplication is formed when moved to the left):

$$\begin{aligned}
\hat{e}_{ij} &= \ln\left(T_{ij}^{-1} T_i^{-1} \exp((-\delta\xi_i)^\wedge) \exp(\delta\xi_j^\wedge) T_j\right)^\vee \\
&= \ln\left(T_{ij}^{-1} T_i^{-1} T_j \exp\left(\left(-\mathrm{Ad}(T_j^{-1})\delta\xi_i\right)^\wedge\right) \exp\left(\left(\mathrm{Ad}(T_j^{-1})\delta\xi_j\right)^\wedge\right)\right)^\vee \\
&\approx \ln\left(T_{ij}^{-1} T_i^{-1} T_j \left[I - (\mathrm{Ad}(T_j^{-1})\delta\xi_i)^\wedge + (\mathrm{Ad}(T_j^{-1})\delta\xi_j)^\wedge\right]\right)^\vee \\
&\approx e_{ij} + \frac{\partial e_{ij}}{\partial \delta\xi_i}\delta\xi_i + \frac{\partial e_{ij}}{\partial \delta\xi_j}\delta\xi_j
\end{aligned}, \tag{9.8}$$

where the two Jacobians are:

$$\frac{\partial e_{ij}}{\partial \delta\xi_i} = -\mathcal{J}_r^{-1}(e_{ij}) \mathrm{Ad}(T_j^{-1}) \tag{9.9}$$

and

$$\frac{\partial e_{ij}}{\partial \delta\xi_j} = \mathcal{J}_r^{-1}(e_{ij}) \mathrm{Ad}(T_j^{-1}). \tag{9.10}$$

If readers find it difficult to understand the derivation of this part, please return to lecture 4 to review the content of SE(3). However, as mentioned earlier, because the left and right Jacobian $\mathcal{J}_r$ forms on $\mathfrak{se}(3)$ are too complicated, we usually approximate them. If the errors are close to zero, we can set them approximately as I or as:

$$\mathcal{J}_r^{-1}(e_{ij}) \approx I + \frac{1}{2}\begin{bmatrix} \phi_e^\wedge & \rho_e^\wedge \\ 0 & \phi_e^\wedge \end{bmatrix}. \tag{9.11}$$

Theoretically speaking, even after optimization, since the observation data given by each edge is not consistent, the error is usually not close to zero, so simply setting the $\mathcal{J}_r$ here as $\boldsymbol{I}$ will have an inevitable loss. Later we will see through practice to see if the theoretical difference is noticeable.

After discussion about the Jacobian derivation, the rest is the same as ordinary graph optimization. In short, all pose vertices and pose edges constitute a graph optimization, which is essentially a least-squares problem. The optimization variable is the pose of each vertex, and the edges come from the pose observation constraints. Let $\mathcal{E}$ be the set of all edges, then the overall objective function is:

$$\min \frac{1}{2} \sum_{i,j \in \mathcal{E}} \boldsymbol{e}_{ij}^T \boldsymbol{\Sigma}_{ij}^{-1} \boldsymbol{e}_{ij}. \tag{9.12}$$

We can still use Gauss Newton's method, the Levenberg-Marquardt method, etc., to solve this problem. Based on previous experience, we can naturally solve this with Ceres or *g2o*. We won't discuss the detailed process of optimization anymore. We have talked about enough in the last lecture.

9.3 Practice: Pose Graph

9.3.1 *Pose Graph Using g2o Built-in Classes*

Now let's demonstrate the use of *g2o* to optimize the pose graph. First of all, please use *g2o_viewer* command to open our pre-generated simulation pose graph, located in "slambook2/ch10/sphere.g2o", as shown in Fig. 9.4.

The pose graph is simulated by the create sphere program that comes with *g2o*. Its ground-truth trajectory is a ball composed of multiple layers from bottom to top. Each layer is a perfect circle. Many circular layers of different sizes form a complete sphere, which contains 2500 pose nodes (Fig. 9.4 upper left). Then, the simulation program generates odometry edges from $t-1$ to t. The edges between layers are also generated, called loop closure edges (loop detection algorithm will be introduced in detail in the next lecture). Then, we add observation noise to each edge and reset the initial value of the node according to the noisy initial value of the odometry edge. In this way, the pose graph data will be affected by accumulative error (Fig. 9.4 bottom right). It looks like part of a sphere locally, but the overall shape is far from that of a sphere. Now we start from the initial values of these noisy edges and nodes, optimize the entire pose graph, and get data that approximates the true value.

Of course, the real robot will certainly not have such a spherical motion trajectory and complete odometry and loop observation data. The advantage of the simulation into a perfect sphere is that we can intuitively see whether the optimization result is correct (as long as to see it is round or not). Readers can click the optimize function in *g2o_viewer* to see the optimization results and convergence process of each step.

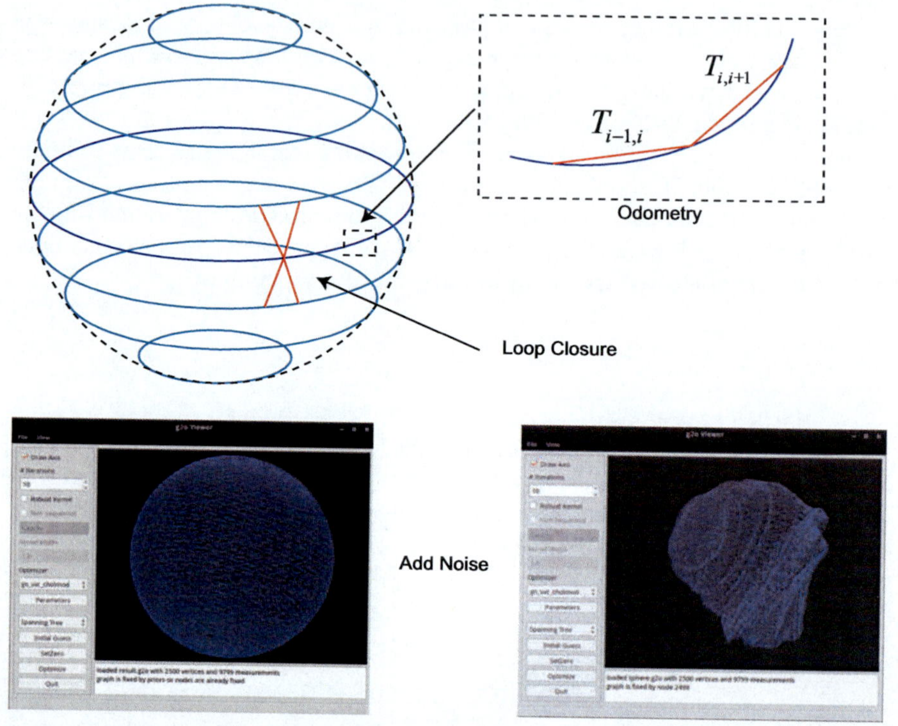

Fig. 9.4 Pose graph generated by simulation. The true value is a perfect sphere. After adding noise to the true value, the simulation data with accumulated error is obtained

On the other hand, sphere.g2o is also a text file. You can open it with a text editor to view its contents. The first half of the file is composed of nodes, and the second half is edges:

```
VERTEX_SE3:QUAT 0 -0.125664 -1.53894e-17 99.9999 0.706662 4.32706e-17
     0.707551 -4.3325e-17
......
EDGE_SE3:QUAT 1524 1574 -0.210399 -0.0101193 -6.28806 -0.00122939
     0.0375067 -2.85291e-05 0.999296 10000 0 0 0 0 0 10000 0 0 0 0 10000
     0 0 0 40000 0 0 40000 0 40000
```

As you can see, the node type is VERTEX_SE3, which expresses a camera pose. *g2o* uses quaternion and translation vector to describe the pose by default, so the meaning of the following fields are ID, $t_x, t_y, t_z, q_x, q_y, q_z, q_w$. The first 3 are translation vector elements, and the last 4 are unit quaternions that represent rotation. Similarly, the edge information is the ID of the two nodes, $t_x, t_y, t_z, q_x, q_y, q_z, q_w$, the upper right corner of the information matrix (because the information matrix is a symmetric matrix, you only need to save half of it). You can see that the information matrix is set to a diagonal matrix.

In order to optimize the pose graph, we can use the default vertices and edges of *g2o*, which are represented by quaternions. Since the simulation data is also generated by *g2o*, optimization by *g2o* itself does not require us to do any more work. Just configure the optimization parameters. The program slambook2/ch10/pose_graph_g2o_SE3.cpp demonstrates how to use the Levenberg-Marquardt method to optimize the pose image and store the result in the result.g2o file.

Listing 9.1 slambook2/ch10/pose_graph_g2o_SE3.cpp

```cpp
#include <iostream>
#include <fstream>
#include <string>

#include <g2o/types/slam3d/types_slam3d.h>
#include <g2o/core/block_solver.h>
#include <g2o/core/optimization_algorithm_levenberg.h>
#include <g2o/solvers/eigen/linear_solver_eigen.h>

using namespace std;

int main(int argc, char **argv) {
  if (argc != 2) {
    cout << "Usage: pose_graph_g2o_SE3 sphere.g2o" << endl;
    return 1;
  }
  ifstream fin(argv[1]);
  if (!fin) {
    cout << "file " << argv[1] << " does not exist." << endl;
    return 1;
  }

  // setting up g2o
  typedef g2o::BlockSolver<g2o::BlockSolverTraits<6, 6>> BlockSolverType;
  typedef g2o::LinearSolverEigen<BlockSolverType::PoseMatrixType> LinearSolverType;
  auto solver = new g2o::OptimizationAlgorithmLevenberg(
    g2o::make_unique<BlockSolverType>(g2o::make_unique<LinearSolverType>()));
  g2o::SparseOptimizer optimizer;
  optimizer.setAlgorithm(solver);
  optimizer.setVerbose(true);

  int vertexCnt = 0, edgeCnt = 0;
  while (!fin.eof()) {
    string name;
    fin >> name;
    if (name == "VERTEX_SE3:QUAT") {
      // built-in SE3 vertex in g2o
      g2o::VertexSE3 *v = new g2o::VertexSE3();
      int index = 0;
      fin >> index;
      v->setId(index);
      v->read(fin);
      optimizer.addVertex(v);
      vertexCnt++;
      if (index == 0)
      v->setFixed(true);
    } else if (name == "EDGE_SE3:QUAT") {
      // SE3-SE3 edges
      g2o::EdgeSE3 *e = new g2o::EdgeSE3();
      int idx1, idx2;
      fin >> idx1 >> idx2;
      e->setId(edgeCnt++);
      e->setVertex(0, optimizer.vertices()[idx1]);
      e->setVertex(1, optimizer.vertices()[idx2]);
      e->read(fin);
      optimizer.addEdge(e);
```

```
    }
    if (!fin.good()) break;
  }

  cout << "read total " << vertexCnt << " vertices, " << edgeCnt << " edges
     ." << endl;

  cout << "optimizing ..." << endl;
  optimizer.initializeOptimization();
  optimizer.optimize(30);

  cout << "saving optimization results ..." << endl;
  optimizer.save("result.g2o");

  return 0;
}
```

We choose the block solver of 6×6, using the Levenberg-Marquardt descent method, and the number of iterations is 30. Call this program to optimize the pose graph:

Listing 9.2 Terminal input:

```
$ build/pose_graph_g2o_SE3 sphere.g2o
read total 2500 vertices, 9799 edges.
optimizing ...
iteration= 0 chi2= 1023011093.851879 edges= 9799 schur= 0 lambda=
     805.622433 levenbergIter= 1
iteration= 1 chi2= 385118688.233188 time= 0.863567 cumTime= 1.71545 edges=
     9799 schur= 0 lambda= 537.081622 levenbergIter= 1
iteration= 2 chi2= 166223726.693659 time= 0.861235 cumTime= 2.57668 edges=
     9799 schur= 0 lambda= 358.054415 levenbergIter= 1
iteration= 3 chi2= 86610874.269316 time= 0.844105 cumTime= 3.42079 edges=
     9799 schur= 0 lambda= 238.702943 levenbergIter= 1
iteration= 4 chi2= 40582782.710190 time= 0.862221 cumTime= 4.28301 edges=
     9799 schur= 0 lambda= 159.135295 levenbergIter= 1
......
iteration= 28 chi2= 45095.174398 time= 0.869451 cumTime= 30.0809 edges=
     9799 schur= 0 lambda= 0.003127 levenbergIter= 1
iteration= 29 chi2= 44811.248504 time= 1.76326 cumTime= 31.8442 edges= 9799
     schur= 0 lambda= 0.003785 levenbergIter= 2
saving optimization results ...
```

Then, open "result.g2o" with *g2o_viewer* to view the results, as shown in Fig. 9.5.

The result was optimized from an irregular shape to a seemingly complete ball. This process is essentially the same as clicking the optimize button on *g2o_viewer*. Below, we implement the optimization using Sophus based on the previous derivation.

9.3.2 *Pose Graph Using Sophus*

Remember how we used Sophus to express the poses? Let's try to use *Sophus* in *g2o* to define our own vertices and edges.

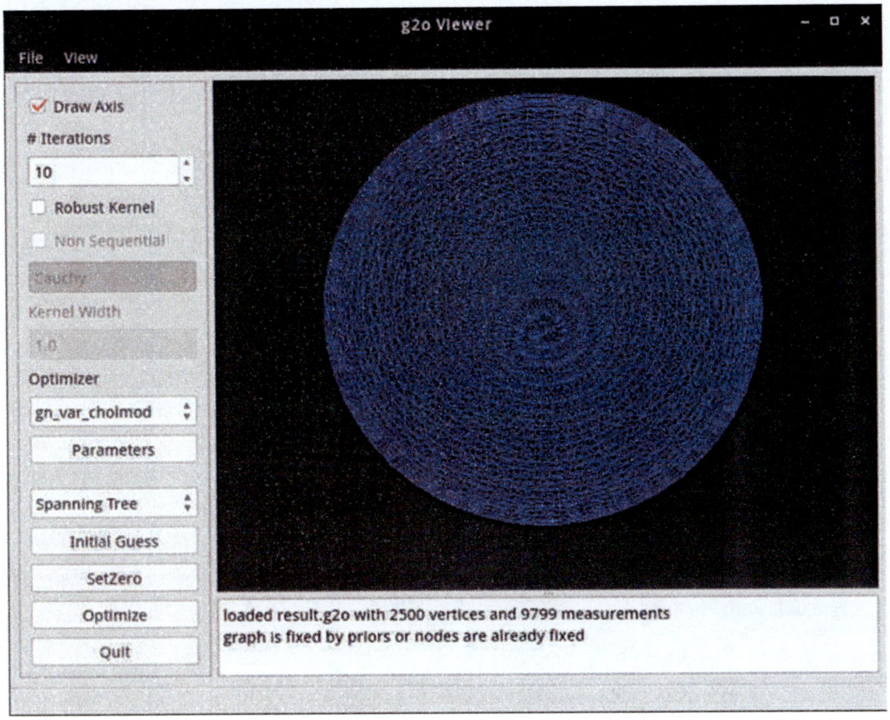

Fig. 9.5 Optimization results using built-in *g2o* classes

Listing 9.3 slambook2/ch10/pose_graph_g2o_lie_algebra.cpp (part)

```
typedef Matrix<double, 6, 6> Matrix6d;

// inverse of J in SE(3)
Matrix6d JRInv(const SE3d &e) {
  Matrix6d J;
  J.block(0, 0, 3, 3) = SO3d::hat(e.so3().log());
  J.block(0, 3, 3, 3) = SO3d::hat(e.translation());
  J.block(3, 0, 3, 3) = Matrix3d::Zero(3, 3);
  J.block(3, 3, 3, 3) = SO3d::hat(e.so3().log());
  J = J * 0.5 + Matrix6d::Identity();
  return J;
}

typedef Matrix<double, 6, 1> Vector6d;
class VertexSE3LieAlgebra : public g2o::BaseVertex<6, SE3d> {
  public:
  EIGEN_MAKE_ALIGNED_OPERATOR_NEW

  virtual bool read(istream &is) override {
    double data[7];
    for (int i = 0; i < 7; i++)
      is >> data[i];
    setEstimate(SE3d(
      Quaterniond(data[6], data[3], data[4], data[5]),
      Vector3d(data[0], data[1], data[2])
    ));
```

```
  }

  virtual bool write(ostream &os) const override {
    os << id() << " ";
    Quaterniond q = _estimate.unit_quaternion();
    os << _estimate.translation().transpose() << " ";
    os << q.coeffs()[0] << " " << q.coeffs()[1] << " " << q.coeffs()[2] << " " << q.coeffs()[3] << endl;
    return true;
  }

  virtual void setToOriginImpl() override {
    _estimate = SE3d();
  }

  // left multiplication for update
  virtual void oplusImpl(const double *update) override {
    Vector6d upd;
    upd << update[0], update[1], update[2], update[3], update[4], update[5];
    _estimate = SE3d::exp(upd) * _estimate;
  }
};

class EdgeSE3LieAlgebra : public g2o::BaseBinaryEdge<6, SE3d, VertexSE3LieAlgebra, VertexSE3LieAlgebra> {
public:
  EIGEN_MAKE_ALIGNED_OPERATOR_NEW

  virtual bool read(istream &is) override {
    double data[7];
    for (int i = 0; i < 7; i++)
      is >> data[i];
    Quaterniond q(data[6], data[3], data[4], data[5]);
    q.normalize();
    setMeasurement(SE3d(q, Vector3d(data[0], data[1], data[2])));
    for (int i = 0; i < information().rows() && is.good(); i++)
      for (int j = i; j < information().cols() && is.good(); j++) {
        is >> information()(i, j);
        if (i != j)
          information()(j, i) = information()(i, j);
      }
    return true;
  }

  virtual bool write(ostream &os) const override {
    VertexSE3LieAlgebra *v1 = static_cast<VertexSE3LieAlgebra *> (_vertices[0]);
    VertexSE3LieAlgebra *v2 = static_cast<VertexSE3LieAlgebra *> (_vertices[1]);
    os << v1->id() << " " << v2->id() << " ";
    SE3d m = _measurement;
    Eigen::Quaterniond q = m.unit_quaternion();
    os << m.translation().transpose() << " ";
    os << q.coeffs()[0] << " " << q.coeffs()[1] << " " << q.coeffs()[2] << " " << q.coeffs()[3] << " ";

    // information matrix
    for (int i = 0; i < information().rows(); i++)
      for (int j = i; j < information().cols(); j++) {
        os << information()(i, j) << " ";
      }
    os << endl;
    return true;
  }

  virtual void computeError() override {
```

```
    SE3d v1 = (static_cast<VertexSE3LieAlgebra *> (_vertices[0]))->estimate();
    SE3d v2 = (static_cast<VertexSE3LieAlgebra *> (_vertices[1]))->estimate();
    _error = (_measurement.inverse() * v1.inverse() * v2).log();
  }

  // Jacobians
  virtual void linearizeOplus() override {
    SE3d v1 = (static_cast<VertexSE3LieAlgebra *> (_vertices[0]))->estimate();
    SE3d v2 = (static_cast<VertexSE3LieAlgebra *> (_vertices[1]))->estimate();
    Matrix6d J = JRInv(SE3d::exp(_error));
    // TODO try to approximate J with I?
    _jacobianOplusXi = -J * v2.inverse().Adj();
    _jacobianOplusXj = J * v2.inverse().Adj();
  }
};
```

In order to realize the storage and reading of *g2o* files, the program implements the read and write functions, and pretend itself as the build-in SE3 vertex of *g2o*, so that *g2o_viewer* can recognize and render it. In fact, apart from using Sophus's Lie algebra representation internally, it looks no different from the outside.

It is worth noting the calculation process of Jacobian here. We have several options: one is not to provide the Jacobian calculation function, let *g2o* automatically calculate the numerical Jacobian. The second is to provide a complete or approximate Jacobian calculation process. Here we use JRInv() function to provide approximate $\mathcal{J}_r^{-1}$. Readers can try to approximate it to $\boldsymbol{I}$, or simply comment out the oplusImpl function to see what the difference is.

Then call *g2o* to optimize the problem:

Listing 9.4 Terminal input:

```
$ build/pose_graph_g2o_lie sphere.g2o
read total 2500 vertices, 9799 edges.
optimizing ...
iteration= 0 chi2= 626657736.014949 time= 0.549125 cumTime= 0.549125 edges=
      9799 schur= 0 lambda= 6706.585223 levenbergIter= 1
iteration= 1 chi2= 233236853.521434 time= 0.510685 cumTime= 1.05981 edges=
      9799 schur= 0 lambda= 2235.528408 levenbergIter= 1
iteration= 2 chi2= 142629876.750105 time= 0.557893 cumTime= 1.6177 edges=
      9799 schur= 0 lambda= 745.176136 levenbergIter= 1
iteration= 3 chi2= 84218288.615592 time= 0.525079 cumTime= 2.14278 edges=
      9799 schur= 0 lambda= 248.392045 levenbergIter= 1
......
```

We found that after 23 iterations, the overall error remains the same, which can actually stop the optimization algorithm. In the previous experiment, the error is still decreasing after 30 iterations.[2] After calling optimization, check "result_lie.g2o" to observe its results, as shown in Fig. 9.6. We cannot see any difference from the naked eye.

---

[2] Please note that although the error here is larger numerically, because we redefine the calculation method of the error when we customize the edge, therefore, the absolute cost value here cannot be directly used for comparison.

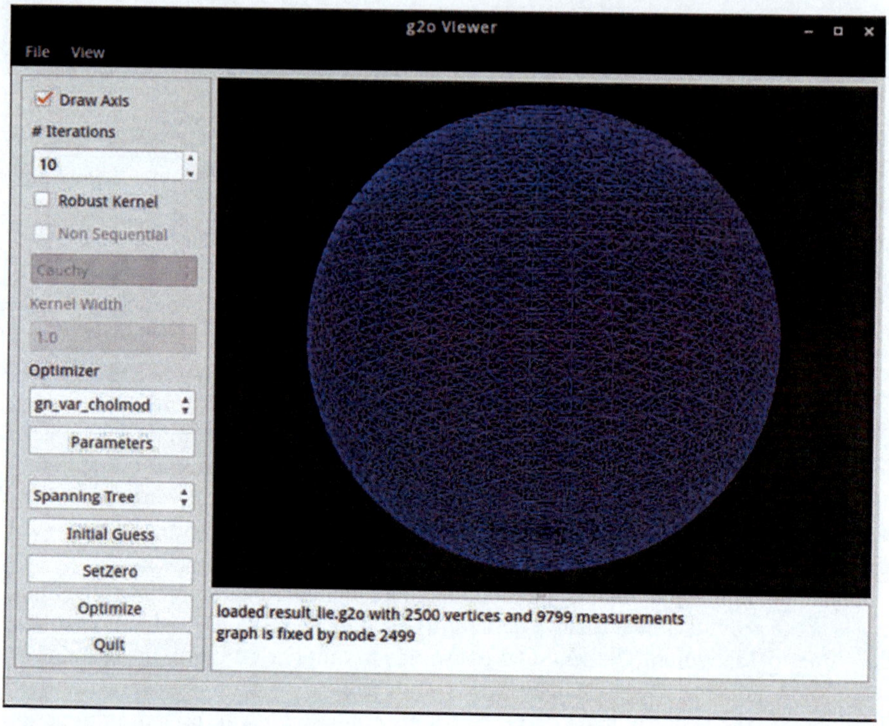

Fig. 9.6 Pose graph with Sophus library

If you press the optimize button in this *g2o_viewer* interface, *g2o* will use its own SE3 vertex to optimize, you can see in the text box below:

```
loaded result_lie.g2o with 2500 vertices and 9799 measurements
graph is fixed by node 2499
# Using CHOLMOD poseDim -1 landMarkDim -1 blockordering 0
Preparing (no marginalization of Landmarks)
iteration= 0 chi2= 44360.509723 time= 0.567504 cumTime= 0.567504 edges=
     9799 schur= 0
iteration= 1 chi2= 44360.471110 time= 0.595993 cumTime= 1.1635 edges= 9799
     schur= 0
iteration= 2 chi2= 44360.471110 time= 0.582909 cumTime= 1.74641 edges= 9799
     schur= 0
```

The overall error is 44360 under the measure of SE3 edge, which is slightly smaller than 44811 in the previous 30 iterations. This shows that after using Lie algebra for optimization, we have obtained better results with fewer iterations.[3] In fact, even if we use the identity matrix to approximate $\mathcal{J}_r^{-1}$, you will converge to a similar value. This is mainly because when the error is close to zero, the Jacobian is very close to the identity matrix.

[3] Because no more experiments have been done, this conclusion is only valid for the "ball" example here.

9.4 Summary

The example of the ball is a more representative case. It has odometry and loop closure similar to the actual ones, just like the pose graph in real SLAM systems. Simultaneously, the "ball" also has a certain computational scale: it has a total of 2,500 pose nodes and nearly 10,000 edges, and we found that optimizing it takes a lot of time (compared to the frontend with strong real-time requirements). On the other hand, the pose graph is generally considered one of the most straightforward graphs. If we do not assume how the robot moves, it is difficult to discuss its sparsity furthermore. The robot may move forward in a straight line to form a ribbon-shaped pose graph, which is, of course, sparse. It may also be a swinging motion from left to right, creating many small loops that need to be optimized (loopy motion). If there is no further information, it seems that we can no longer use the solution structure like BA.

Since PTAM [97] was proposed, people have realized that backend optimization does not need to respond to the frontend image data in real-time. People tend to separate the frontend and the backend, running in two separate threads, historically called tracking and mapping, although so-called, the mapping part mainly refers to the backend optimization content. In engineering, the frontend needs to respond to the video in real-time, such as 30 frames per second. At the same time, the backend optimization can run slowly, as long as a result is returned to the frontend when the optimization is completed. Therefore, we usually do not put high-speed requirements on backend optimization.

Exercises

1. If the error in the pose graph is defined as $\Delta \boldsymbol{\xi}_{ij} = \boldsymbol{\xi}_i \circ \boldsymbol{\xi}_j^{-1}$, please derive the Jacobians of the left perturbation model.
2. Derive the Jacobians using right perturbation model.
3. Implement a Ceres version for the "ball" example.
4. Implement a gtsam version for the "ball" example and compare the efficiency.
5. *Read iSAM's paper [98, 99] and learn how to make incremental optimization.

Chapter 10
Loop Closure

Check for updates

> **Goal of Study**
> 1. Understand why loop closure is needed in SLAM.
> 2. Understand the principles of the bag-of-words model.
> 3. Use DBoW3 to detect similar images.

In this lecture, we will introduce another main module in SLAM: loop closure detection. We know that the primary purpose of SLAM (frontend, backend) is to estimate camera movement. However, the loop closure module is quite different from the previous content, so it is usually considered an independent module. We will introduce the loop detection method in visual SLAM: the bag of words model. And then, we carry out an experiment on the DBoW library so that readers can get a more intuitive understanding.

10.1 Loop Closure and Detection

10.1.1 Why Loop Closure Is Needed

We have already introduced the frontend and the backend: the frontend provides short-time trajectory/landmarks estimation and the map's initial value. The backend is responsible for optimizing all these data. However, if we only consider the adjacent keyframes like VO, then the errors will inevitably accumulate with time so that the entire SLAM will suffer from the accumulative error. The result of long-term estimation will not be reliable. In other words, we cannot construct *globally consistent* trajectories and maps.

Let's take an example. In the map-building stage of autonomous driving, we usually designate the collection vehicle to circle several times in a given area to cover all the collection areas. Suppose we extract the features at the frontend, then

X. Gao and T. Zhang, *Introduction to Visual SLAM*,
https://doi.org/10.1007/978-981-16-4939-4_10

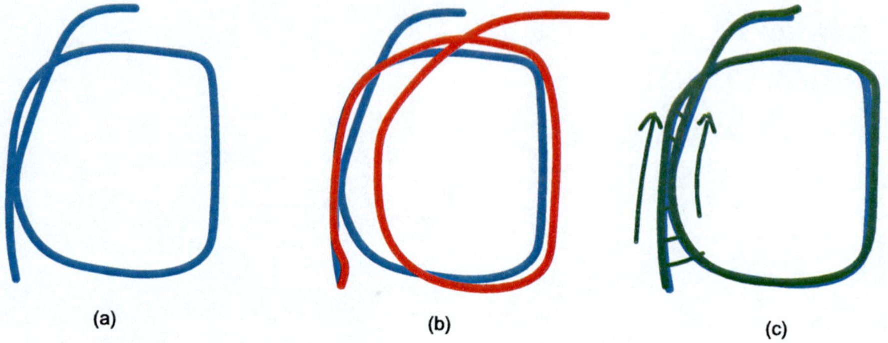

Fig. 10.1 Accumulated drift. **a** Real trajectory. **b** Accumulated error if we only consider adjacent keyframes. **c** Add loop closure to reduce the accumulated drift

ignore the feature points, and use a pose graph to optimize the entire trajectory at the backend, as shown in Fig. 10.1a. Since the frontend gives only sequential pose constraints, for example, it may be $\mathbf{x}_1 - \mathbf{x}_2$, $\mathbf{x}_2 - \mathbf{x}_3$, etc. However, due to the error in the estimation of $\mathbf{x}_1$, $\mathbf{x}_2$ is determined according to $\mathbf{x}_1$, $\mathbf{x}_3$ is again determined by $\mathbf{x}_2$. By analogy, errors will be accumulated, making the result of backend optimization look like what is shown in Fig. 10.1b, which gradually tends to be inaccurate. In this application scenario, we should use the loop closure information to determine that the vehicle reaches the same place when we pass the same road.

Although the backend can estimate the maximum posterior error, it depends on the BA or the pose graph's structure. If there are only adjacent keyframe constraints, we cannot do much, nor can we eliminate the accumulated error. However, the loop closure detection module can provide constraints on longer time other than just adjacent frames: for example, the pose constraint between $\mathbf{x}_1 - \mathbf{x}_{100}$. Why do we want to constrain those keyframes? Because we noticed that the camera returns to a previously visited scene, and it acquires similar data in history. The key to loop detection is how to detect this fact effectively. If we can successfully see this, we can provide more valid constraints for the backend pose graph to get a better estimate, especially a globally consistent estimate. Since the pose graph can be regarded as a point-spring system, loop detection is equivalent to adding additional springs to the graph, which improves the system's stability. The reader can also intuitively imagine that the loop edge "pulls" the edge with accumulated error back to the correct position-if the loop itself is right.

Loop detection is of great significance to SLAM systems. It is related to the correctness of our estimated trajectory and map in a long time. On the other hand, since loop detection provides the correlation between current data and historical data, we can also use loop detection for relocation. Relocation is also beneficial in most applications. For example, if we record a track for a scene in advance and build a map, we can let the robot follow this track for navigation, and relocation can help us determine our position on this track. Therefore, the loop detection improves the accuracy and robustness of the entire SLAM system. In some cases, we call a system with only a frontend and a local backend as *VO* and call a system with loop closing and a global backend as *SLAM*.

10.1.2 How to Close the Loops

Let's consider how to implement loop detection. There are several different ways of thinking about this problem, including theoretical and engineering.

The simplest way is to perform feature matching on any image pairs and determine which of them are related according to the number of correct matches. This is indeed a simple and effective idea. The disadvantage is that we blindly assume that "any two images may have loops," which makes the number of detection too large: for N possible loops, we have to detect C_N^2 times. It has the complexity of $O(N^2)$, which grows too fast as the trajectory becomes longer and is not practical in most real-time systems. Another simple way is to extract historical data and perform loop detection randomly. For example, randomly select five frames among n frames and compare them with the current frame. This approach can maintain a constant time calculation, but when the number of frames N increases, the probability of drawing a loop is significantly reduced, making the detection efficiency low.

The simple ideas mentioned above are too coarse. Although random detection is indeed useful in some implementations like [100], we at least hope that there is a prediction of "there may be a loop somewhere" so that the detection is not so blind. Such approaches can be roughly divided into two ideas: odometry-based or appearance-based. Based on the geometric relationship, when we find that the current camera moves close to a certain position before, we detect whether they have a loop relationship [101]. This is naturally an intuitive idea, but it is hard to estimate the accumulated drift amount unless we have global position measurements like GPS. Therefore, this approach has a logical problem because loop detection aims to eliminate accumulated errors. But the odometry-based approach assumes that "the accumulated error is small so that the loop can be detected." If the assumption does not hold, such methods cannot work when the cumulative error is large [12].

The other method is based on appearance. It has nothing to do with the estimation of the frontend or the backend and only determines the loop detection relationship based on the two images' similarity. This approach eliminates accumulated errors and makes the loop detection module a relatively independent module in the SLAM system (of course, the frontend can provide the extracted feature points). Since it was proposed in the early 21st century, the appearance-based loop detection method can effectively work in different scenarios and has become the mainstream method in visual SLAM and applied to the actual system [88, 96, 102].

In the appearance-based loop detection algorithm, the core problem is how to calculate the similarity between images. For example, for image **A** and image **B**, we need to design a method to calculate the similarity score between them: $s(\mathbf{A}, \mathbf{B})$. Of course, this score will take a value in a certain interval, and when it is greater than a certain amount, we think that there is a loop. Readers may have questions: Is it difficult to calculate the similarity between two images? For example, intuitively, an image can be expressed as a matrix, so how about subtracting two images directly and then taking a certain norm?

$$s(\mathbf{A}, \mathbf{B}) = \|\mathbf{A} - \mathbf{B}\|. \tag{10.1}$$

Why don't we do this?

1. As mentioned earlier, the pixel grayscale is an unstable measurement value, which is severely affected by the ambient light and camera exposure. Assuming that the camera is not moving and we turn on an electric light, the image will be brighter overall. In this way, even for the same data, we will get a significant difference value.
2. On the other hand, when the camera's viewing angle changes a little, even if each object's luminosity does not change, their pixels will be transformed in the image, resulting in a large difference in value.

Due to the existence of these two situations, in practice, even for very similar images, $\mathbf{A} - \mathbf{B}$ will often get an (unrealistic) enormous value. So we say that this function cannot reflect the similar relationship between images. It involves a definition of "good" and "bad." We have to ask, what kind of function can better reflect the similar relationship, and what kind of function is not good enough? From here, two concepts can be drawn: *perceptual aliasing* and *perceptual variability*. Let's discuss it in more detail now.

10.1.3 Precision and Recall

From a human point of view (at least we think), we can feel that "the two images are similar" or "the two photos were taken from the same place" with high accuracy. But since we have not yet grasped the human brain's working principle, we cannot clearly describe how we accomplish this. From a program point of view, we hope that algorithms can reach judgments consistent with humans or facts. When we feel that the two images were taken from the same place, we expect the loop detection algorithm to result as "this is a loop." Conversely, if we think that the two images were taken from different places, the program should also judge that "this is not a loop." Readers with a machine learning background should feel how similar this passage is to machine learning. Of course, the judgment of the algorithm is not always consistent with our human thinking, so there may be four situations in Table 10.1:

The term negative/positive come from medical terms. False-positive is also called perceptual bias, and false negative is called perceptual variation (see Fig. 10.2). For the convenience of writing, we use the abbreviation TP for true-positive, and FN for false-negative, etc. Since we want the algorithm to be consistent with human judgment, we hope that TP and TN should be as high as possible, and FP and FN

Table 10.1 Classification of the loop detection results

Algorithm\Fact	Is loop	Not loop
Is loop	True positive	False positive
Not loop	False negative	True negative

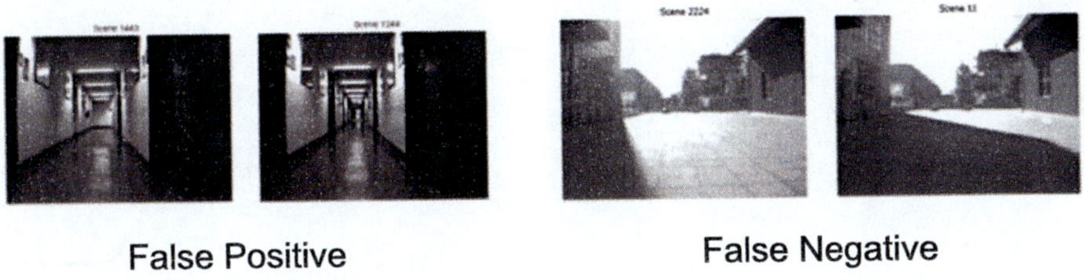

Fig. 10.2 Example of false-positive and false-negative scenes. Left: the images look the same but actually taken from different spaces. Right: the images are from the same place, but the appearance is different

should be as low as possible. Therefore, for a particular algorithm, we can count the number of occurrences of TP, TN, FP, and FN on a certain dataset and calculate two statistics: *accuracy rate* and *recall rate* (precision & recall)

$$\text{Precision} = \text{TP}/(\text{TP} + \text{FP}), \quad \text{Recall} = \text{TP}/(\text{TP} + \text{FN}). \tag{10.2}$$

Literally, the accuracy rate describes the probability that all the loops extracted by the algorithm are indeed true loops. The recall rate refers to the probability of loops being detected from all real loops. Why shall we take these two statistics instead of one? Because they are usually a pair of contradictions.

An algorithm often has many setting parameters. For example, when a certain threshold is raised, the algorithm may become more "strict". It detects fewer loops and improves accuracy. But at the same time, because the number of detection has decreased, many real loops may be missed, resulting in a decline in the recall rate. Conversely, if we choose a more relaxed configuration, the number of detected loops will increase, resulting in a higher recall rate. But there may be incorrectly detected loops so that the accuracy rate will decrease.

In order to evaluate the quality of the algorithm, we will test its P and R values under various configurations and then make a precision-recall curve (see Fig. 10.3). When using the recall rate on the horizontal axis and the accuracy rate on the vertical axis, we will care about the degree to which the entire curve deviates to the upper right, the recall rate at 100% accuracy, or the accuracy at 50% recall rate, as evaluation indicators. However, please note that we usually cannot say that algorithm A is better than algorithm B in general. We may say that A has a good recall rate when the accuracy rate is high, while B can guarantee a good accuracy rate when the recall rate is 70%, and so on.

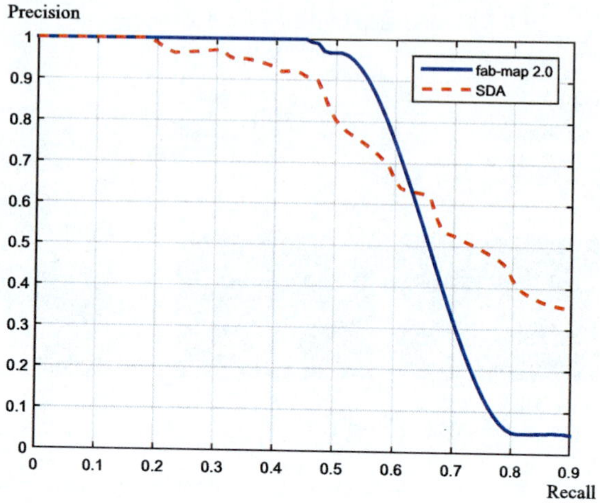

Fig. 10.3 Example of the precision-recall curve [103]. As the recall rate increases, the detection conditions become looser, and the accuracy rate decreases. A good algorithm can still guarantee a better accuracy rate in a high recall rate

It is worth mentioning that we have higher requirements for accuracy in SLAM while being relatively tolerant to the recall. The false-positive loops will add fundamentally wrong edges to the backend pose graph, sometimes causing the optimization algorithm to give completely wrong results. Imagine if the SLAM mistakenly treats all the desks as the same one. What will happen to the created map? You may see that the corridor is not straight, the walls are staggered, and finally, the entire map is invalid. In contrast, if the recall rate is lower, some loops are probably not detected, and the map may be affected by some accumulated errors, but the other loops may eliminate them. Therefore, when choosing the loop detection algorithm, we are more inclined to set the parameters more strictly or add the step of *loop verification* after the detection.

So, back to the previous question, why not use $\mathbf{A} - \mathbf{B}$ to calculate similarity? We will find that its accuracy and recall are inferior to most current methods, and there may be many false-positive or false-negative cases, so it is "not good." So, which method is better?

10.2 Bag of Words

Since directly subtracting two images is not good enough, then we need a more reliable method. Recalling the previous lectures' content, we may have an intuitive idea: Why not use VO feature points to detect the loops? We may match the feature points of the two images just like VO. Furthermore, according to the feature matching, we can also calculate the motion between the two images. Of course, there are some problems with this approach. For example, feature matching will be time-consuming, feature description may be unstable when the illumination changes, etc. But it is very close to the bag of words we will introduce. Let's talk about the bag of words first and then discuss the implementation details.

The purpose of Bag-of-Words (BoW) is to describe an image in terms of "what kinds of features are there on the image." For example, we say a person and a car in one photo; and two people and a dog in another photo. According to this description, the similarity of the two images can be measured. To be more specific, we need to do the following things:

1. Determine the concepts of people, cars, and dogs-corresponding to the word. Many words are put together to form a dictionary.
2. Detect which predefined words in the dictionary appear in an image-we use the appearance of words (histogram) to describe the entire image. In this way, we convert an image into a vector description.
3. Compute the similarity by the histogram in the second step.

Let's give an example. Assume we get a dictionary in some way. There are many words recorded in the dictionary, and each word has a certain meaning. For example, *person*, *car*, and *dog* are all words recorded in the dictionary. We might as well write them as w_1, w_2, w_3. Then, for any image A, according to the words they contain, it can be written as:

$$A = 1 \cdot w_1 + 1 \cdot w_2 + 0 \cdot w_3. \tag{10.3}$$

Since the dictionary is fixed, we may use the vector $[1, 1, 0]^{\mathrm{T}}$ to express the meaning of A. Through dictionaries and words, only one vector can describe the entire image. This vector represents the information of "whether the image contains a certain type of feature," which is more stable than the pure gray value. And because the description vector is only about the existence of the words rather than their order, it has nothing to do with the object's spatial position and arrangement order. Therefore, when the camera moves a little, as long as the object is still appearing in the field of vision, we can guarantee that the description vector does not change.[1] Based on this feature, we call it Bag-of-Words instead of List-of-Words. The emphasis is on the presence or absence of Words, regardless of their order. Therefore, it can be said that a dictionary is similar to a collection of words.

Going back to the above example, in the same way, the vector $[2, 0, 1]^{\mathrm{T}}$ can describe the image B. If you only consider "whether it appears" without considering the quantity, it can also be $[1, 0, 1]^{\mathrm{T}}$. At this time, this vector is binary. Therefore, by designing a particular calculation method based on these two vectors, the similarity between the images can be determined. Of course, if there are still some different ways to calculate the difference between two vectors, for example, for $\mathbf{a}, \mathbf{b} \in \mathbb{R}^W$, we can calculate:

$$s(\mathbf{a}, \mathbf{b}) = 1 - \frac{1}{W} \|\mathbf{a} - \mathbf{b}\|_1, \tag{10.4}$$

where we take the $\mathcal{L}_1$ norm, which is the sum of the absolute values of the elements. Please note that when the two vectors are exactly the same, we will get 1; when the

[1] Although this property sometimes brings some problems, for example, is the face with the eyes under the mouth still a human face?

two vectors are completely different (where **a** is 0, **b** is 1), we will get 0. This defines the similarity of the two description vectors and establishes the similarity between the images.

Yes, what's next?

1. Where does the dictionary come from?
2. If we can calculate the similarity score between two images, is it enough to close the loop?

Next, we first introduce the dictionary's generation method and then present how to use the dictionary to calculate the similarity between two images.

10.3 Train the Dictionary

10.3.1 The Structure of Dictionary

According to the previous introduction, the dictionary is composed of many words, and each word represents a concept. A word is different from a single feature point. It is not extracted from a single image but a combination of a specific type of feature. Therefore, the dictionary generation problem is similar to a clustering problem.

Clustering is a widespread approach in unsupervised machine learning, which lets the machine find the data structure by itself. Training BoW's dictionary is also one of them. First, suppose we have extracted feature points from many, let's say N, images. Now, we want to find a dictionary with k words. Each word can be regarded as a collection of local adjacent feature points. We can solve this with the classic K-means (K-means) algorithm [104].

K-means is a straightforward and effective method. It is widely used in unsupervised learning, and we briefly introduce its principles below. Let's say there are N data, and you want to classify them into k categories, then using K-means to do it mainly includes the following steps:

1. Randomly select k centers: $c_1, \ldots, c_k$.
2. Compute the distance between each data sample to the cluster centers. Assign the closest cluster to this sample.
3. Re-compute the centers of the clusters.
4. If the centers converge, exit. Otherwise, return to step 2.

The K-means approach is simple and effective, but there are some problems, such as the need to specify the number of clusters, randomly select the center point so that each clustering result is different, and some efficiency issues. Later, researchers also developed algorithms such as hierarchical clustering, K-means++ [105] to make up for its shortcomings, but these are all later improvements. We will not discuss them in detail. In short, according to K-means, we can cluster a large number of feature points that have been extracted into a dictionary containing k words. The question now becomes how to find the dictionary's corresponding word based on a specific feature point in the image.

There is still a simple idea: just compare each word and choose the most similar word-this is, of course, a practical approach. However, considering the versatility of the dictionary,[2] we usually use a larger-scale dictionary to ensure that the image features in the current environment have all appeared in the dictionary or have similar expressions. If you think it's not a hassle to compare ten words one by one, how about ten thousand? How about one hundred thousand?

If readers have learned data structure before, this $O(n)$ search algorithm is obviously not what we want. If the dictionary is sorted, then the binary search can improve search efficiency and reach the logarithmic level of complexity. In practice, we may use more complex data structures, such as the Chou–Liu tree [106] in Fabmap [107–109]. But we don't want to write this book as a collection of complex details, so we introduce another simpler and more practical tree structure [110].

In the document of [110], a k-d tree is used to express the dictionary. Its idea is straightforward (as shown in Fig. 10.4), similar to hierarchical clustering, and a direct extension of k-means. Suppose we have N feature points, and we want to build a tree with a depth of d and k forks, then the approach is as follows[3]:

1. At the root node, use k-means to cluster all samples into k classes (in practice, k-means++ is used to ensure clustering uniformity). This gives the first layer.
2. For each node in the first layer, gather the samples belonging to that node into the k class to get the next layer.
3. Repeat the second step until the leaf layer. The leaf layer is the so-called *words*.

We are still building words at the leaf level in the end, and the intermediate nodes in the tree structure are only used for quick search. Such a tree with k branch and depth d can hold k^d words. On the other hand, when looking for a word corresponding to a given feature, you only need to compare it with each intermediate node's cluster center (a total of d times) to find the last word, ensuring the logarithmic level search efficiency.

10.3.2 Practice: Creating the Dictionary

Now we've talked about dictionary generation. Let's demonstrate it in practice. The previous VO part used a lot of ORB feature descriptions, so here is how to generate and use the ORB dictionary.

---

[2] Would you call a page of paper with only ten words a dictionary? I believe that most people's dictionaries are quite heavy.

[3] We used k and d to express the branches of the tree, and the depth, which may remind you about the k-d tree [111]. Although the practices are different, the meanings they express are indeed the same.

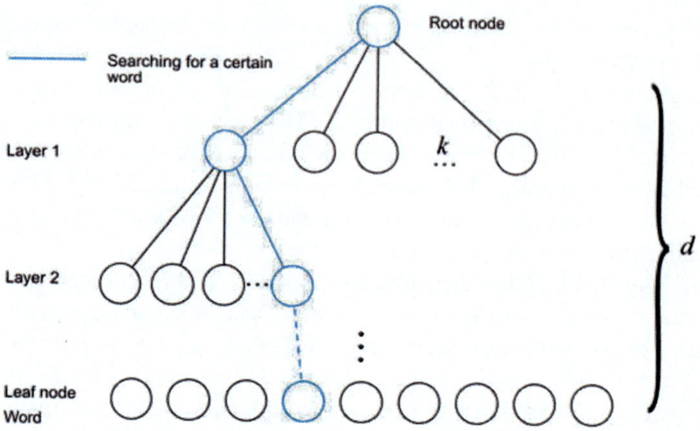

Fig. 10.4 Schematic view of K-d tree dictionary. When training the dictionary, K-means clustering is used layer by layer. When searching for words based on known features, we can also compare them layer by layer to find the corresponding words

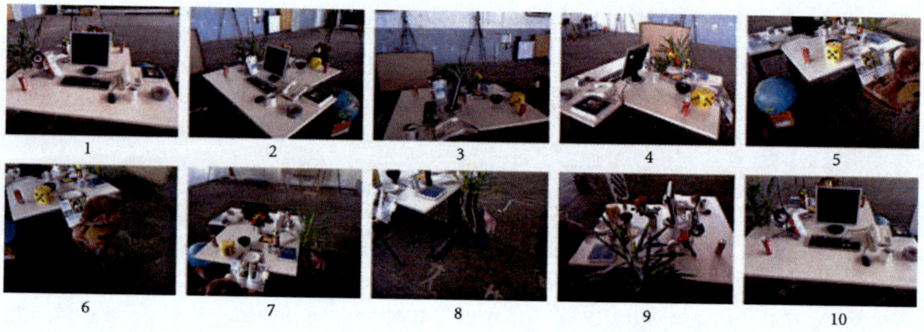

Fig. 10.5 The 10 images used in the demonstration experiment

In this experiment, we select 10 images in the TUM dataset (located in slambook2/ch11/data, as shown in Fig. 10.5), which come from a set of real camera motion trajectories. It can be seen that the first image and the last image were obviously taken from the same place. Now we have to see if the program can detect this. According to the bag-of-words model, we first generate the dictionary corresponding to these ten images.

What needs to be stated is that the BoW dictionary is often generated from a larger dataset and preferably from a place similar to the target environment. We usually use larger-scale dictionaries-the larger the dictionary, the more abundant it is, and the easier it is to find the words corresponding to the current image, but it should not exceed our computing power and memory. The author does not plan to store a large dictionary file on GitHub, so we temporarily train a small dictionary from ten images. If readers want to further pursue better results, please download more data and train larger dictionaries so that the program will be practical. You can also use a dictionary trained by others, but please pay attention to whether the feature types used in the dictionary are consistent.

Let's start training the dictionary. First, please install the BoW library used by this program. We use DBoW3[4]: https://github.com/rmsalinas/DBow3. Readers can also find it in the 3rdparty folder of the code in this book. It is also a cmake project. Please follow the cmake process to compile and install it.

Now let's do the training work:

Listing 10.1 slambook2/ch11/feature_training.cpp

```
int main(int argc, char **argv) {
  // read the image
  cout << "reading images... " << endl;
  vector<Mat> images;
  for (int i = 0; i < 10; i++) {
    string path = "./data/" + to_string(i + 1) + ".png";
    images.push_back(imread(path));
  }
  // detect ORB features
  cout << "detecting ORB features ... " << endl;
  Ptr<Feature2D> detector = ORB::create();
  vector<Mat> descriptors;
  for (Mat &image:images) {
    vector<KeyPoint> keypoints;
    Mat descriptor;
    detector->detectAndCompute(image, Mat(), keypoints, descriptor);
    descriptors.push_back(descriptor);
  }

  // create vocabulary
  cout << "creating vocabulary ... " << endl;
  DBoW3::Vocabulary vocab;
  vocab.create(descriptors);
  cout << "vocabulary info: " << vocab << endl;
  vocab.save("vocabulary.yml.gz");
  cout << "done" << endl;

  return 0;
}
```

The use of DBoW3 is very easy. We extract ORB features from 10 target images, store them in a vector container, and then call DBoW3's dictionary generation interface. In the constructor of the DBoW3::Vocabulary object, we can specify the number and depth of the tree branches, but the default constructor is used here, which is $k = 10, d = 5$. This is a small-scale dictionary with a maximum of 100,000 words. We also use the default parameters for image features, that is, 500 feature points per image. Finally, we store the dictionary as a compressed file.

Run this program, you will see the following dictionary information output:

Listing 10.2 Terminal output:

```
$ build/feature_training
reading images...
detecting ORB features ...
creating vocabulary ...
vocabulary info: Vocabulary: k = 10, L = 5, Weighting = tf-idf, Scoring =
    L1-norm, Number of words = 4983
done
```

[4] The reason for choosing it is that it has good compatibility with OpenCV3, and it is easy to compile and use.

We see that the number of branches k is 10, and the depth L is 5,[5] the number of words is 4983, which does not reach the maximum capacity. But what are the remaining weighting and scoring? Literally, the weighting is the weight values. Scoring seems to refer to the similarity score, but how is the score calculated?

10.4 Calculate the Similarity

10.4.1 Theoretical Part

Let's discuss the problem of similarity calculation. After having a dictionary, given any feature f_i, as long as you look up layer by layer in the dictionary tree, you can finally find the corresponding word w_j. If the dictionary is large enough, we can think f_i and w_j comes from the same class of objects (although there is no theoretical guarantee, but only in the sense of clustering). Then, suppose that N features are extracted from an image. After finding the words corresponding to these N features, we are equivalent to having the image's distribution in the word list or histogram. Intuitively (or ideally), it is equivalent to saying, "There are a person and a car in this picture". According to Bag-of-Words, it may be considered a bag.

Note that in this approach, we treat all words equally-if there is a word, the weight is one. If there is not, the weight is zero. Is this good? Considering that the importance of different words in distinguishing is not the same. For example, words such as "of" and "is" may appear in many sentences, and we cannot indicate the type of sentence based on them. But if there are words such as "document" and "football", it will be more effective in distinguishing sentences. Some, it can be said that they provide more information. In summary, we hope to evaluate the distinguishability or importance of words and give them different weights to achieve better results.

TF-IDF (Term Frequency-Inverse Document Frequency) [112, 113] is a weighting method commonly used in text retrieval and is also used in BoW models. The TF's idea is that if a word often appears in an image, its distinguishing degree is high. On the other hand, IDF's idea is that the lower the frequency of a word in the dictionary, the higher the degree of discrimination when classifying images.

We can calculate IDF when building a dictionary: Count the ratio of the number of features in a leaf node w_i to the number of all features as the IDF part. Assuming that the number of all features is n and the number of w_i is n_i, then the IDF of the word is:

$$\mathrm{IDF}_i = \log \frac{n}{n_i}. \tag{10.5}$$

On the other hand, the TF part refers to a certain feature's frequency in a single image. Assuming that the word w_i appears n_i times in the image A, and the total number of words appears n, then TF is:

---

[5] Here, L is the same as the d mentioned above.

$$\mathrm{TF}_i = \frac{n_i}{n}. \tag{10.6}$$

Therefore, the weight of w_i is equal to the product of TF times IDF:

$$\eta_i = \mathrm{TF}_i \times \mathrm{IDF}_i. \tag{10.7}$$

After considering the weights, for a certain image A, its feature points can correspond to many words, forming its Bag-of-Words:

$$A = \{(w_1, \eta_1), (w_2, \eta_2), \ldots, (w_N, \eta_N)\} \triangleq \mathbf{v}_A. \tag{10.8}$$

Since similar features may fall into the same class, there will be many zeros in the actual $\mathbf{v}_A$. In any case, we use a single vector $\mathbf{v}_A$ to describe an image A through the bag of words. This vector $\mathbf{v}_A$ is a sparse vector whose non-zero parts indicate which words are contained in the image A, and the values of these parts are the value of TF-IDF.

The next question is: Given $\mathbf{v}_A$ and $\mathbf{v}_B$, how to calculate the difference between them? This problem is the same as the way the norm is defined. There are several solutions, such as the L_1 norm form mentioned in the document [114]:

$$s\left(\mathbf{v}_A - \mathbf{v}_B\right) = 2\sum_{i=1}^{N} |\mathbf{v}_{Ai}| + |\mathbf{v}_{Bi}| - |\mathbf{v}_{Ai} - \mathbf{v}_{Bi}|. \tag{10.9}$$

Of course, many other ways are waiting for you to explore. Here we just give an example as a demonstration. So far, we have explained how to calculate the similarity between random images through the bag of words model. Let's practice it through the program.

10.4.2 Practice Part

In the practice part of the previous section, we have generated a dictionary for ten images. This time we use this dictionary to create Bag-of-Words and compare their differences to see how they differ from reality.

Listing 10.3 slambook/ch12/loop_closure.cpp

```
int main(int argc, char **argv) {
  // read the images and database
  cout << "reading database" << endl;
  DBoW3::Vocabulary vocab("./vocabulary.yml.gz");
  // DBoW3::Vocabulary vocab("./vocab_larger.yml.gz"); // use large vocab
      if you want:
  if (vocab.empty()) {
    cerr << "Vocabulary does not exist." << endl;
    return 1;
  }
  cout << "reading images... " << endl;
```

```
  vector<Mat> images;
  for (int i = 0; i < 10; i++) {
    string path = "./data/" + to_string(i + 1) + ".png";
    images.push_back(imread(path));
  }

  // NOTE: in this case we are comparing images with a vocabulary
      generated by themselves, this may lead to overfit.
  // detect ORB features
  cout << "detecting ORB features ... " << endl;
  Ptr<Feature2D> detector = ORB::create();
  vector<Mat> descriptors;
  for (Mat &image:images) {
    vector<KeyPoint> keypoints;
    Mat descriptor;
    detector->detectAndCompute(image, Mat(), keypoints, descriptor);
    descriptors.push_back(descriptor);
  }

  // we can compare the images directly or we can compare one image to a
      database
  // images :
  cout << "comparing images with images " << endl;
  for (int i = 0; i < images.size(); i++) {
    DBoW3::BowVector v1;
    vocab.transform(descriptors[i], v1);
    for (int j = i; j < images.size(); j++) {
      DBoW3::BowVector v2;
      vocab.transform(descriptors[j], v2);
      double score = vocab.score(v1, v2);
      cout << "image " << i << " vs image " << j << " : " << score << endl;
    }
    cout << endl;
  }

  // or with database
  cout << "comparing images with database " << endl;
  DBoW3::Database db(vocab, false, 0);
  for (int i = 0; i < descriptors.size(); i++)
  db.add(descriptors[i]);
  cout << "database info: " << db << endl;
  for (int i = 0; i < descriptors.size(); i++) {
    DBoW3::QueryResults ret;
    db.query(descriptors[i], ret, 4); // max result=4
    cout << "searching for image " << i << " returns " << ret << endl <<
        endl;
  }
  cout << "done." << endl;
}
```

This program demonstrates two comparison methods: the direct comparison between images and comparison between images and databases-even though they are similar. Besides, we output the Bag-of-Words description vector corresponding to each image, and readers can see them from the output data.

Listing 10.4 Terminal output:

```
$ build/feature_training
reading database
reading images...
detecting ORB features ...
comparing images with images
desp 0 size: 500
transform image 0 into BoW vector: size = 455
```

```
key value pair = <1, 0.00155622>, <3, 0.00222645>, <12, 0.00222645>, <13,
    0.00222645>, <14, 0.00222645>, <22, 0.00222645>, <33, 0.00222645>,
    <37, 0.00155622>, <38, 0.00222645>, <39, 0.00222645>, <43,
    0.00222645>, <57, 0.00155622> ......
```

As you can see, the BoW description vector contains each word's ID and weight, which constitute the entire sparse vector. When we compare two vectors, DBoW3 will calculate a score for us. The calculation method is defined by the previous dictionary construction:

Listing 10.5 Terminal output:

```
image 0 vs image 0 : 1
image 0 vs image 1 : 0.0234552
image 0 vs image 2 : 0.0225237
image 0 vs image 3 : 0.0254611
image 0 vs image 4 : 0.0253451
image 0 vs image 5 : 0.0272257
image 0 vs image 6 : 0.0217745
image 0 vs image 7 : 0.0231948
image 0 vs image 8 : 0.0311284
image 0 vs image 9 : 0.0525447
```

When querying the database, DBoW sorts the above scores and gives the most similar results:

Listing 10.6 Terminal output:

```
searching for image 0 returns 4 results:
<EntryId: 0, Score: 1>
<EntryId: 9, Score: 0.0525447>
<EntryId: 8, Score: 0.0311284>
<EntryId: 5, Score: 0.0272257>

searching for image 1 returns 4 results:
<EntryId: 1, Score: 1>
<EntryId: 2, Score: 0.0339641>
<EntryId: 8, Score: 0.0299387>
<EntryId: 3, Score: 0.0256668>

searching for image 2 returns 4 results:
<EntryId: 2, Score: 1>
<EntryId: 7, Score: 0.036092>
<EntryId: 9, Score: 0.0348702>
<EntryId: 1, Score: 0.0339641>

searching for image 3 returns 4 results:
<EntryId: 3, Score: 1>
<EntryId: 9, Score: 0.0357317>
<EntryId: 8, Score: 0.0278496>
<EntryId: 5, Score: 0.0270168>

searching for image 4 returns 4 results:
<EntryId: 4, Score: 1>
<EntryId: 5, Score: 0.0493492>
<EntryId: 0, Score: 0.0253451>
<EntryId: 6, Score: 0.0253017>

searching for image 5 returns 4 results:
<EntryId: 5, Score: 1>
<EntryId: 4, Score: 0.0493492>
<EntryId: 9, Score: 0.028996>
<EntryId: 6, Score: 0.0277584>

searching for image 6 returns 4 results:
```

```
<EntryId: 6, Score: 1>
<EntryId: 8, Score: 0.0306241>
<EntryId: 5, Score: 0.0277584>
<EntryId: 3, Score: 0.0267135>

searching for image 7 returns 4 results:
<EntryId: 7, Score: 1>
<EntryId: 2, Score: 0.036092>
<EntryId: 1, Score: 0.0239091>
<EntryId: 0, Score: 0.0231948>

searching for image 8 returns 4 results:
<EntryId: 8, Score: 1>
<EntryId: 9, Score: 0.0329149>
<EntryId: 0, Score: 0.0311284>
<EntryId: 6, Score: 0.0306241>

searching for image 9 returns 4 results:
<EntryId: 9, Score: 1>
<EntryId: 0, Score: 0.0525447>
<EntryId: 3, Score: 0.0357317>
<EntryId: 2, Score: 0.0348702>
```

Readers can view all the output to see how different images differ from similar image scores. We see that the apparent similarity of Figs. 1 and 10 (subscripts are 0 and 9, respectively in C++), the similarity score is about 0.0525, while the other images are about 0.02.

In the demonstration experiment, we see that the scores of similar images 1 and 10 are significantly higher than other image pairs. But, they are not as evident in terms of value as we thought. Ordinarily, if we compare ourselves with a similarity of 100%, then we (from a human perspective) think that Figs. 1 and 10 also have at least 70–80% similarity, while other images maybe 20–30%. However, the experimental results are about 2% of irrelevant images and about 5% of similar images, which seems not as obvious as we thought. Is this the result we want to see?

10.5 Discussion About the Experiment

10.5.1 Increasing the Dictionary Scale

In the field of machine learning, if there is no error in the code and the result is not satisfactory, we first doubt whether the network structure is large enough, whether the number of layers is deep enough, whether the data samples are enough, etc. This is because the principle of "a good model is no match for bad data" (On the one hand, it is also because of the lack of deeper theoretical analysis). Although we are now studying SLAM, we will first wonder when this happens: Is the dictionary too small? After all, we generated a dictionary only from ten images and then calculated the image similarity based on this dictionary.

Another dictionary at *slambook2/ch11/vocab_larger.yml.gz* is a slightly larger dictionary we generated. It is generated for all images of the same data sequence, which has about 2,900 images. The dictionary's size is still taken as $k = 10, d = 5$, that is, up to 10,000 words. Readers can use the gen_vocab_large.cpp file in the same directory to train the dictionary by themselves. Please note that to train a large dictionary, you may need a larger memory machine and wait patiently for a while. We slightly modify the program in the previous section and use a larger dictionary to detect image similarity:

Listing 10.7 Terminal output:

```
comparing images with database
database info: Database: Entries = 10, Using direct index = no. Vocabulary
    : k = 10, L = 5, Weighting = tf-idf, Scoring = L1-norm, Number of
    words = 99566
searching for image 0 returns 4 results:
<EntryId: 0, Score: 1>
<EntryId: 9, Score: 0.0320906>
<EntryId: 8, Score: 0.0103268>
<EntryId: 4, Score: 0.0066729>

searching for image 1 returns 4 results:
<EntryId: 1, Score: 1>
<EntryId: 2, Score: 0.0238409>
<EntryId: 8, Score: 0.00814409>
<EntryId: 3, Score: 0.00697527>

searching for image 2 returns 4 results:
<EntryId: 2, Score: 1>
<EntryId: 1, Score: 0.0238409>
<EntryId: 5, Score: 0.00897928>
<EntryId: 8, Score: 0.00893477>

searching for image 3 returns 4 results:
<EntryId: 3, Score: 1>
<EntryId: 5, Score: 0.0107005>
<EntryId: 8, Score: 0.00870392>
<EntryId: 6, Score: 0.00720695>

searching for image 4 returns 4 results:
<EntryId: 4, Score: 1>
<EntryId: 6, Score: 0.0069998>
<EntryId: 0, Score: 0.0066729>
<EntryId: 5, Score: 0.0062834>

searching for image 5 returns 4 results:
<EntryId: 5, Score: 1>
<EntryId: 3, Score: 0.0107005>
<EntryId: 2, Score: 0.00897928>
<EntryId: 4, Score: 0.0062834>

searching for image 6 returns 4 results:
<EntryId: 6, Score: 1>
<EntryId: 7, Score: 0.00915307>
<EntryId: 3, Score: 0.00720695>
<EntryId: 4, Score: 0.0069998>

searching for image 7 returns 4 results:
<EntryId: 7, Score: 1>
<EntryId: 6, Score: 0.00915307>
<EntryId: 8, Score: 0.00814517>
<EntryId: 1, Score: 0.00538609>

searching for image 8 returns 4 results:
<EntryId: 8, Score: 1>
<EntryId: 0, Score: 0.0103268>
<EntryId: 2, Score: 0.00893477>
```

```
<EntryId: 3, Score: 0.00870392>

searching for image 9 returns 4 results:
<EntryId: 9, Score: 1>
<EntryId: 0, Score: 0.0320906>
<EntryId: 8, Score: 0.00636511>
<EntryId: 1, Score: 0.00587605>
```

It can be seen that when the size of the dictionary increases, the similarity of irrelevant images decreases significantly. Similar images, such as images 1 and 10, although the absolute scores have dropped slightly, the relative scores have become more considerably larger than other images. This shows that increasing the dictionary training samples is beneficial. Similarly, readers can try to use larger-scale dictionaries to see how the results will change.

10.5.2 Similarity Score Processing

We can give a similarity score for any two images, but just using the absolute value of this score is not necessarily very helpful. For example, the appearance of some environments is very similar in nature. Offices often have many tables and chairs of the same style. But in other environments, there are lots of differences from place to place. Considering this situation, we will take a *prior similarity* $s\left(\mathbf{v}_t, \mathbf{v}_{t-\Delta t}\right)$, which means the similarity between the keyframe at a certain moment and the keyframe at the previous moment. Then, other scores are normalized with reference to this value:

$$s\left(\mathbf{v}_t, \mathbf{v}_{t_j}\right)' = s\left(\mathbf{v}_t, \mathbf{v}_{t_j}\right) / s\left(\mathbf{v}_t, \mathbf{v}_{t-\Delta t}\right). \tag{10.10}$$

From this perspective, we say: If the similarity between the current frame and a previous keyframe exceeds 3 times the similarity between the current frame and the previous keyframe, it is considered that there may be a loop. This step avoids introducing an absolute similarity threshold so that the algorithm can adapt to more environments.

10.5.3 Processing the Keyframes

When detecting loops, we must consider the selection of keyframes. If the keyframes are selected too close, the similarity between the two keyframes will be too high. In contrast, it is not easy to detect loops in the historical data. For example, the detection result is often that the nth frame is the most similar to the $n-2$th frame and the $n-3$th frame. This result seems to be too trivial and of little significance. Therefore, in practice, the frames used for loop closure detection are better to be sparse, different from each other and cover the entire environment.

On the other hand, if a loop is successfully detected, for example, the first frame and the nth frame. The $n + 1$th frame and the $n + 2$th frame will likely form a loop with the first frame too. However, confirming that there is a loop between the 1st frame and the nth frame is helpful for trajectory optimization. But the help of $n + 1$th and $n + 2$th frame with the first frame will not be that great. Because we have already used the previous information to eliminate the accumulated error, and more loops will not bring more information. Therefore, we may group the "similar" loops into one so that the algorithm does not repeatedly detect the same type of loops.

10.5.4 Validation of the Detected Loops

The BoW's loop detection algorithm completely relies on the appearance without using any geometric information, which causes images with a similar appearance to be easily regarded as loops. Moreover, since the bag of words does not care about the order of words, it only cares about the expression of words, which is more likely to cause perceptual deviation. Therefore, after the loop closure detection, we usually have an extra verification step [96, 115].

There are many verification methods. One is to set up a buffering mechanism for loops. It is considered that a single detected loop is not enough to constitute a useful constraint, and a loop that has been detected for a while is regarded as a correct loop. This can be seen as a time consistency check. Another method is spatial consistency detection, which is to perform feature matching on the two frames detected by the loop to estimate the camera's movement. Then, we can put the previous pose graph's tendency to check whether there is a big difference from the previous estimate. In short, the verification part is usually necessary, but how to implement it is a matter of opinion.

10.5.5 Relationship with Machine Learning

As can be seen from the previous discussion, loop detection and machine learning are inextricably linked. Loop detection itself is very much like a classification problem. The difference with traditional pattern recognition is that the number of categories in the loop is enormous. The samples of each category are very small. When the robot moves, the image changes, and new categories are created. Categories are treated as continuous variables rather than discrete variables; loop detection, equivalent to two images falling into the same category, is rarely seen. From another perspective, loop detection is also equivalent to learning the concept of *similarity* or *metric learning*. Since humans can determine whether images are similar, it is possible for machines to learn such concepts.

From the BoW model, it is a typical unsupervised machine learning process. The dictionary is equivalent to clustering feature descriptors, and a tree is just a data structure for quick search of the clustered classes. Since it is clustering, combined with the knowledge in machine learning, we can at least ask:

1. Is it possible to use the learned features instead of manually designed features such as SURF and ORB to detect the loops?
2. Is there a better way to do the clustering than the straightforward K-d tree and K-means?

Combined with the current development of machine learning, the learning of binary descriptors and unsupervised clustering are all very promising problems that can be solved in the deep learning framework. We have also seen the use of machine learning for loop detection. Although the bag-of-words method is still the mainstream at present, the author believes that in the future, deep learning methods are promising to defeat these artificially designed, "traditional" machine learning methods [103, 116, 117]. After all, the bag-of-words method is obviously inferior to neural networks in object recognition, and loop detection is a very similar problem. For example, the improved form of the BoW model, VLAD, has a CNN-based implementation [118, 119], and there are also some grids after training, which can be returned from the image to collect the camera's pose [120], these may become new loop detection algorithm.

Exercises

1. Please write a small program to calculate the PR curve. It may be easier to use MATLAB or Python because they are good at drawing pictures.
2. A dataset with manually marked loops is required to verify the loop detection result, such as [107]. However, it is very inconvenient to manually mark the loop. We will consider calculating the loop based on the standard trajectory. That is, if the poses of two frames in the trajectory are very similar, they are considered to be loops. Please calculate the loop in a data set based on the standard trajectory given by the TUM dataset. Are these looped images really similar?
3. Learn the usage of DBoW3 or DBoW2 library, look for a few pictures by yourself, and see if you can correctly detect the loops.
4. Survey the common methods of measuring image similarity. Which of them are commonly used?
5. What is the principle of the Chow–Liu tree? How is it used to build a dictionary and loop detection?
6. Read the literature [121]. In addition to the bag-of-words model, what other methods are used for loop detection?

Chapter 11
Dense Reconstruction

Goal of Study

1. Learn how to estimate the dense depth in monocular SLAM.
2. Implement the dense depth estimation in monocular SLAM.
3. Learn some of the commonly used map forms in RGB-D reconstruction.

In this lecture, we start to introduce the algorithm of the mapping part. In the front and backends, we focus on simultaneously estimating the camera motion trajectory and the feature points' spatial position. However, in real applications, in addition to localizing the camera, there are many other requirements. For example, consider the SLAM on the robot. We hope that the map can be used for localization, navigation, obstacle avoidance, and interaction. The feature point map obviously cannot meet all these needs. Therefore, in this lecture, we will discuss various forms of maps and point out the shortcomings in current visual SLAM maps.

11.1 Brief Introduction

Mapping should be one of the two goals of SLAM-because SLAM is called simultaneous localization and mapping. But until now, we have been discussing the localization problems for a long time, including localization through feature points, direct method, and backend optimization. Does this imply that mapping is not that important in SLAM, so we didn't discuss it until this lecture?

The answer is negative. In fact, in the classic SLAM model, a map is just a collection of all landmarks. Once the location of the landmarks is determined, it can be said that we have completed the mapping. Therefore, the aforementioned visual odometry or bundle adjustment model is proposed to model the landmarks' positions and optimize them. From this perspective, we have discussed the issue of mapping. So why do we need to list the map separately?

X. Gao and T. Zhang, *Introduction to Visual SLAM*,
https://doi.org/10.1007/978-981-16-4939-4_11

This is because people have different needs for mapping. As a kind of underlying technology, SLAM is often used to provide information for upper-layer applications. If the upper layer is a robot, then the application developer may want to use SLAM as a global localization and navigation module on the map. For example, a sweeper needs to complete the sweeping work, hoping to calculate a path covering the entire map. If the upper layer is an augmented reality device, the developer may wish to superimpose the virtual object on the real thing. It may also need to handle the occlusion relationship between the virtual object and the real object.

We found that the requirements for *localization* at the application level are similar. They hope that SLAM provides the real-time spatial pose information of the subject carrying the camera. For maps, there are many different requirements. From the point of view of visual SLAM, *mapping* is simultaneously done with *localization*. But from the perspective of the application, *mapping* obviously has many other requirements. Regarding the usage of the map, we can roughly summarize it as follows:

1. **Localization**. Localization is a basic function of the map. In the previous part of visual odometry, we discussed how to use local maps to achieve localization. In the loop detection part, we also saw that we can also determine the robot's position through re-location with the descriptors. Furthermore, we also hope to save the map so that the robot can still locate on the map after the next startup. We only need to model the map once instead of doing a complete SLAM every time the robot is started.
2. **Navigation**. Navigation refers to the process in which a robot can plan a path on a map, find a path between any two map points, and then control its movement to a target point. In this process, we need to know at least which places on the map are not passable and which places are passable. This is beyond the capability of sparse feature point maps, and we must have another map form. We will say later that this must be at least a dense map.
3. **Obstacle avoidance**. Obstacle avoidance is also a problem often encountered by robots. It is similar to navigation but pays more attention to the handling of local and dynamic obstacles. Similarly, we cannot judge whether a feature point is an obstacle, so a dense map is also needed here.
4. **Reconstruction**. Sometimes, we hope to use SLAM to obtain the reconstruction of the surrounding environment. This kind of map is mainly used for demonstration, so we hope it looks more comfortable and beautiful. We can also use the map for communication so that others can remotely watch the 3D objects or scenes we reconstructed-such as 3D video calls or online shopping. This kind of map is also dense, and we also have some additional requirements for its appearance. We may not be satisfied with the dense point cloud reconstruction but hope to build a textured plane, just like the three-dimensional scene in a video game.
5. **Interaction**. Interaction mainly refers to the interaction between people and the map. For example, in augmented reality, we will place virtual objects in the room and interact with them. For example, we will click on the virtual web browser to watch a video or throw objects on the wall, hoping that they will have a virtual physical collision. On the other hand, there will also be interactions with people

and maps in robot applications. For example, the robot may receive the command "take the newspaper on the table". In addition to the traditional map, the robot needs to know where the "table", what is called "above", and what is called "newspaper". This requires robots to have a higher level of knowledge of maps-also known as semantic maps.

Figure 11.1 visually explains the relationship between the various map types discussed above and their uses. Our previous discussion was basically focused on the "sparse feature map" part and did not discuss dense maps. The so-called dense map not only models the part of interest, i.e., the feature points, but also models all the seen objects. For a table, the sparse map may only model the four corners of the table, while the dense map will model the entire desktop. From the localization perspective, a map with only four corners can also be used to locate the camera. But since we cannot infer the spatial structure between these points from the four corners, it is impossible to complete navigation with only four corners., Obstacle avoidance and other tasks that require dense maps to complete.

As can be seen from the above discussion, the dense map occupies a significant position. So, the remaining question is: Can a dense map be established through visual SLAM? If so, how to build it?

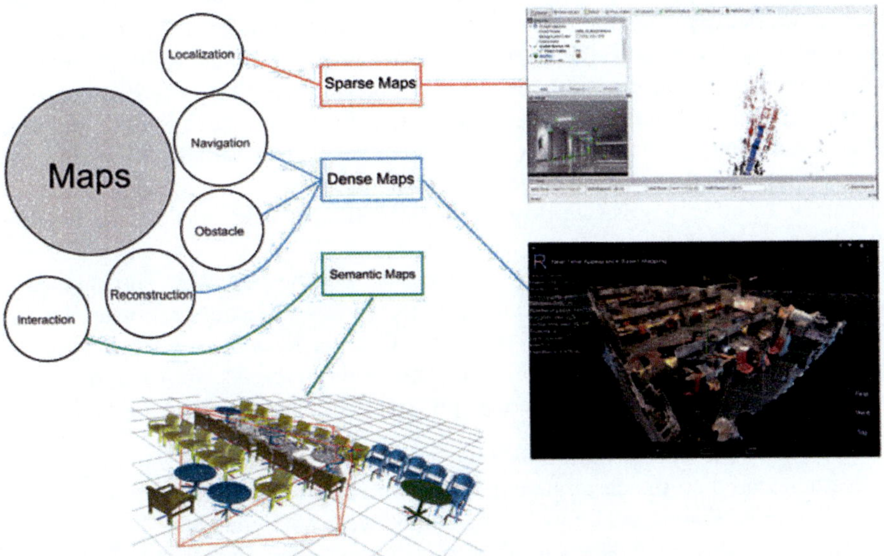

Fig. 11.1 Different maps from [88, 122, 123]

11.2 Monocular Dense Reconstruction

11.2.1 Stereo Vision

The dense reconstruction of visual SLAM is an important topic of this lecture. Cameras have long been considered as bearing-only sensors. The pixels in a single image can only provide the angle between the object and the camera's imaging plane and the brightness collected by the object, but not the distance (range). In dense reconstruction, we need to know each pixel's distance (or most of the pixels). There are roughly the following solutions for this:

1. Use monocular cameras and estimate the depth using triangulation after motion.
2. Use stereo cameras by its disparity (similar for more than two eyes).
3. Use the depth sensor in RGB-D cameras to directly get the depth.

The first two methods are sometimes called stereo vision, and the moving monocular is also called moving view stereo (MVS). Compared with the depth measured directly by RGB-D, the acquisition of depth by monocular and binocular is often fragile. We need to spend a lot of calculations and finally get some unreliable depth estimation. Of course, RGB-D also has some limits on the range, application range, and illumination conditions, but compared to monocular and binocular results, using RGB-D for dense reconstruction is often a more common choice. The advantage of monocular and binocular is that in outdoor and large scenes where RGB-D is not yet well applied, depth information can still be estimated through stereo vision.

Having said that, in this section, we will lead readers to realize a single-purpose dense estimation and experience why it is thankless. Let's start with the simplest case: estimate the depth of an image based on a video sequence with a given camera trajectory. In other words, we do not consider SLAM, first consider the slightly simpler mapping problem.

Assuming that there is a video sequence, we get the trajectory corresponding to each frame through some magic (of course, it is also probably estimated by the frontend of the visual odometry). We use the first image as the reference frame to calculate the depth (or distance) of each pixel in the reference frame. First of all, please remember how we completed the process in the feature matching section:

1. First, we extract features from the image and calculate the matching between the features based on the descriptor. In other words, through the feature, we track a certain spatial point and know its position between each image.
2. Then, since we cannot determine the feature point's position with only one image, we must estimate its depth through observations under different viewing angles. The principle is the triangulation mentioned above.

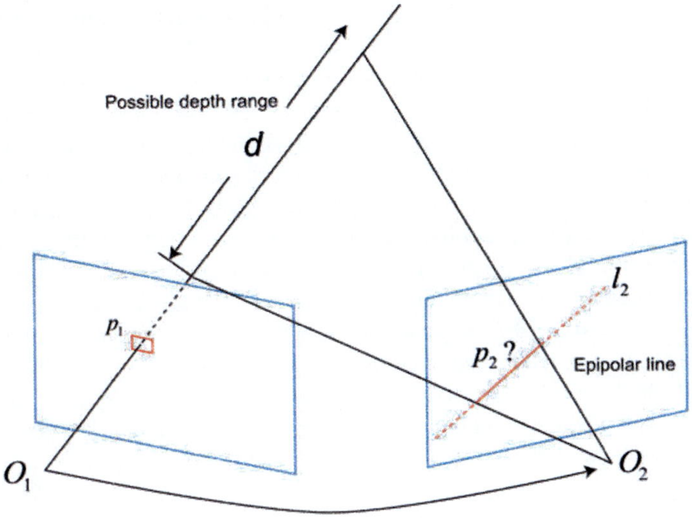

Fig. 11.2 Epipolar line search

11.2.2 Epipolar Line Search and Block Matching

Let's first discuss the geometric relationship produced by observing the same point from different perspectives. This is very similar to the epipolar geometry discussed in Sect. 6.3. Please see Fig. 11.2. The camera on the left has observed a certain pixel $\mathbf{p}_1$. Since this is a monocular camera, we have no way of knowing its depth, so assuming that the depth may be within a certain area, it may be said that it is between a certain minimum and infinity: $(d_{\min}, +\infty)$. Therefore, the spatial points corresponding to the pixel are distributed on a certain line segment (a ray in this example). Seeing from the camera on the right, we may find that this line segment's projection also forms a line on the image plane, which we know is called the epipolar line. When the movement between the two cameras is known, this epipolar line can be determined.[1] Then the question is: Which point on the epipolar line is the $\mathbf{p}_1$ point we just saw?

Again, in the feature matching method, we find the position of $\mathbf{p}_2$ through features. However, now we don't have a descriptor, so we can only search for points similar to $\mathbf{p}_1$ on the epipolar line. To be more specific, we may walk along one end of the epipolar line in the second image to the other and compare each pixel's similarity one by one with $\mathbf{p}_1$. From the point of view of directly comparing pixels, this approach is the same as the direct method.

When discussing the direct method, we know that comparing a single pixel's brightness value is not necessarily stable and reliable. One obvious thing is: if there are many similarities to $\mathbf{p}_1$ in the epipolar line, how can we determine which one is true? This seems to return to the question we mentioned in the loop detection: how

[1] On the contrary, if the movement is not known, the epipolar line cannot be determined.

to determine two images' similarity (or two points)? Loop detection is solved by the bag of words, but we have to find another way because there are no features.

An intuitive idea is: Since one pixel's brightness is not distinguishable, is it possible to compare pixel blocks? We take a small block of size $w \times w$ around $\mathbf{p}_1$, and then take many small blocks of the same size on the epipolar line for comparison. It should improve the discrimination to a certain extent. This is the so-called block match method. Note that in this process, this comparison is meaningful only if the gray value of the entire small block remains unchanged between different images. Therefore, the algorithm's assumption has changed from the grayscale invariance of one pixel to image blocks. To a certain extent, it has become more assertive.

Okay, now we have taken the small blocks around $\mathbf{p}_1$, and many small blocks on the epipolar line. We denote the small blocks around $\mathbf{p}_1$ as $\mathbf{A} \in \mathbb{R}^{w \times w}$, and denote the n small blocks on the epipolar line into $\mathbf{B}_i, i = 1, \ldots, n$. So, how to calculate the difference between a small block and another block? There are several different calculation methods:

1. SAD (Sum of Absolute Difference):

$$S(\mathbf{A}, \mathbf{B})_{\mathrm{SAD}} = \sum_{i,j} |\mathbf{A}(i, j) - \mathbf{B}(i, j)|. \tag{11.1}$$

2. SSD (Sum of Squared Distance), not solid state drive:

$$S(\mathbf{A}, \mathbf{B})_{\mathrm{SSD}} = \sum_{i,j} (\mathbf{A}(i, j) - \mathbf{B}(i, j))^2. \tag{11.2}$$

3. NCC (Normalized Cross Correlation):

$$S(\mathbf{A}, \mathbf{B})_{\mathrm{NCC}} = \frac{\sum\limits_{i,j} \mathbf{A}(i, j)\mathbf{B}(i, j)}{\sqrt{\sum\limits_{i,j} \mathbf{A}(i, j)^2 \sum\limits_{i,j} \mathbf{B}(i, j)^2}}. \tag{11.3}$$

Please note that since correlation is used here, correlation close to 0 means that the two images are not similar, and close to 1 means similar. The first two distances are reversed. Close to 0 means similarity, while a larger value means different.

Like many situations we have encountered, these calculation methods often have a contradiction between accuracy and efficiency. Methods with good accuracy often require complex calculations, while simple and fast algorithms often do not work well. This requires us to make a choice in actual engineering. In addition to these simple versions, we can remove each small block's mean first, which is called zero-mean SSD, zero mean NCC, and so on. After removing the mean value, we allow situations like "a small piece of **B** is brighter than **A** as a whole, but still very

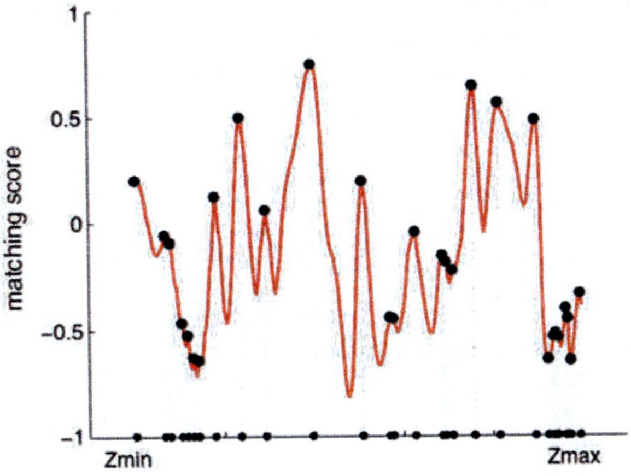

Fig. 11.3 Matching score along with the epipolar line [126]

similar",[2] so it is more robust than before. If readers are interested in other block matching measurement methods, it is recommended to read the literature [124, 125] as supplementary material.

We have now calculated the similarity measure between $\mathbf{A}$ and $\mathbf{B}_i$ on the epipolar line. For the convenience of description, suppose we use NCC, then we will get an NCC distribution along the epipolar line. This distribution's shape depends heavily on the image data, as shown in Fig. 11.3. In a long search distance, we usually get a non-convex function: this distribution has many peaks, but there must be only one true corresponding point. In this case, we tend to use probability distributions to describe depth values rather than using a single value to describe the depth. Therefore, our question turns to update the depth distribution when we continue to search for different images with epipolar lines. This is the so-called *depth filter*.

11.2.3 Gaussian Depth Filters

The estimation of pixel depth can also be modeled as a state estimation problem, so there are naturally two ways of solving the problem: filter and nonlinear optimization. Although nonlinear optimization is literally better, in the case of strong real-time requirements such as SLAM, considering that the frontend has already occupied many calculations, the filter method with less calculation is usually used in the mapping. This is also the purpose of the depth filter discussion in this section.

There are several different ways of assuming the distribution of depth. Under relatively simple assumptions, we can assume that the depth value obeys the Gaussian distribution and gets a Kalman-like method (but in fact, it is just a normalized production, as we will see later). On the other hand, in [68, 126] and other documents,

[2] The overall lighter may be caused by the ambient light or the camera exposure parameters increase.

the assumption of uniform-Gaussian mixture distribution is also used to derive a more complex depth filter. Based on the principle of simplicity and ease of use, we first introduce and demonstrate the depth filter under the assumption of Gaussian distribution and then take the filter with uniform-Gaussian mixture distribution as an exercise.

Assume the depth d of a certain pixel satisfy:

$$P(d) = N(\mu, \sigma^2). \tag{11.4}$$

And whenever new data arrives, we will observe its depth. Similarly, suppose this observation is also a Gaussian distribution:

$$P(d_{\text{obs}}) = N(\mu_{\text{obs}}, \sigma_{\text{obs}}^2). \tag{11.5}$$

Therefore, our question is how to use the observed information to update the original distribution of d. This is exactly an information fusion problem. According to Appendix A, we know that the normalized product of two Gaussian distributions is still a Gaussian distribution. Suppose the distribution of d after fusion is $N(\mu_{\text{fuse}}, \sigma_{\text{fuse}}^2)$, then according to the product of the Gaussian distribution, there are:

$$\mu_{\text{fuse}} = \frac{\sigma_{\text{obs}}^2 \mu + \sigma^2 \mu_{\text{obs}}}{\sigma^2 + \sigma_{\text{obs}}^2}, \quad \sigma_{\text{fuse}}^2 = \frac{\sigma^2 \sigma_{\text{obs}}^2}{\sigma^2 + \sigma_{\text{obs}}^2}. \tag{11.6}$$

Since we only have observation equations and no motion equations, the depth here only uses the information fusion part, and there is no need to predict and update like the complete Kalman filter. Of course, you can consider it as if the motion equation is fixed for the depth value. It can be seen that the fusion equation is indeed relatively simple and easy to understand, but the question remains: how to determine the distribution of the depth we observe? That is, how to calculate $\mu_{\text{obs}}, \sigma_{\text{obs}}$?

Regarding of μ_{obs} and σ_{obs}, there are also some different processing methods. For example, [75] considers the sum of geometric uncertainty and photometric uncertainty, while [126] only considers geometric uncertainty. For now, we only consider the uncertainty caused by geometric relations. Suppose we have determined the projection position of a pixel through epipolar search and block matching. We know that the block matching's accuracy is about one pixel. So, how does this uncertainty affect the estimated depth?

Take Fig. 11.4 as an example. Consider in an epipolar searching, we find the pixel $\mathbf{p}_2$ corresponding to $\mathbf{p}_1$, and then compute the depth value of $\mathbf{p}_1$. Let the 3D point corresponding to $\mathbf{p}_1$ is $\mathbf{P}$. We denote that $\mathbf{O}_1\mathbf{P}$ is $\mathbf{p}$, $\mathbf{O}_1\mathbf{O}_2$ is the camera's translation $\mathbf{t}$, and $\mathbf{O}_2\mathbf{P}$ is $\mathbf{a}$. And, let the two bottom angles of this triangle be α and β. Now, consider that there is one pixel error on the epipolar line l_2, so that the angle of β becomes β', and $\mathbf{p}_2$ also changes to $\mathbf{p}_2'$. Let the upper corner be γ. We want to ask, if $\mathbf{p}_2\mathbf{p}_2'$ is one pixel, then how long is $\mathbf{p}'$ and $\mathbf{p}$?

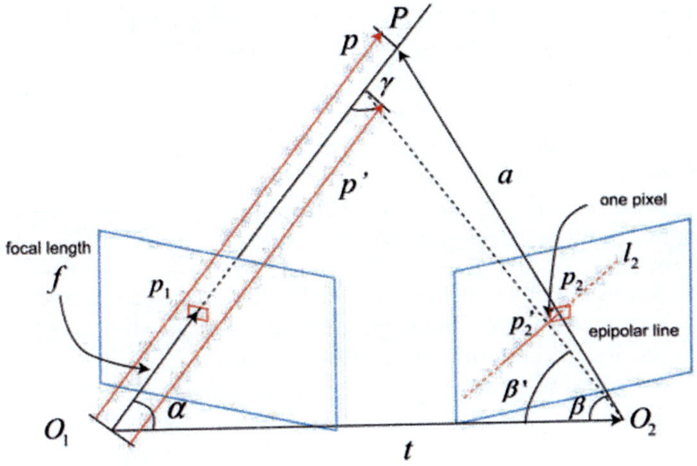

Fig. 11.4 Uncertainty analysis

This is a typical geometric problem. Let's list the geometric relationship between these quantities. Obviously:

$$\begin{aligned} \mathbf{a} &= \mathbf{p} - \mathbf{t} \\ \alpha &= \arccos \langle \mathbf{p}, \mathbf{t} \rangle \\ \beta &= \arccos \langle \mathbf{a}, -\mathbf{t} \rangle . \end{aligned} \tag{11.7}$$

Perturbing $\mathbf{p}_2$ by one pixel will cause β to produce a change, which becomes β'. According to the geometric relationship, there are:

$$\begin{aligned} \beta' &= \arccos \left\langle \mathbf{O}_2\mathbf{p}_2', -\mathbf{t} \right\rangle \\ \gamma &= \pi - \alpha - \beta'. \end{aligned} \tag{11.8}$$

Therefore, according to the law of sine, $\mathbf{p}'$ can be obtained as:

$$\|\mathbf{p}'\| = \|\mathbf{t}\| \frac{\sin \beta'}{\sin \gamma}. \tag{11.9}$$

Thus, we computed the depth uncertainty caused by the uncertainty of a single pixel. If we assume that the block matching of the epipolar search has only one pixel error, then we can set:

$$\sigma_{\text{obs}} = \|\mathbf{p}\| - \|\mathbf{p}'\|. \tag{11.10}$$

Of course, if the uncertainty of epipolar search is greater than one pixel, we can also amplify this uncertainty according to this derivation. The depth fusion process has been introduced before. In engineering, when the uncertainty is less than a certain threshold, it can be considered that the depth data has converged.

In summary, we give a complete process of estimating the pixel depth:

1. Assume that the depth of all pixels meets an initial Gaussian distribution.
2. When a new image is generated, the projected point's location is determined through epipolar search and block matching.
3. Calculate the depth and uncertainty of the triangle based on the geometric relationship.
4. Fuse the current observation into the last estimate. If it converges, stop the calculation, otherwise return to step 2.

These steps constitute a feasible depth estimation method. Please note that the depth value mentioned here is the length of O_1P, which is slightly different from the depth we mentioned in the pinhole camera model. The depth in a pinhole camera refers to the z value of the pixel. We will demonstrate the results of the algorithm in the practical part.

11.3 Practice: Monocular Dense Reconstruction

The sample program in this section will use the test dataset of REMODE [127, 128]. It provides a total of 200 monocular top-view images collected by a drone, and it also provides the ground-truth pose of each image. Based on these data, we can estimate each pixel's depth value of the first frame, which is exactly the dense monocular reconstruction.

http://rpg.ifi.uzh.ch/datasets/remode_test_data.zip. You can use a web browser or download software to download. After decompression, all images from 0 to 200 can be found in test_data/Images, and a text file in the test_data directory records the poses.

```
scene_000.png 1.086410 4.766730 -1.449960 0.789455 0.051299 -0.000779 0.611661
scene_001.png 1.086390 4.766370 -1.449530 0.789180 0.051881 -0.001131 0.611966
scene_002.png 1.086120 4.765520 -1.449090 0.788982 0.052159 -0.000735 0.612198
......
```

Figure 11.5 shows images from several moments. It can be seen that the scene is mainly composed of the ground, a table, and small objects on the table. If the depth estimate is roughly correct, we can at least see the difference between the table's depth and the ground. Below, we follow the previous explanation to write the dense depth estimation program. The program is written in C language style and placed in a single file for ease of understanding. This program is a bit long compared with previous demos. We will focus on a few important functions. Please read the rest of the content from the source code on GitHub.

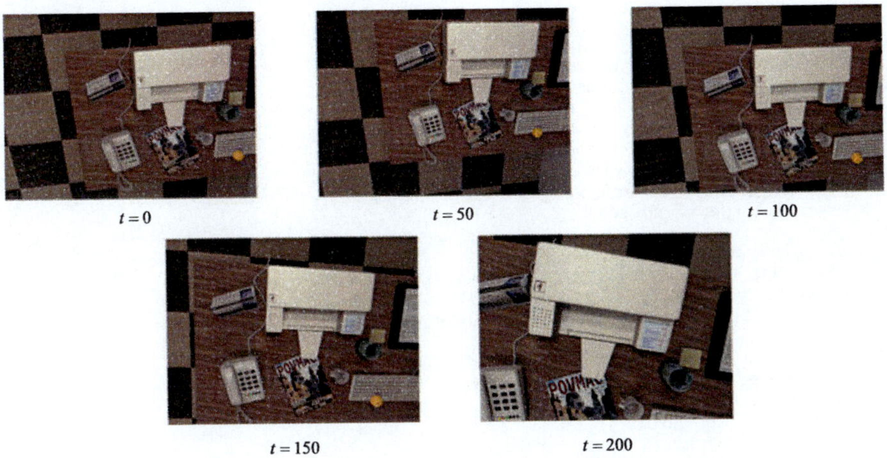

Fig. 11.5 Sample images from the dataset

Listing 11.1 slambook2/ch12/dense_monocular/dense_mapping.cpp (part)

```
/***********************************************
* This program demonstrates the dense depth estimation of a monocular
    camera under a known trajectory
* use epipolar search + NCC matching method, which corresponds to section
    12.2 of the book
* Please note that this program is not perfect, you can improve it by
    yourself.
***********************************************/

//
// parameters
const int boarder = 20; // image boarder
const int width = 640; // image width
const int height = 480; // image height
const double fx = 481.2f; // camera intrinsicss
const double fy = -480.0f;
const double cx = 319.5f;
const double cy = 239.5f;
const int ncc_window_size = 3; // half window size of NCC
const int ncc_area = (2 * ncc_window_size + 1) * (2 * ncc_window_size + 1);
    // area of NCC
const double min_cov = 0.1; // converge criteria: minimal covariance
const double max_cov = 10; // disconverge criteria: maximal covariance

//
// important functions
/**
* update depth using new images
* @param ref refernce image
* @param curr current image
* @param T_C_R matrix from ref to cur
* @param depth depth estimation
* @param depth_cov covariance of depth
* @return true if success
*/
bool update(
  const Mat &ref, const Mat &curr, const SE3d &T_C_R,
  Mat &depth, Mat &depth_cov2);

/**
```

```
* epipolar search
* @param ref refernce image
* @param curr current image
* @param T_C_R matrix from ref to cur
* @param pt_ref point in ref
* @param depth_mu mean of depth
* @param depth_cov cov of depth
* @param pt_curr point in current
* @param epipolar_direction
* @return true if success
*/
bool epipolarSearch(
  const Mat &ref, const Mat &curr, const SE3d &T_C_R,
  const Vector2d &pt_ref, const double &depth_mu, const double &depth_cov,
  Vector2d &pt_curr, Vector2d &epipolar_direction);

/**
* update depth filter
* @param pt_ref point in ref
* @param pt_curr point in cur
* @param T_C_R matrix from ref to cur
* @param epipolar_direction
* @param depth mean of depth
* @param depth_cov2 cov of depth
* @return true if success
*/
bool updateDepthFilter(
  const Vector2d &pt_ref, const Vector2d &pt_curr, const SE3d &T_C_R,
  const Vector2d &epipolar_direction, Mat &depth, Mat &depth_cov2);

/**
* NCC computation
* @param ref reference image
* @param curr current image
* @param pt_ref reference pixel
* @param pt_curr current pixel
* @return NCC score
*/
double NCC(const Mat &ref, const Mat &curr, const Vector2d &pt_ref, const
     Vector2d &pt_curr);

// bilinear interpolation
inline double getBilinearInterpolatedValue(const Mat &img, const Vector2d &
     pt) {
  uchar *d = &img.data[int(pt(1, 0)) * img.step + int(pt(0, 0))];
  double xx = pt(0, 0) - floor(pt(0, 0));
  double yy = pt(1, 0) - floor(pt(1, 0));
  return ((1 - xx) * (1 - yy) * double(d[0]) +
  xx * (1 - yy) * double(d[1]) +
  (1 - xx) * yy * double(d[img.step]) +
  xx * yy * double(d[img.step + 1])) / 255.0;
}

int main(int argc, char **argv) {
  if (argc != 2) {
    cout << "Usage: dense_mapping path_to_test_dataset" << endl;
    return -1;
  }

  // read data
  vector<string> color_image_files;
  vector<SE3d> poses_TWC;
  Mat ref_depth;
  bool ret = readDatasetFiles(argv[1], color_image_files, poses_TWC,
       ref_depth);
  if (ret == false) {
    cout << "Reading image files failed!" << endl;
```

```
    return -1;
  }
  cout << "read total " << color_image_files.size() << " files." << endl;

  // first image
  Mat ref = imread(color_image_files[0], 0); // gray-scale image
  SE3d pose_ref_TWC = poses_TWC[0];
  double init_depth = 3.0; // initial depth
  double init_cov2 = 3.0; // initial covariance
  Mat depth(height, width, CV_64F, init_depth); // depth image
  Mat depth_cov2(height, width, CV_64F, init_cov2); // depth cov image

  for (int index = 1; index < color_image_files.size(); index++) {
    cout << "*** loop " << index << " ***" << endl;
    Mat curr = imread(color_image_files[index], 0);
    if (curr.data == nullptr) continue;
    SE3d pose_curr_TWC = poses_TWC[index];
    SE3d pose_T_C_R = pose_curr_TWC.inverse() * pose_ref_TWC; // T_C_W *
         T_W_R = T_C_R
    update(ref, curr, pose_T_C_R, depth, depth_cov2);
    evaludateDepth(ref_depth, depth);
    plotDepth(ref_depth, depth);
    imshow("image", curr);
    waitKey(1);
  }

  cout << "estimation returns, saving depth map ..." << endl;
  imwrite("depth.png", depth);
  cout << "done." << endl;

  return 0;
}


bool update(const Mat &ref, const Mat &curr, const SE3d &T_C_R, Mat &depth,
      Mat &depth_cov2) {
  for (int x = boarder; x < width - boarder; x++)
  for (int y = boarder; y < height - boarder; y++) {
    if (depth_cov2.ptr<double>(y)[x] < min_cov || depth_cov2.ptr<double>(y)[
         x] > max_cov) {
      // converge or abort
      continue;
    }
    // search match of (x,y) along the epipolar line
    Vector2d pt_curr;
    Vector2d epipolar_direction;
    bool ret = epipolarSearch(
    ref, curr, T_C_R, Vector2d(x, y), depth.ptr<double>(y)[x], sqrt(
         depth_cov2.ptr<double>(y)[x]), pt_curr, epipolar_direction);

    if (ret == false) // failed
      continue;

    // un-comment this to display the match result
    // showEpipolarMatch(ref, curr, Vector2d(x, y), pt_curr);

    // update if succeed
    updateDepthFilter(Vector2d(x, y), pt_curr, T_C_R, epipolar_direction,
         depth, depth_cov2);
  }
}


bool epipolarSearch(
  const Mat &ref, const Mat &curr,
  const SE3d &T_C_R, const Vector2d &pt_ref,
  const double &depth_mu, const double &depth_cov,
```

```
    Vector2d &pt_curr, Vector2d &epipolar_direction) {
    Vector3d f_ref = px2cam(pt_ref);
    f_ref.normalize();
    Vector3d P_ref = f_ref * depth_mu; // reference vector

    Vector2d px_mean_curr = cam2px(T_C_R * P_ref); // pixel according to mean depth
    double d_min = depth_mu - 3 * depth_cov, d_max = depth_mu + 3 * depth_cov;
    if (d_min < 0.1) d_min = 0.1;
    Vector2d px_min_curr = cam2px(T_C_R * (f_ref * d_min)); // pixel of minimal depth
    Vector2d px_max_curr = cam2px(T_C_R * (f_ref * d_max)); // pixel of maximal depth

    Vector2d epipolar_line = px_max_curr - px_min_curr; // epipolar line
    epipolar_direction = epipolar_line; // normalized
    epipolar_direction.normalize();
    double half_length = 0.5 * epipolar_line.norm();
    if (half_length > 100) half_length = 100; // we don'e want to search too much

    // un-comment this to show the epipolar line
    // showEpipolarLine( ref, curr, pt_ref, px_min_curr, px_max_curr );

    // epipolar search
    double best_ncc = -1.0;
    Vector2d best_px_curr;
    for (double l = -half_length; l <= half_length; l += 0.7) { // l+=sqrt(2)
      Vector2d px_curr = px_mean_curr + l * epipolar_direction;
      if (!inside(px_curr))
        continue;
      // compute NCC score
      double ncc = NCC(ref, curr, pt_ref, px_curr);
      if (ncc > best_ncc) {
        best_ncc = ncc;
        best_px_curr = px_curr;
      }
    }
    if (best_ncc < 0.85f) // only trust NCC with high scores
      return false;
    pt_curr = best_px_curr;
    return true;
}

double NCC(
    const Mat &ref, const Mat &curr,
    const Vector2d &pt_ref, const Vector2d &pt_curr) {
    // zero-mean NCC
    // compute the mean
    double mean_ref = 0, mean_curr = 0;
    vector<double> values_ref, values_curr;
    for (int x = -ncc_window_size; x <= ncc_window_size; x++)
    for (int y = -ncc_window_size; y <= ncc_window_size; y++) {
      double value_ref = double(ref.ptr<uchar>(int(y + pt_ref(1, 0)))[int(x + pt_ref(0, 0))]) / 255.0;
      mean_ref += value_ref;

      double value_curr = getBilinearInterpolatedValue(curr, pt_curr + Vector2d(x, y));
      mean_curr += value_curr;

      values_ref.push_back(value_ref);
      values_curr.push_back(value_curr);
    }

    mean_ref /= ncc_area;
```

```
  mean_curr /= ncc_area;

  // compute Zero mean NCC
  double numerator = 0, demoniator1 = 0, demoniator2 = 0;
  for (int i = 0; i < values_ref.size(); i++) {
    double n = (values_ref[i] - mean_ref) * (values_curr[i] - mean_curr);
    numerator += n;
    demoniator1 += (values_ref[i] - mean_ref) * (values_ref[i] - mean_ref);
    demoniator2 += (values_curr[i] - mean_curr) * (values_curr[i] -
         mean_curr);
  }
  return numerator / sqrt(demoniator1 * demoniator2 + 1e-10);
}

bool updateDepthFilter(
  const Vector2d &pt_ref, const Vector2d &pt_curr, const SE3d &T_C_R,
  const Vector2d &epipolar_direction, Mat &depth, Mat &depth_cov2) {
  // anybody still reading?
  // tri-angulation
  SE3d T_R_C = T_C_R.inverse();
  Vector3d f_ref = px2cam(pt_ref);
  f_ref.normalize();
  Vector3d f_curr = px2cam(pt_curr);
  f_curr.normalize();

  // equation:
  // d_ref * f_ref = d_cur * ( R_RC * f_cur ) + t_RC
  // f2 = R_RC * f_cur
  // convert to this:
  // => [ f_ref^T f_ref, -f_ref^T f2 ] [d_ref] [f_ref^T t]
  // [ f_cur^T f_ref, -f2^T f2 ] [d_cur] = [f2^T t ]
  Vector3d t = T_R_C.translation();
  Vector3d f2 = T_R_C.so3() * f_curr;
  Vector2d b = Vector2d(t.dot(f_ref), t.dot(f2));
  Matrix2d A;
  A(0, 0) = f_ref.dot(f_ref);
  A(0, 1) = -f_ref.dot(f2);
  A(1, 0) = -A(0, 1);
  A(1, 1) = -f2.dot(f2);
  Vector2d ans = A.inverse() * b;
  Vector3d xm = ans[0] * f_ref; // result in ref
  Vector3d xn = t + ans[1] * f2; // result in cur
  Vector3d p_esti = (xm + xn) / 2.0; // take average as p
  double depth_estimation = p_esti.norm(); // depth

  // compute the covariance
  Vector3d p = f_ref * depth_estimation;
  Vector3d a = p - t;
  double t_norm = t.norm();
  double a_norm = a.norm();
  double alpha = acos(f_ref.dot(t) / t_norm);
  double beta = acos(-a.dot(t) / (a_norm * t_norm));
  Vector3d f_curr_prime = px2cam(pt_curr + epipolar_direction);
  f_curr_prime.normalize();
  double beta_prime = acos(f_curr_prime.dot(-t) / t_norm);
  double gamma = M_PI - alpha - beta_prime;
  double p_prime = t_norm * sin(beta_prime) / sin(gamma);
  double d_cov = p_prime - depth_estimation;
  double d_cov2 = d_cov * d_cov;

  // Gaussian fusion
  double mu = depth.ptr<double>(int(pt_ref(1, 0)))[int(pt_ref(0, 0))];
  double sigma2 = depth_cov2.ptr<double>(int(pt_ref(1, 0)))[int(pt_ref(0,
       0))];
```

```
    double mu_fuse = (d_cov2 * mu + sigma2 * depth_estimation) / (sigma2 +
        d_cov2);
    double sigma_fuse2 = (sigma2 * d_cov2) / (sigma2 + d_cov2);

    depth.ptr<double>(int(pt_ref(1, 0)))[int(pt_ref(0, 0))] = mu_fuse;
    depth_cov2.ptr<double>(int(pt_ref(1, 0)))[int(pt_ref(0, 0))] =
        sigma_fuse2;

    return true;
}
```

We omit functions such as drawing and reading data and only show the part related to depth calculation. If the reader understands the previous section's content, I believe it is not difficult to understand the source code here. Nevertheless, we will briefly explain several key functions:

1. The main function (not listed here) is very simple. It is only responsible for reading the image from the dataset and then handing it over to the update function to update the depth map.
2. In the update function, we traverse each pixel of the reference frame, first look for an epipolar match in the current frame. If it can match, use the epipolar match to update the estimation of the depth map.
3. The principle of epipolar search is roughly the same as the one introduced in the previous section. Still, some details have been added to the implementation. Because the depth value is assumed to obey the Gaussian distribution, we take the mean value as the center and take $\pm 3\sigma$ as the radius and then look for the epipolar line's projection in the current frame. Then, traverse the pixels on this epipolar line (the step size is approximately 0.7 of $\sqrt{2}/2$), and find the point with the highest NCC as the matching point. If the highest NCC is also lower than the threshold (here taken as 0.85), the match is considered failed.
4. The calculation of NCC uses the zero-mean version, that is, for the image block $\mathbf{A}, \mathbf{B}$, take:

$$\mathrm{NCC}_z(\mathbf{A}, \mathbf{B}) = \frac{\sum\limits_{i,j} \left(\mathbf{A}(i,j) - \bar{\mathbf{A}}(i,j)\right)\left(\mathbf{B}(i,j) - \bar{\mathbf{B}}(i,j)\right)}{\sqrt{\sum\limits_{i,j} \left(\mathbf{A}(i,j) - \bar{\mathbf{A}}(i,j)\right)^2 \sum\limits_{i,j} \left(\mathbf{B}(i,j) - \bar{\mathbf{B}}(i,j)\right)^2}}. \tag{11.11}$$

5. The triangulation part is consistent with Sect. 6.5, and the calculation of uncertainty is consistent with the Gaussian fusion method and the previous section.

Although the program is a bit long, I believe readers can understand it according to the above tips. Let's take a look at the results.

Experimental Results

After compiling this program, run it in the dataset directory[3]:

[3] Please note that the dense depth estimation is time-consuming. If your computer is older, please wait patiently for a while.

Listing 11.2 Terminal ouptut:

```
$ build/dense_mapping~/dataset/test_data
read total 202 files.
*** loop 1 ***
*** loop 2 ***
......
```

The program's output is relatively concise, only showing the number of iterations, the current image, and the depth map. Regarding the depth map, we show the depth image by multiplying it by 0.4-that is, the depth of the pure white point (the value is 1.0) is about 2.5 m. The darker the color, the smaller the depth value, and the closer the object to us. If you have run the program, you should find that depth estimation is a dynamic process that gradually converges from an uncertain initial value to a stable value. Our initial value used a distribution with a mean and variance of 3.0. Of course, you can also modify the initial distribution to see how it affects the results.

From Fig. 11.6 it can be found that when the number of iterations exceeds a specific value, the depth map becomes stable, and no more changes are made to the new data. The stabilized depth map shows that the difference between the floor and the table can be roughly seen, and the depth of the objects is close to the table. Some of the estimations are correct, but there are also many wrong estimates. They appear as inconsistencies between the depth pixel and the surrounding data, which seems to

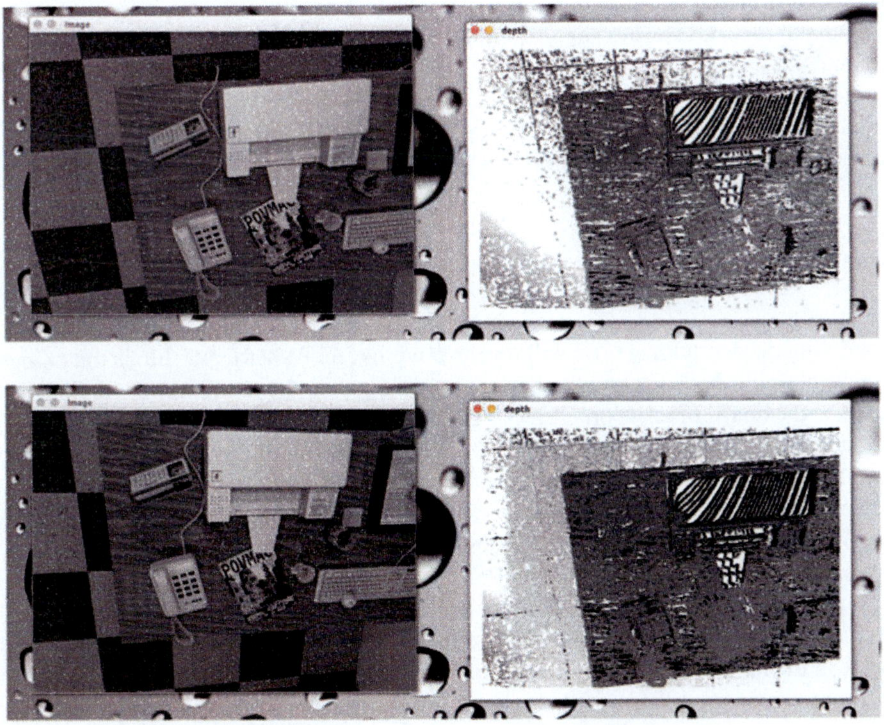

Fig. 11.6 Snapshots of running the depth filter after 10 and 30 iterations

be too large or too small estimates. The wrong places are often located at the edges because the number of times seen is not enough, so they are not correctly estimated. In summary, we think that most of the depth map is correct, but it did not achieve the desired effect. We will analyze the causes of these situations in the next section and discuss what can be improved.

11.3.1 Discussion

In the previous section, we demonstrated the dense mapping of a mobile monocular camera and estimated each pixel's depth of the reference frame. Our code is relatively simple and straightforward, without using any tricks. This is a common situation that simple methods are always not effective in real engineering.

In the previous section, we demonstrated the dense mapping of a mobile monocular camera and estimated each pixel's depth of the reference frame. Our code is relatively simple and straightforward, without using any tricks. This is a common situation that simple methods are always not effective in real engineering. Due to the complexity of real data, programs that can work in a real environment often require careful consideration and many engineering tricks, making lots of practical code extremely complicated. They are difficult to explain to beginners, so we have to use a less effective but relatively easy read and write implementation. Of course, we can put forward several suggestions for improving the demo program, but we do not intend to present the modified (very complicated) program directly to the reader.

Below we conduct a preliminary analysis of the results of the experiment in the previous section. We will analyze the results of the demonstration experiment from the perspective of computer vision and filters.

11.3.2 Pixel Gradients

Observing the depth image, we will find an obvious fact. Whether the block matching is correct or not depends on whether the image block is distinguishable. Obviously, suppose the image block is only a piece of black or white, lacking visual information. In that case, we may erroneously match it with some surrounding pixels. For example, the printer surface in the demo program is uniformly white. It is very easy to cause mismatches, so the depth information on the printer's surface is mostly incorrect. The sample program's spatial surface has obviously undesirable striped depth estimates, and according to our intuitive imagination, The surface of the printer must be smooth.

This involves a problem that has been seen once in the direct method chapter. When performing block matching (and calculation of NCC), we must assume that the small block is unchanged, and then we compare it with other blocks. Of course, blocks with noticeable gradients will have good discrimination and will not easily cause mismatches. For pixels with inconspicuous gradients, since there is no discrimination in block matching, it is difficult for us to effectively estimate its depth. Conversely, the depth information we have evident gradients will be relatively accurate, such as magazines, phones, and other objects with obvious texture on the desktop. Therefore, the demo program reflects a widespread stereo vision problem: dependence on the texture. This problem is also extremely common in binocular vision, which shows that the reconstruction quality of stereo vision is very dependent on the environmental texture.

Our demo program deliberately uses a good-textured environment, such as a checkerboard-like floor, a wood-grained desktop, etc., so we can get a seemingly good result. However, in practice, places with uniform brightness such as walls and smooth surfaces will often appear, affecting our depth estimation. From a certain perspective, the problem cannot be improved or solved on the current algorithm flow (block matching) if we only care about the neighborhood around a certain block.

Further discussing the pixel gradient problem, we will also find the connection between the pixel gradient and the epipolar line. The literature [75] has discussed their relationship in detail, but it is also intuitively reflected in our demo program.

Taking Fig. 11.7 as an example, we will give two extreme cases: the pixel gradient is parallel to the epipolar direction and orthogonal to the epipolar direction. Let's look at the orthogonal situation first. In this case, even if the blocks have noticeable gradients, when we do block matching along the epipolar line, we will find that the matching degree is the same, so no effective matching can be obtained. Conversely, in the parallel case, we can accurately determine where the highest matching point appears. In reality, the gradient and the epipolar line are probably somewhere in between: they are neither completely orthogonal nor completely parallel. When the angle between the pixel gradient and the epipolar line is large, the uncertainty of the epipolar line matching is large. When the angle is small, the uncertainty of the matching becomes smaller. We uniformly treat these conditions in the demo program as one pixel error, which is not fair enough. A more accurate uncertainty model should be used after considering the angle between the epipolar line and the pixel gradient. Specific adjustments and improvements are left as exercises.

11.3.3 Inverse Depth Filter

From another perspective, we might as well ask: Is it appropriate to assume that the pixel depth is a Gaussian distribution? This is related to a *parameterization* problem.

In the previous content, we often use the world coordinates x, y, z to describe a point, which is one of many parameterization methods. We may think that the three quantities x, y, z are random, and they obey the 3D Gaussian distribution. However, this lecture uses the image coordinates u, v and the depth value d to describe a certain spatial point. We think that u, v do not move, and d obeys (one-dimensional) Gaussian distribution, which is another form of parameterization. Then we have to ask: Is there any difference between these two parameterized forms?

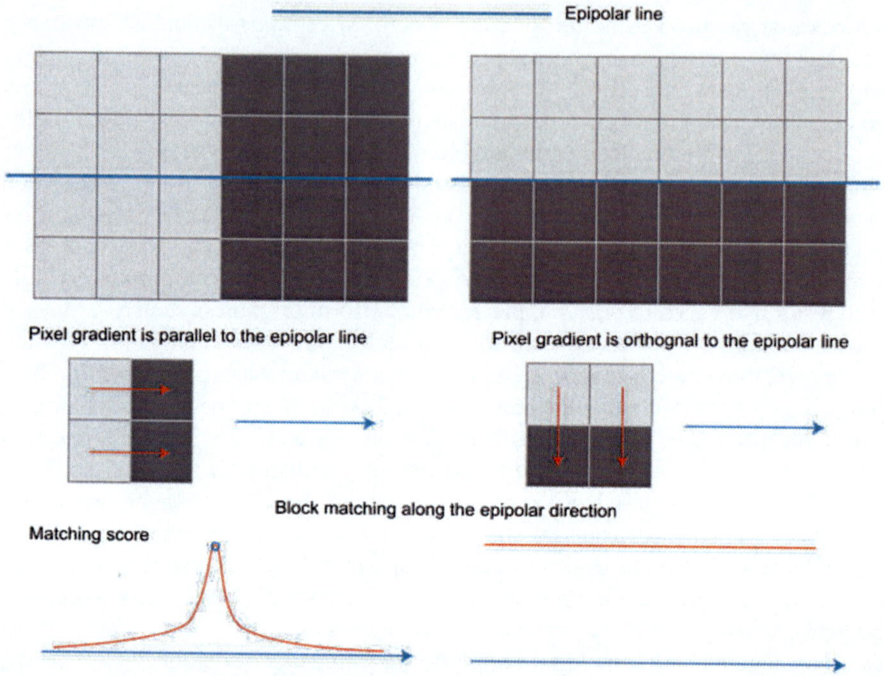

Fig. 11.7 Relationship of pixel gradients and the epipolar direction

Different parameterized forms are used to describe the same quantity, that is, a 3D spatial point. Considering that when we see a specific point on the camera, its image coordinates u, v are relatively accurate. The uncertainty of the depth value d is very uncertain. If the world coordinates x, y, z are used to describe this point, then according to the camera's current pose, there will be an apparent correlation between the three quantities x, y, z. The non-diagonal elements are of the covariance matrix will not be zero. And if a point is parameterized with u, v, d, then its u, v and d are approximately independent. We can even think that u, v are independent. Then the covariance matrix is roughly diagonal, which is more concise.

Inverse depth is a widely used parameterization technique [129, 130] that has appeared in SLAM research in recent years. In the demo program, we assume that the depth value satisfies the Gaussian distribution: $d \sim N(\mu, \sigma^2)$. But is it reasonable to do so? Does the depth really approximate a Gaussian distribution? If we think more deeply, there are indeed some problems with the normal distribution of depth:

1. What we actually want to express is: the depth of this scene is about 5~10 m, there may be some further points, but the close distance will definitely not be less than the camera focal length (or the depth will not be less than 0). This distribution does not form a symmetrical shape like the Gaussian distribution. Its tail may be slightly longer, and the negative area is zero.

2. There may be points that are very far away in some outdoor applications or even at infinity. It is difficult to cover these points in our initial value, and there will be some numerical difficulties in describing the large depth values with Gaussian distribution.

Thus, the inverse depth came into being. People found in the simulation that the hypothesis that the inverse depth forms a Gaussian distribution is more effective [130]. Later, in practical applications, the inverse depth also has better numerical stability, which gradually becomes a general and standard practice in the existing SLAM systems [68, 69, 88].

It is not complicated to change the demonstration program from positive depth to inverse depth. Just change d to the inverse depth d^{-1} in the previous depth's derivation. We also leave this change as an exercise for readers to complete.

11.3.4 Pre-Transform the Image

Before block matching, it is also a common preprocessing method to do a transformation from image to image. This is because we assume that image patches remain unchanged when the camera is moving, and this assumption can be held when the camera is shifted (the example data set is basically such an example). But when the camera rotates significantly, it is no more effective. In particular, when the camera rotates around the optical center, an image block that is black on the bottom may become a black on top, causing the correlation to directly become a negative number (although it is still the same block).

To prevent this situation, we usually need to consider the motion between the reference frame with the current frame. A point $\mathbf{P}_R$ in the reference frame has the following relationship with the 3D point $\mathbf{P}_W$:

$$d_R \mathbf{P}_R = \mathbf{K}\left(\mathbf{R}_{RW}\mathbf{P}_W + \mathbf{t}_{RW}\right). \tag{11.12}$$

Similarly, for the current frame, there is a projection of $\mathbf{P}_W$ on it, denoted as $\mathbf{P}_C$:

$$d_C \mathbf{P}_C = \mathbf{K}\left(\mathbf{R}_{CW}\mathbf{P}_W + \mathbf{t}_{CW}\right). \tag{11.13}$$

Substituting and eliminating $\mathbf{P}_W$, the pixel relationship between the two images is obtained:

$$d_C \mathbf{P}_C = d_R \mathbf{K}\mathbf{R}_{CW}\mathbf{R}_{RW}^T\mathbf{K}^{-1}\mathbf{P}_R + \mathbf{K}\mathbf{t}_{CW} - \mathbf{K}\mathbf{R}_{CW}\mathbf{R}_{RW}^T\mathbf{K}\mathbf{t}_{RW}. \tag{11.14}$$

When we know $d_R, \mathbf{P}_R$, we can calculate the projection position of $\mathbf{P}_C$. At this time, give the two components of $\mathbf{P}_R$ an increment $\mathrm{d}u, \mathrm{d}v$, then the increment of $\mathbf{P}_C$ can be obtained as $\mathrm{d}u_c, \mathrm{d}v_c$. In this way, a linear relationship between the coordinate transformation of the reference frame and the current frame image in a local range is calculated to form an affine transformation:

$$\begin{bmatrix} \mathrm{d}u_c \\ \mathrm{d}v_c \end{bmatrix} = \begin{bmatrix} \frac{\mathrm{d}u_c}{\mathrm{d}u} & \frac{\mathrm{d}u_c}{\mathrm{d}v} \\ \frac{\mathrm{d}v_c}{\mathrm{d}u} & \frac{\mathrm{d}v_c}{\mathrm{d}v} \end{bmatrix} \begin{bmatrix} \mathrm{d}u \\ \mathrm{d}v \end{bmatrix} \tag{11.15}$$

According to the affine transformation matrix, we can transform the current frame's pixels (or reference frame) and then perform block matching to obtain a better effect on rotation.

11.3.5 Parallel Computing

In the experiment, we have also seen that the dense depth map estimation is very time-consuming. This is because the estimated points have changed from the original hundreds of feature points to hundreds of thousands of pixels. Even now, the mainstream CPUs are impossible to do such a large calculation in real-time. However, the problem also has another nature: the depth estimates of these hundreds of thousands of pixels are independent. This makes parallelization possible.

In the sample program, we traverse all the pixels in a double loop and perform epipolar searches one by one. When we use the CPU, this process is carried out sequentially. The calculation of the next pixel must wait for the previous pixel. However, there is no need to wait because the calculation is independent. So we can use multiple threads to calculate each pixel separately and then collect the results. Theoretically, if we have 300,000 threads, the calculation time for this problem is the same as calculating one pixel.

The parallel computing architecture of GPU is very suitable for such problems. Therefore, in dense reconstruction, the GPU is often used for parallel acceleration. Of course, this book will not involve GPU programming, so we only point out the possibility of using GPU acceleration here, and the specific practice is left to the reader as a verification. According to some similar work, the dense depth estimation using GPU can be real-time on mainstream GPUs.

11.3.6 Other Improvements

In fact, we can also propose many improvements to this example, such as:

1. If each pixel is completely calculated independently, there may be cases where one pixel's depth is small, and the next one is large. We don't have any smooth constraints on the depth map. However, we can assume that the adjacent depth will not change too much, thus adding a spatial regularization term to the depth estimation.

2. We did not explicitly deal with the case of outliers. Due to various factors such as occlusion, lighting, motion blur, etc., it is impossible to maintain a successful match for every pixel. As long as the NCC is greater than a specific value in the demonstration program, it is considered a successful match, and the mismatch is not considered. There are also several ways to handle mismatches. For example, the depth filter under the uniform-Gaussian mixture distribution is proposed in [126] explicitly distinguishes the inlier from the outlier and performs probabilistic modeling, which can better process the outliers. However, this type of filter theory is more complicated, and this book does not want to involve too much. Please read the original paper if you are interested.

As can be seen from the above discussion, there are many possible improvements. If we carefully improve every step, we can hope to get a good dense mapping algorithm in the end. However, as we discussed, there are problems with theoretical difficulties, such as the dependence on the texture and the correlation between the pixel gradient and the epipolar direction (the orthogonal case). These problems are difficult to solve by only adjusting the code or parameters. So, until now, although binoculars and mobile monoculars can build dense maps, we usually think that they rely too much on environmental textures and lighting and are not robust enough.

11.4 Dense RGB-D Mapping

In addition to using monocular and binocular for dense reconstruction, RGB-D cameras are a better choice within the application scope. The depth estimation problem discussed in detail in the last lecture can be obtained by hardware measurement in RGB-D cameras without consuming many computing resources. In addition, the structured light or time-of-flight principle of RGB-D ensures that the depth data is independent of texture. Even when facing a solid-colored object, we can measure its depth as long as it can reflect light. This is also a major advantage of RGB-D sensors.

It is relatively easy to use RGB-D for dense mapping. However, depending on the map format, there are also several different mainstream mapping methods. The most intuitive and straightforward way is to convert the RGB-D data into a point cloud based on the estimated camera pose and then stitch them into a global point cloud map composed of discrete points. On this basis, if we have further requirements for the appearance and want to estimate the object's surface, we can use the triangular mesh and the surface (surfel) to build the map. On the other hand, if you want to know the map's obstacle information and navigate the map, you can also create an occupancy map through voxels.

We seem to have introduced many new concepts. Please don't worry, we will investigate them one by one slowly. For some suitable experiments, we will also provide several demonstration programs as usual. Since there is not much theoretical knowledge involved in RGB-D mapping, the following sections will directly introduce the practical part. GPU mapping is beyond this book's scope, so we will briefly explain its principles and not demonstrate them.

11.4.1 Practice: RGB-D Point Cloud Mapping

First, let's explain the simplest point cloud map. The so-called point cloud is a map represented by a set of discrete points. The most basic point contains three-dimensional coordinates of x, y, z and also color information of r, g, b. Since the RGB-D camera provides a color map and a depth map, it is easy to calculate the RGB-D point cloud based on the camera's internal parameters. If the camera's pose is obtained, then we can directly merge the keyframes into a global point cloud. In the Sect. 4.4.2 of this book, an example of merging point clouds through camera internal and external parameters was given. However, that example is mainly for the reader to understand the camera parameters. In the real mapping, we will also add some filtering processing to the point cloud to obtain a better visual effect. This program mainly uses two kinds of filters: the outer point removal filter and the voxel grid filter. The code of the sample program is as follows:

Listing 11.3 slambook/ch12/dense_RGBD/pointcloud_mapping.cpp (part)

```
int main(int argc, char **argv) {
  vector<cv::Mat> colorImgs, depthImgs;
  vector<Eigen::Isometry3d> poses;

  ifstream fin("./data/pose.txt");
  if (!fin) {
    cerr << "cannot find pose file" << endl;
    return 1;
  }

  for (int i = 0; i < 5; i++) {
    boost::format fmt("./data/%s/%d.%s"); // we use boost::format to read
        image files
    colorImgs.push_back(cv::imread((fmt % "color" % (i + 1) % "png").str()))
    ;
    depthImgs.push_back(cv::imread((fmt % "depth" % (i + 1) % "png").str(),
        -1)); // use -1 to read the unchanged data

    double data[7] = {0};
    for (int i = 0; i < 7; i++) {
      fin >> data[i];
    }
    Eigen::Quaterniond q(data[6], data[3], data[4], data[5]);
    Eigen::Isometry3d T(q);
    T.pretranslate(Eigen::Vector3d(data[0], data[1], data[2]));
    poses.push_back(T);
  }

  // merge the point clouds
  // intrinsics
  double cx = 319.5;
  double cy = 239.5;
  double fx = 481.2;
  double fy = -480.0;
  double depthScale = 5000.0;

```

```
  cout << "convering image to point cloud ..." << endl;

  // use XYZRGB as our format
  typedef pcl::PointXYZRGB PointT;
  typedef pcl::PointCloud<PointT> PointCloud;

  PointCloud::Ptr pointCloud(new PointCloud);
  for (int i = 0; i < 5; i++) {
    PointCloud::Ptr current(new PointCloud);
    cout << "converting " << i + 1 << endl;
    cv::Mat color = colorImgs[i];
    cv::Mat depth = depthImgs[i];
    Eigen::Isometry3d T = poses[i];
    for (int v = 0; v < color.rows; v++)
    for (int u = 0; u < color.cols; u++) {
      unsigned int d = depth.ptr<unsigned short>(v)[u]; // depth data
      if (d == 0) continue; // 0 means invalid reading
      Eigen::Vector3d point;
      point[2] = double(d) / depthScale;
      point[0] = (u - cx) * point[2] / fx;
      point[1] = (v - cy) * point[2] / fy;
      Eigen::Vector3d pointWorld = T * point;

      PointT p;
      p.x = pointWorld[0];
      p.y = pointWorld[1];
      p.z = pointWorld[2];
      p.b = color.data[v * color.step + u * color.channels()];
      p.g = color.data[v * color.step + u * color.channels() + 1];
      p.r = color.data[v * color.step + u * color.channels() + 2];
      current->points.push_back(p);
    }
    // depth filter and statistical removal
    PointCloud::Ptr tmp(new PointCloud);
    pcl::StatisticalOutlierRemoval<PointT> statistical_filter;
    statistical_filter.setMeanK(50);
    statistical_filter.setStddevMulThresh(1.0);
    statistical_filter.setInputCloud(current);
    statistical_filter.filter(*tmp);
    (*pointCloud) += *tmp;
  }

  pointCloud->is_dense = false;
  cout << "we have " << pointCloud->size() << " points." << endl;

  // voxel filter
  pcl::VoxelGrid<PointT> voxel_filter;
  double resolution = 0.03;
  voxel_filter.setLeafSize(resolution, resolution, resolution); //
      resolution
  PointCloud::Ptr tmp(new PointCloud);
  voxel_filter.setInputCloud(pointCloud);
  voxel_filter.filter(*tmp);
  tmp->swap(*pointCloud);

  cout << "Now we have " << pointCloud->size() << " points after voxel
      filtering." << endl;

  pcl::io::savePCDFileBinary("map.pcd", *pointCloud);
  return 0;
}
```

This code needs to install the point cloud library. In Ubuntu 18.04, just one command is enough:

Listing 11.4 Terminal input:

```
sudo apt-get install libpcl-dev pcl-tools
```

The code does not change much compared with the lecture 4. The main differences are:

1. When generating the point cloud of each frame, we remove the points with invalid depth values. This is mainly because of the effective range of Kinect. The depth value after exceeding the range will have a large error or return a zero.
2. Use the statistical filter method to remove outliers. This filter counts the distance distribution between each point and the nearest N points and removes those with extremely large distances. In this way, we keep those sticky points and remove isolated noise points.
3. Finally, the voxel filter is used for downsampling. There will be many redundant points in the overlapping area due to multiple viewing angles. This will take up a lot of memory space in vain. Voxel filtering ensures that there is only one point in a specific size cube (or voxel), which is equivalent to downsampling the 3D space, saving a lot of storage space.

In the second edition of the book, we use the ICL-NUIM dataset [131] as an example. This data set is a synthetic RGB-D dataset, which allows us to get noise-free depth data to facilitate experiments. We store five images and depth maps and the corresponding camera poses in the data/ directory. In the voxel filter, we set the resolution to 0.03, which means that there is only one point reserved in each $0.03\times0.03\times0.03$ grid. This is a relatively high resolution. We can see from the program output that the number of points has been significantly reduced (from 1.3 million points to 30,000 points, only 2% storage space) but still keeps a similar visual effect.

Run the program in the dense_RGBD directory:

Listing 11.5 Terminal input:

```
./build/pointcloud_mapping
```

The point cloud file map.pcd can be obtained in the same directory. Then, open the pcd with the pcl_viewer tool. We can see the content, as shown in Fig. 11.8.

The point cloud map provides us with a relatively basic visual map, allowing us to roughly understand what the environment looks like. It is stored in three dimensions so we can quickly browse all corners of the scene and even roam in the scene. A significant advantage of point clouds is that they can be efficiently generated directly from RGB-D images without additional processing. Its filtering operation is also very intuitive, and the processing efficiency is acceptable. However, the point clouds maps are still very fundamental. Let's see if point cloud maps can meet the requirements mentioned earlier.

1. Localization requirements: depends on the implementation of the frontend visual odometry. If it is based on feature points, the point cloud map cannot be directly used for localization because there is no feature information stored in the point

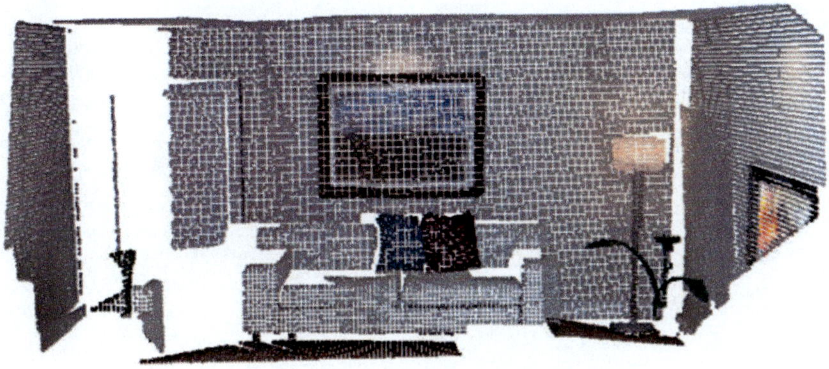

Fig. 11.8 Point cloud mapping using five image pairs in ICL-NUIM

cloud. If the frontend uses ICP for point cloud alignment, then you can perform ICP for the local point cloud to the global point cloud to estimate the camera's pose. However, this requires the global point cloud to have better accuracy. We only merge the point clouds without any optimization in this demo, so it is not enough.

2. Navigation and obstacle avoidance: point clouds cannot be used directly for navigation and obstacle avoidance. A pure point cloud may be disturbed by dynamic objects and cannot represent the occupancy information. We usually need post-processes based on raw point clouds to obtain a map format that is more suitable for navigation and obstacle avoidance.
3. Visualization and interaction: point clouds have basic visualization and interaction capabilities. We can see the appearance of the scene, and we can also walk through the scene. From the perspective of visualization, since the point cloud only contains discrete points and no surface information (such as normals), it does not conform to people's visualization habits. For example, the object of a point cloud map is the same from the front and the back, and you can see what is behind it through the object: these are not compatible with our daily experience.

In summary, we say that the raw point cloud map is *basic* or *primary*, which means that it is closer to the raw data read by the sensor. It has some essential functions, but it is usually used for debugging and basic display, inconvenient in most applications. If we want the map to have more advanced functions, the point cloud map is a good starting point. For example, for the navigation function, we can start from the point cloud to construct an occupancy grid map for the navigation algorithm to query whether a point can pass. Another example is the Poisson reconstruction [132] method commonly used in SfM, which can reconstruct the meshes from the point cloud to obtain the surface information. In addition to Poisson reconstruction, surfel

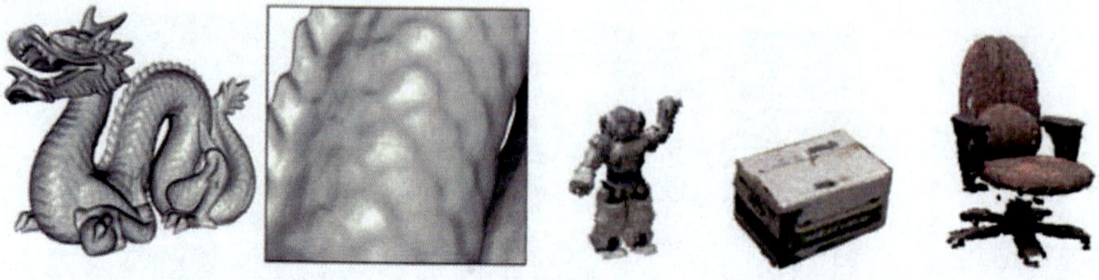

Fig. 11.9 Reconstruction results of Poisson and surfel model

is also a way to express the surface of an object. Using facets as the basic unit of the map, it can build a visually satisfactory map [133].

Figure 11.9 shows an example of Poisson reconstruction and surfel. It can be seen that their visual effects are significantly better, and they can all be constructed through point clouds. Most of the map formats obtained from point cloud conversion are provided in the PCL library, and interested readers can further explore the PCL's contents. As this book serves as introductory material, it does not introduce every map form in detail.

11.4.2 Building Meshes from Point Cloud

Reconstructing the mesh from the point cloud is also relatively easy. Let's demonstrate how to build the mesh based on the point cloud file just now. The general idea is: first calculate the point cloud's normal and then calculate the grid from the normal.

Listing 11.6 slambook2/ch12/dense_RGBD/surfel_mapping.cpp

```
#include <pcl/point_cloud.h>
#include <pcl/point_types.h>
#include <pcl/io/pcd_io.h>
#include <pcl/visualization/pcl_visualizer.h>
#include <pcl/kdtree/kdtree_flann.h>
#include <pcl/surface/surfel_smoothing.h>
#include <pcl/surface/mls.h>
#include <pcl/surface/gp3.h>
#include <pcl/surface/impl/mls.hpp>

// typedefs
typedef pcl::PointXYZRGB PointT;
typedef pcl::PointCloud<PointT> PointCloud;
typedef pcl::PointCloud<PointT>::Ptr PointCloudPtr;
typedef pcl::PointXYZRGBNormal SurfelT;
typedef pcl::PointCloud<SurfelT> SurfelCloud;
typedef pcl::PointCloud<SurfelT>::Ptr SurfelCloudPtr;

SurfelCloudPtr reconstructSurface(
const PointCloudPtr &input, float radius, int polynomial_order) {
  pcl::MovingLeastSquares<PointT, SurfelT> mls;
  pcl::search::KdTree<PointT>::Ptr tree(new pcl::search::KdTree<PointT>);
  mls.setSearchMethod(tree);
```

```
  mls.setSearchRadius(radius);
  mls.setComputeNormals(true);
  mls.setSqrGaussParam(radius * radius);
  mls.setPolynomialFit(polynomial_order > 1);
  mls.setPolynomialOrder(polynomial_order);
  mls.setInputCloud(input);
  SurfelCloudPtr output(new SurfelCloud);
  mls.process(*output);
  return (output);
}

pcl::PolygonMeshPtr triangulateMesh(const SurfelCloudPtr &surfels) {
  // Create search tree*
  pcl::search::KdTree<SurfelT>::Ptr tree(new pcl::search::KdTree<SurfelT>);
  tree->setInputCloud(surfels);

  // Initialize objects
  pcl::GreedyProjectionTriangulation<SurfelT> gp3;
  pcl::PolygonMeshPtr triangles(new pcl::PolygonMesh);

  // Set the maximum distance between connected points (maximum edge length)
  gp3.setSearchRadius(0.05);

  // Set typical values for the parameters
  gp3.setMu(2.5);
  gp3.setMaximumNearestNeighbors(100);
  gp3.setMaximumSurfaceAngle(M_PI / 4); // 45 degrees
  gp3.setMinimumAngle(M_PI / 18); // 10 degrees
  gp3.setMaximumAngle(2 * M_PI / 3); // 120 degrees
  gp3.setNormalConsistency(true);

  // Get result
  gp3.setInputCloud(surfels);
  gp3.setSearchMethod(tree);
  gp3.reconstruct(*triangles);

  return triangles;
}

int main(int argc, char **argv) {
  // Load the points
  PointCloudPtr cloud(new PointCloud);
  if (argc == 0 || pcl::io::loadPCDFile(argv[1], *cloud)) {
    cout << "failed to load point cloud!";
    return 1;
  }
  cout << "point cloud loaded, points: " << cloud->points.size() << endl;

  // Compute surface elements
  cout << "computing normals ... " << endl;
  double mls_radius = 0.05, polynomial_order = 2;
  auto surfels = reconstructSurface(cloud, mls_radius, polynomial_order);

  // Compute a greedy surface triangulation
  cout << "computing mesh ... " << endl;
  pcl::PolygonMeshPtr mesh = triangulateMesh(surfels);

  cout << "display mesh ... " << endl;
  pcl::visualization::PCLVisualizer vis;
  vis.addPolylineFromPolygonMesh(*mesh, "mesh frame");
  vis.addPolygonMesh(*mesh, "mesh");
  vis.resetCamera();
  vis.spin();
}
```

Fig. 11.10 Building meshes from point clouds

This program demonstrates the process of calculating normals and meshes. Use:

Listing 11.7 Terminal input:

```
./build/surfel_mapping map.pcd
```

to convert the point cloud into a grid map, as shown in Fig. 11.10. It can be seen that after the mesh is reconstructed, the normals, texture, and other information can be constructed from the point cloud without surface information. For the point cloud reconstruction algorithms (moving least-square and greedy projection) demonstrated in this section, readers can find them in the literature [134, 135], which are classic algorithms in this area.

11.4.3 Octo-Mapping

The following section introduces a map format that is commonly used in navigation and has better compression performance: the *octree map*. In the point cloud map, although we have a three-dimensional structure and voxel filtering to adjust the resolution, the point cloud has several obvious defects:

- The point cloud map is usually very large, so the pcd file size will be very large. An image of 640pixel×480pixel will generate 300,000 spatial points and require a lot of storage space. Even after some filtering, the pcd file size is still unacceptable in large-scale environments. And the annoying thing is that its huge size is not necessary for map usage. Point cloud maps provide a lot of unnecessary details. The folds on the carpet and the shadows in the wall, we do not particularly care about these things. Putting them on the map is a waste of space. In the navigation task, we only want to know if it is passable or not. Of course, reducing the resolution will save space, but it also decreases the map's quality. Is there any way to compress and store the map and discard some duplicate information?

- The point cloud map cannot handle moving objects. Our approach only adds points, and there is no mechanism like removing points when they disappear. In the real environment, the ubiquity of moving objects makes point cloud maps not practical enough.

We will introduce next is a flexible, compressed, and updateable map format: Octo-map [136]. I'm not doing an advertisement.

We know that it is common to model the 3D space as many small cubes (or voxels). If we cut each face of a small cube into two pieces on average, this small cube will become eight smaller pieces of the same size. This step can be repeated continuously until the final size reaches the highest accuracy of modeling. In this process, dividing a small square into eight of the same size is regarded as expanding from one node into eight child nodes, then the whole process of subdividing from the largest space to the smallest space is an octo-tree.

As shown in Fig. 11.11, the left side shows a large cube continuously divided into eight pieces evenly until it becomes the smallest one. Therefore, the entire large cube can be regarded as the root node, and the smallest block can be regarded as the leaf node. Therefore, when we move up one level in the octree, the map's volume can be expanded to eight times the original. We might as well do a simple calculation: if the size of the leaf node is $1\,\mathrm{cm}^3$, then when we limit the octree to 10 levels, the total volume that can be modeled is about $8^{10}\ \mathrm{cm}^3 = 1,073\,\mathrm{m}^3$, which is enough to model a room. The volume and depth have an exponential relationship. When we use deeper depth, the modeled volume will grow very fast.

Readers may be wondering that we also limit a voxel to only one point? Why do we say that the point cloud takes up space while the octree saves space? This is

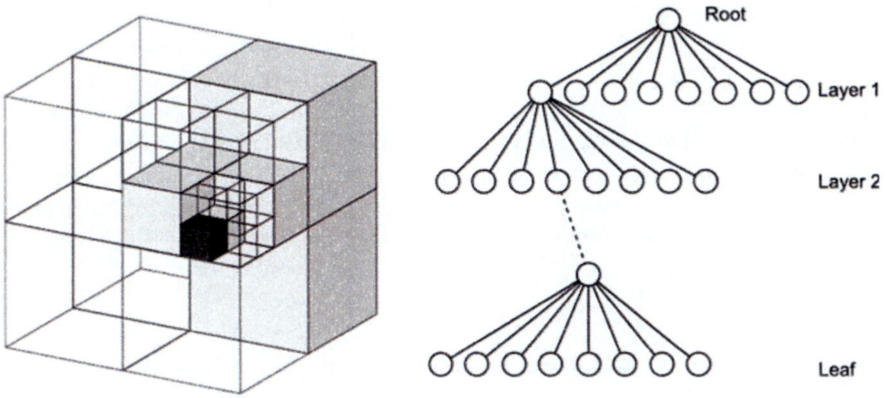

Fig. 11.11 The structure of an octo-tree

because, in the octree, we store information about whether it is occupied or not in the node. However, the difference is that when all the child nodes of a block are occupied or not, then we don't need to expand this node. For example, when the initial map is blank, we only need a root node instead of a complete tree. When adding information, since occupied objects and black spaces are often connected together, most octree nodes do not need to be expanded to the leaf level. So, octree saves a lot of storage space than point cloud.

As mentioned earlier, the nodes of the octree store information about whether it is occupied. From the point cloud level, we can naturally use 0 for blank and 1 for occupied. This 0–1 representation can be stored in one bit, saving space, but it seems a bit too simple. Due to noise's influence, we may see a certain point as 0 for a while and 1 for a while. Or 0 for most time and 1 for a small amount of time. In addition to the two cases of yes and no, there is an unknown state in navigation. We usually want to choose the probability to express whether a node is occupied. For example, use a floating-point number $x \in [0, 1]$ to express. This x takes 0.5 at the beginning to describe the unknown state. If we keep observing that it is occupied, we increase its value. On the contrary, if we keep observing that it is blank, we decrease it.

In this way, we dynamically model the obstacle information in the map. However, the current method has a small problem: if x is kept increasing or decreasing, it may go outside the range of [0, 1], causing inconvenience in processing. So we do not directly use the probability but use log-odds to describe it. Let $y \in \mathbb{R}$ be the logarithmic value and x be the probability of 0~1, then the transformation between them is described by the *logit* transformation:

$$y = \text{logit}(x) = \log\left(\frac{x}{1-x}\right). \tag{11.16}$$

The inverse transform is:

$$x = \text{logit}^{-1}(y) = \frac{\exp(y)}{\exp(y)+1}. \tag{11.17}$$

It can be seen that when y changes from $-\infty$ to $+\infty$, x changes from 0 to 1 accordingly. When y takes 0, x takes 0.5. Therefore, we might as well store y to express whether the node is occupied. When the occupation is continuously observed, let y increase by one value; otherwise, let y decrease. When querying the probability, use the inverse logit transformation to convert y to probability. In mathematical terms, suppose a certain node is n, and the observed data is z. Then the logarithm value of the probability of a node from the beginning to the moment t is $L(n|z_{1:t})$, and the time $t+1$ is:

$$L(n|z_{1:t+1}) = L(n|z_{1:t-1}) + L(n|z_t). \tag{11.18}$$

If written in probabilistic form instead of logarithmic form of probability, it would be a bit more complicated:

$$P(n|z_{1:T}) = \left[1 + \frac{1 - P(n|z_T)}{P(n|z_T)} \frac{1 - P(n|z_{1:T-1})}{P(n|z_{1:T-1})} \frac{P(n)}{1 - P(n)}\right]^{-1}. \tag{11.19}$$

With logarithmic probability, we can update the entire octree map based on RGB-D data. Suppose we observe a specific pixel with depth d in the RGB-D image, it means: (1) There observed point is occupied; (2) The line from the camera center to the observed point is free. With this information, the octree map can be updated, and dynamic objects can also be handled.

11.4.4 Practice: Octo-mapping

Let's demonstrate the process of octo-mapping through the program. Please install the octomap library first. After 18.04, octomap and the corresponding visualization tool octovis have been integrated into the apt library and can be installed by the following command:

Listing 11.8 Terminal input:

```
sudo apt-get install liboctomap-dev octovis
```

We will directly demonstrate how to generate an octree map from the previous five images and then draw it with octovis.

Listing 11.9 slambook/ch13/dense_RGBD/octomap_mapping.cpp (part)

```
// octomap tree
octomap::OcTree tree(0.01); // resolution=0.01

for (int i = 0; i < 5; i++) {
  cout << "Converting " << i + 1 << endl;
  cv::Mat color = colorImgs[i];
  cv::Mat depth = depthImgs[i];
  Eigen::Isometry3d T = poses[i];

  octomap::Pointcloud cloud; // the point cloud in octomap

  for (int v = 0; v < color.rows; v++)
  for (int u = 0; u < color.cols; u++) {
    unsigned int d = depth.ptr<unsigned short>(v)[u];
    if (d == 0) continue;
    Eigen::Vector3d point;
    point[2] = double(d) / depthScale;
    point[0] = (u - cx) * point[2] / fx;
    point[1] = (v - cy) * point[2] / fy;
    Eigen::Vector3d pointWorld = T * point;
    cloud.push_back(pointWorld[0], pointWorld[1], pointWorld[2]);
  }

  // save into octo tree
  tree.insertPointCloud(cloud, octomap::point3d(T(0, 3), T(1, 3), T(2, 3)))
      ;
}

// update and save into a bt file
tree.updateInnerOccupancy();
cout << "saving octomap ... " << endl;
tree.writeBinary("octomap.bt");
```

octomap (resolution=0.05) octomap (resolution=0.1)

Fig. 11.12 The display results of the octree map at different resolutions

We used octomap::OcTree to build the entire map. In fact, octomap provides many kinds of octrees: some with RGB maps and some with occupancy information. You can also define which variables each node needs to carry. For simplicity, we used the most basic octree map without color information.

A point cloud structure is provided inside Ocotmap. It is slightly simpler than the point cloud of PCL and only carries the point's spatial position information. According to the RGB-D image and camera pose information, we first transfer the coordinates of the point to the world coordinates, then put it into the point cloud of octomap, and finally give it to the octree map. After that, the octomap will update the internal occupation probability according to the projection information introduced before and finally save it as a compressed octree map. We save the generated map as *octomap.bt* file. Now, call the octovis to open the map file, and you can see the map.

Figure 11.12 shows the result of the map we built. Since we did not add color information to the map, it will be gray. Press the "1" key to color it according to the height information. Readers can explore the octovis interface by themselves, including map viewing, rotation, zooming, etc.

There is an octree depth limit bar on the right, where you can adjust the map's resolution. Since the default depth is 16 layers, the 16th layer is the highest resolution displayed here, which is blocks with 0.05 m. If we reduce the depth by one layer, the octree leaf nodes are raised by one layer, and the resolution doubles to 0.1 m. As you can see, we can easily adjust the map resolution to suit different occasions.

Octomap also has some places that can be explored. For example, we can easily query the occupation probability of any point to design a navigation method in the map [137]. Readers can also compare the file sizes of point cloud maps and octree maps. The disk file of the point cloud map generated in the previous section is about 6.9 MB, while the octomap is only 56 KB, which is less than one percent of the point cloud map, which can effectively model larger scenes.

11.5 *TSDF and RGB-D Fusion Series

At the end of this lecture, we introduce a research direction that is very similar to SLAM but slightly different: real-time 3D reconstruction. This section involves GPU programming and does not provide examples, so it is used as optional reading material.

In the previous map model, localization is regarded as the main body. The mesh or octomap is used as a post-processing step. This framework has become mainstream because the localization algorithm can meet the real-time requirements, and the processing of the map can be processed at the keyframe without real-time response. Localization is usually lightweight, especially when using sparse features or sparse direct methods; accordingly, the maps' expression and storage are heavyweight. The large-scale mapping and its computational requirements are not conducive to real-time processing. Dense maps can only be calculated at the keyframe level.

However, in the current practice, we have not optimized the dense map. For example, when the same chair is observed in two images, we only stitch the point clouds at the two locations based on the two images' poses to generate a map. Since pose estimation is usually error-prone, this direct stitching is often not accurate enough. For example, the point clouds of the same chair cannot be stitched perfectly. At this time, two observations of the same chair will appear on the map-this phenomenon is sometimes vividly called *ghost shadow*.

This phenomenon is obviously not what we want. We hope that the reconstruction result is smooth and complete. Under this kind of thinking, there has been a practice of taking the map as the main body and localization in a secondary position, which is the real-time 3D reconstruction. Since 3D reconstruction takes the reconstruction of an accurate map as the main goal, it usually needs GPU for acceleration. In contrast, SLAM is developing towards lightweight and miniaturization. Some solutions even abandon the mapping and loop detection part and only retain the visual odometry. The real-time reconstruction is developing towards the rebuilding of large-scale dynamic scenes.

Since the emergence of RGB-D sensors, real-time reconstruction using RGB-D images has formed an important development direction, such as Kinect Fusion [138], Dynamic Fusion [139], Elastic Fusion [140], Fusion4D [141], Volumn Deform [142] and other achievements. Among them, Kinect Fusion has completed the basic model reconstruction, but it is limited to small scenes; the follow-up work is to expand it to large, dynamic, and even deformed scenes. We regard them as real-time reconstruction work, but it is impossible to discuss each's working principles in detail due to the large variety. Figure 11.13 shows part of the reconstruction results. You can see that these modeling results are very fine, much more delicate than simple point clouds.

We will introduce the classic TSDF map as a representative. TSDF is the abbreviation of Truncated Signed Distance Function. Although it seems inappropriate to call a function a map, we will temporarily call it a TSDF map, TSDF reconstruction, etc., as long as there is no deviation in understanding.

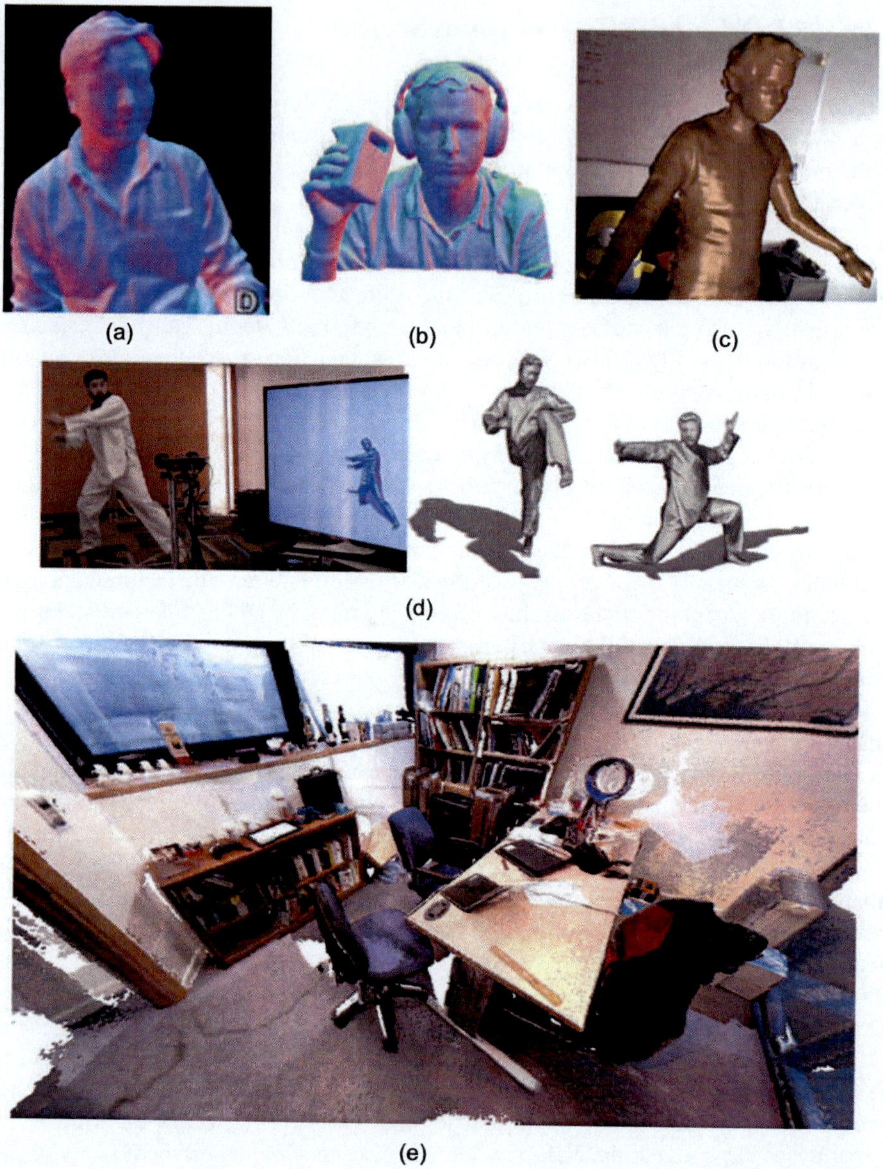

Fig. 11.13 Reconstruction by the RGB-D fusions: **a** Kinect Fusion; **b** Dynamic Fusion; **c** Volumn Deform; **d** Fusion4D; **e** Elastic Fusion

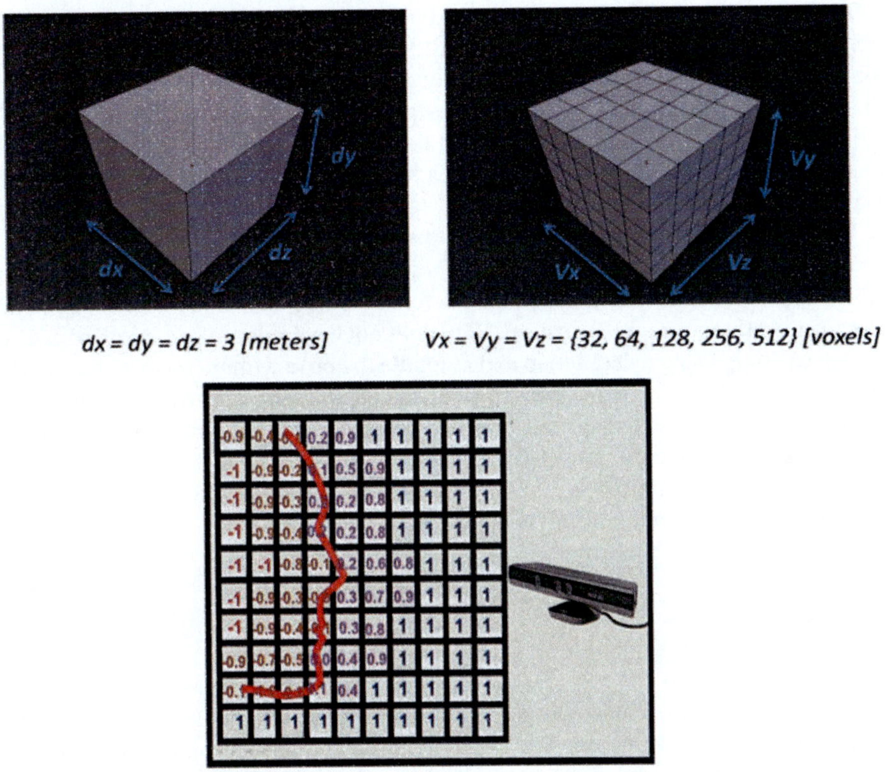

The truncated distance value formed when the camera observes the surface of the object

Fig. 11.14 Truncation Signed Distance Function

Similar to the octree, the TSDF map is also a grid format (or square) map, as shown in Fig. 11.14. First, we select the three-dimensional space to be modeled, such as $3 \times 3 \times 3\,\mathrm{m}^3$, divide this space into many small blocks according to a certain resolution, and store the information inside each small block. The difference is that the entire TSDF map is stored in GPU memory instead of CPU memory. Using the GPU's parallel feature, we can compute and update each voxel in parallel instead of having to serialize as the CPU traverses the memory area.

In each TSDF voxel, the distance between the small block and the closest object's surface is stored. If the block is in front of the object's surface, it has a positive value; conversely, if it is behind the surface, it has a negative value. Since the object's surface is usually a thin layer, the block values are taken as 1 and -1 if they are too small or too large. The distance after truncation is obtained in this way, which is the so-called TSDF. So by definition, the place where TSDF is 0 is the surface itself–or, due to the existence of numerical errors, the place where TSDF changes from negative to positive is the surface itself. In the lower part of Fig. 11.14, we see a surface similar to a human face appearing where the TSDF changes sign.

TSDF also has two localization and map building problems, which are very similar to SLAM, but the specific form is slightly different from the previous lectures in this book. Here, the localization problem mainly refers to comparing the current RGB-D image with the TSDF map in the GPU to estimate the camera pose. The problem of mapping is how to update the TSDF map based on the estimated camera pose. In the traditional approach, we also perform a bilateral Bayesian filter on the RGB-D image to remove the depth map's noise.

The localization of TSDF is similar to the ICP described earlier. Due to GPU's parallel computation, we can perform ICP calculation on the entire depth map and TSDF map without having to calculate the feature points first like traditional visual odometry. At the same time, because TSDF does not have color information, it means that we can only use the depth map and compute the pose without using color maps, which to some extent get rid of the dependence of texture or illumination conditions. The RGB-D reconstruction is more robust.[4] On the other hand, the mapping part is also a process of updating the values in the TSDF in parallel, making the estimated surface smoother and more reliable. Since we do not introduce GPU-related content too much, the specific method will not be elaborated. Please refer to relevant literature for details.

11.6 Summary

This lecture introduces some common map types, especially dense map forms. We see that dense maps can be constructed based on monocular or binocular cameras, while RGB-D sensors are often easier and more stable. The map in this lecture focuses on the measurement map, and the topological map form is quite different from the SLAM research, so it is not discussed in detail.

Exercises

1. Prove (11.6).
2. Change the dense depth estimation in this lecture to semi-dense. You can first filter out the places with obvious gradients
3. *Change the monocular dense reconstruction code demonstrated in this lecture from positive depth to inverse depth, and add affine transformation. Does your experiment improve the results?
4. Can you demonstrate how to navigate or plan a path in an octree?
5. Read [138] to discuss how the TSDF map performs pose estimation and update. What are the similarities and differences between it and the localization mapping algorithm we talked about before?
6. *Study the principle and realization of uniform-Gaussian mixture filter.

[4] But having said that, it is more dependent on the depth map in this way.

Chapter 12
Practice: Stereo Visual Odometry

Goal of Study

1. Implement a stereo visual SLAM from scratch.
2. Understand the problems that are prone to occur in VO and how to fix them.

This lecture is the concluding part of the book. We will use the knowledge we learned before to actually write a visual odometry program. You will manage local robot trajectories and landmarks and experience how a software framework is composed. During the operation, we will encounter many practical problems: how to continuously track the image, control the scale of BA, and so on. To make the program run stably, we need to deal with the above situations, which will bring about many useful discussions on engineering realization.

12.1 Why Do We Have a Separate Engineering Chapter?

Knowing the principles of bricks and cement does not mean that you can build a grand palace.

In the Minecraft game, all the players have are some blocks with different colors and textures. All the player needs to do is place these blocks on a plane and stack them together. Understanding a block is also extremely simple, but most beginners can only make simple matchbox houses when really starting to build the architec-

X. Gao and T. Zhang, *Introduction to Visual SLAM*,
https://doi.org/10.1007/978-981-16-4939-4_12

Fig. 12.1 Great work normally starts from simple things

tures. Experienced and creative players can use these simple blocks to build houses, gardens, terraces, and pavilions, even the big cities (Fig. 12.1).[1]

In SLAM, we believe that engineering implementation and understanding of algorithm principles should be at least the same important, or even more, as the algorithm principles. The algorithms are like blocks in the building. We can discuss their properties clearly, but just understanding the basic units will not enable you to build an entire building. They require a lot of trials, time, and experience. We encourage readers to work in a more practical direction. Of course, this is often very complicated. Just like in Minecraft, you need to master the structure of various columns, walls, and roofs, wall carvings, and calculation of geometric angles. There are far more contents than discussing the nature of each block.

The same is true for the specific implementation of SLAM. A practical program has a lot of engineering design and skills (tricks), and it is necessary to discuss how to deal with problems after each step. In principle, each SLAM implementation is different. Most of the time, we cannot say which implementation method is the best. However, we usually encounter some common problems like managing map points, dealing with mismatches, selecting keyframes, and so on. We hope that readers can have some intuitive feelings about these possible problems. We think this feeling is vital.

Therefore, out of the emphasis on practice, in this lecture, we will lead the readers through the process of building a SLAM framework. Just like the architecture, we have to discuss trivial but essential issues such as column spacing and facade aspect ratio. SLAM engineering is complicated. Even if we only keep the core part, it will take up a lot of space and make this book too verbose. However, please note

[1] The lower left is my practice work. The bottom right is from the work of the Epicwork team: "Old Summer Palace." I used to study in Epicwork for a while. The creativity of the young people and even the children there left a deep impression on me.

that although the final completed project may be complicated, the process of being simple to complex is worthy of detailed discussion. Therefore, we have to start from a simple data structure first, make simple visual odometry, and then slowly add some additional functions.

The code for this lecture is in *slambook2/ch13*. We will implement a simplified version of stereo VO and then see its running effect in the Kitti dataset. This VO consists of a frontend of optical flow tracking and a backend of sliding window BA. Why choose stereo VO? One reason is that the stereo vision is relatively simple to implement, and initialization can be completed in a single frame. The second is that the stereo camera has 3D observation, and the realization effect will be better than that of the monocular.

12.2 Framework

We are discussing a project realization, and the project usually has the concept of a framework. Most Linux libraries will classify and store algorithm code files according to modules. For example, the header files will be placed in the header file directory, and the source code will be placed in the source code directory. Also, there may be configuration files, test files, third-party libraries, and so on. Now we come to classify our files according to the common practice of small algorithm libraries:

1. *bin* stores the compiled binary file;
2. *include/myslam* stores the header files of the SLAM module.
3. *src* stores the source code files, mainly .cpp files.
4. *test* stores the files used for testing, which are also .cpp files.
5. *config* stores the configuration files.
6. *cmake_modules* saves the cmake files of third-party libraries, which are used by libraries such as *g2o*.

In this way, we have determined the location of the code file. Now let's discuss the basic data structure involved in VO.

12.2.1 Data Structure

Before writing code, we should clarify what we want to write. A long time ago, there was a classic sentence that *a program is data structure plus algorithm*. For VO, we have to ask: What kind of data does VO need to process? What are the critical algorithms involved? What is the relationship between them?

- The most basic unit we deal with is the image. In stereo VO, that is a pair of images. We might as well call it a *frame*.
- We will detect visual *features* on the frame. These features are many 2D pixels.

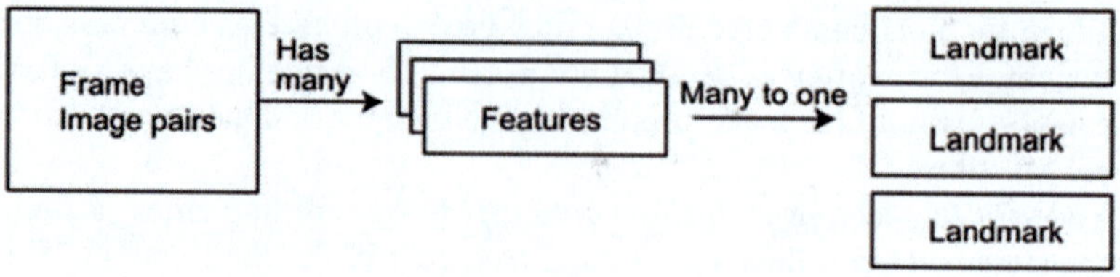

Fig. 12.2 Basic data structure and their relationships

- We look for the association of features between images. If we see a feature multiple times, we use the triangulation method to calculate its 3D position, which forms the *landmarks or map points*.

Obviously, *images*, *features*, and *landmarks* are the most basic elements in our system. The relationship between them is shown in Fig. 12.2. In the following description, we use terms landmarks and map points for points in 3D space, and their semantics are the same.

12.2.2 Pipeline

Next, we have to ask, which algorithms are responsible for feature extraction, which algorithms are accountable for triangulation, and which algorithms to deal with optimization problems? According to the previous contents of this book, the SLAM system consists of front and backends. The frontend is responsible for calculating the feature matching of adjacent images, and the backend is responsible for optimizing the entire problem. In a typical implementation, they should run in separate threads. The frontend ensures real-time performance, and the backend optimizes keyframes to ensure good results. So overall, our program has two essential modules:

- **Frontend**. We get an image frame from the sensor, and the frontend is responsible for extracting the features in the image. It then performs optical flow tracking or feature matching with the previous frame and calculates the frame's position based on the optical flow result. If necessary, new feature points should be added and triangulated. The result of the frontend processing will be used as the initial value of the backend optimization.
- **Backend**. The backend is a slower thread. It gets the processed keyframes and landmark points, optimizes them, and then returns the optimized results. The backend should control the optimization problem's scale within a certain range and cannot keep growing over time.

We can determine the framework of the entire algorithm through this analysis and then draw it in a pipeline diagram, such as Fig. 12.3. We put a map module between the front and backends to handle the data flow between them. Since the front and

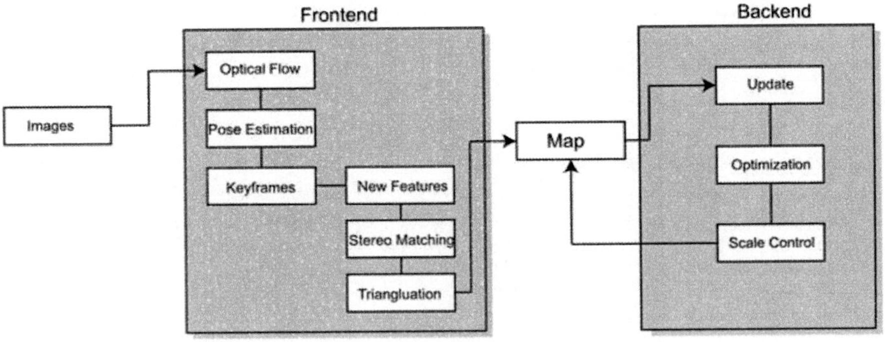

Fig. 12.3 Processing pipeline

backends process data in separate threads, the interaction process should be: (1) The frontend adds new data to the map after finding a new keyframe. (2) When the backend detects that the map has new data, it runs an optimization routine and then resets the map scale. The old keyframes and map points are removed if necessary.

In this way, we have determined the general system flow, which will help the subsequent coding realization. Of course, in addition to the core algorithm, we also need some small peripheral modules to make the system more convenient, such as:

- We should have a camera class to manage the intrinsic and extrinsics as well as the projection functions.
- We need a configuration file management class to facilitate reading content from configuration files. Some critical parameters can be store in the configuration file for quick debugging;
- Because the algorithm runs on the Kitti dataset, we need to read the image data according to Kitti's storage format, which should also be handled by a separate class.
- We need a visualization module to observe the running status of the system. Otherwise, we have to scratch our heads against a series of numeric values.

Although these modules are not the core parts, they are indispensable for implementation. Due to space limitations, we leave the peripheral code to readers.

12.3 Implementation

12.3.1 Implement the Basic Data Structure

Let's first implement the frame, feature, and landmark classes. It is usually recommended to set them as structs without defining complex private variables and interfaces. Considering that these data may be accessed and modified by multiple threads, we need to set thread locks in those parts.

The frame struct is:

Listing 12.1 slambook2/ch13/include/myslam/frame.h

```
struct Frame {
public:
  EIGEN_MAKE_ALIGNED_OPERATOR_NEW;
  typedef std::shared_ptr<Frame> Ptr;

  unsigned long id_ = 0; // id of this frame
  unsigned long keyframe_id_ = 0; // id of keyframe
  bool is_keyframe_ = false; // keyframe?
  double time_stamp_; // timestamp, not used in Kitti
  SE3 pose_;      // pose defined as Tcw
  std::mutex pose_mutex_; // Pose data mutex
  cv::Mat left_img_, right_img_; // stereo images

  // extracted features in left image
  std::vector<std::shared_ptr<Feature>> features_left_;
  // corresponding features in right image, set to nullptr if no
       corresponding
  std::vector<std::shared_ptr<Feature>> features_right_;

public: // data members
  Frame() {}

  Frame(long id, double time_stamp, const SE3 &pose, const Mat &left,
    const Mat &right);

  // set and get pose, thread safe
  SE3 Pose() {
    std::unique_lock<std::mutex> lck(pose_mutex_);
    return pose_;
  }

  void SetPose(const SE3 &pose) {
    std::unique_lock<std::mutex> lck(pose_mutex_);
    pose_ = pose;
  }

  /// Set keyframe and keyframe_id
  void SetKeyFrame();

  /// create new frame and allocate id
  static std::shared_ptr<Frame> CreateFrame();
};
```

We define the frame struct to contain id, pose, image, and features in the left and right images. Among them, the pose will be set or accessed by the front and backends simultaneously, so we define its set and get functions and lock the data in them. Meanwhile, a frame can be constructed by a static function, and we can automatically set its id in the static create function.

Next, the feature struct:

Listing 12.2 slambook2/ch13/include/myslam/feature.h

```
struct Feature {
  EIGEN_MAKE_ALIGNED_OPERATOR_NEW;
  typedef std::shared_ptr<Feature> Ptr;

  std::weak_ptr<Frame> frame_; // the frame that takes this feature
  cv::KeyPoint position_; // 2D pixel position
  std::weak_ptr<MapPoint> map_point_; // assigned map point

```

```
  bool is_outlier_ = false; // is outlier?
  bool is_on_left_image_ = true; // is detected on the left image?

  Feature() {}

  Feature(std::shared_ptr<Frame> frame, const cv::KeyPoint &kp)
    : frame_(frame), position_(kp) {}
};
```

The feature's main information is its 2D position and several flags describing whether it is an abnormal point and whether it is extracted in the left camera. We can access the host frame and its corresponding map point through a feature object. However, the real ownership of frame and map point objects belongs to the map. In order to avoid the circular reference generated by shared_ptr, the weak_ptr is used here.[2]

Finally the map point, or the landmark:

Listing 12.3 slambook2/ch13/include/myslam/mappoint.h

```
struct MapPoint {
  EIGEN_MAKE_ALIGNED_OPERATOR_NEW;
  typedef std::shared_ptr<MapPoint> Ptr;
  unsigned long id_ = 0; // ID
  bool is_outlier_ = false;
  Vec3 pos_ = Vec3::Zero(); // Position in world
  std::mutex data_mutex_;
  int observed_times_ = 0; // being observed by feature matching algo.
  std::list<std::weak_ptr<Feature>> observations_;

  MapPoint() {}

  MapPoint(long id, Vec3 position);

  Vec3 Pos() {
    std::unique_lock<std::mutex> lck(data_mutex_);
    return pos_;
  }

  void SetPos(const Vec3 &pos) {
    std::unique_lock<std::mutex> lck(data_mutex_);
    pos_ = pos;
  };

  void AddObservation(std::shared_ptr<Feature> feature) {
    std::unique_lock<std::mutex> lck(data_mutex_);
    observations_.push_back(feature);
    observed_times_++;
  }

  void RemoveObservation(std::shared_ptr<Feature> feat);

  std::list<std::weak_ptr<Feature>> GetObs() {
    std::unique_lock<std::mutex> lck(data_mutex_);
    return observations_;
  }

  // factory function
  static MapPoint::Ptr CreateNewMappoint();
};
```

[2] In short, the frame holds the shared_ptr of feature, so we should avoid the feature holding frame's shared_ptr again. Otherwise, the two structs refer to each other, which will cause the smart pointer to fail to be automatically destructed.

The most important thing about MapPoint is its 3D position, which is the pos_ variable, also needs to be locked. Its observation_ variable records the features that observed this map point. Because the feature may be judged as an outlier, it needs to be locked when the observation part is changed.

So far, we have realized the basic data structure. In the framework, we let the map class actually hold these frame and map point objects, so we also need to define a map class:

Listing 12.4 slambook2/ch13/include/myslam/map.h

```cpp
class Map {
public:
  EIGEN_MAKE_ALIGNED_OPERATOR_NEW;
  typedef std::shared_ptr<Map> Ptr;
  typedef std::unordered_map<unsigned long, MapPoint::Ptr> LandmarksType;
  typedef std::unordered_map<unsigned long, Frame::Ptr> KeyframesType;

  Map() {}

  void InsertKeyFrame(Frame::Ptr frame);

  void InsertMapPoint(MapPoint::Ptr map_point);

  LandmarksType GetAllMapPoints() {
    std::unique_lock<std::mutex> lck(data_mutex_);
    return landmarks_;
  }

  KeyframesType GetAllKeyFrames() {
    std::unique_lock<std::mutex> lck(data_mutex_);
    return keyframes_;
  }

  LandmarksType GetActiveMapPoints() {
    std::unique_lock<std::mutex> lck(data_mutex_);
    return active_landmarks_;
  }

  KeyframesType GetActiveKeyFrames() {
    std::unique_lock<std::mutex> lck(data_mutex_);
    return active_keyframes_;
  }

  void CleanMap();

private:

  void RemoveOldKeyframe();

  std::mutex data_mutex_;
  LandmarksType landmarks_; // all landmarks
  LandmarksType active_landmarks_; // active landmarks
  KeyframesType keyframes_; // all keyframes
  KeyframesType active_keyframes_; // active keyframes

  Frame::Ptr current_frame_ = nullptr;

  // settings
  int num_active_keyframes_ = 7;
};
```

The map stores all keyframes and corresponding landmarks in a hash form and maintains an activated keyframe and map point set. Here the concept of activation is what we call the sliding window before. The backend will optimize the activated keyframes and landmark points and fix the rest to control the optimization scale. Of course, the activation strategy is defined by ourselves. The simple activation strategy is to remove the oldest keyframe and keep the latest keyframes in time. We only keep the latest 7 keyframes in this demo.

12.3.2 Implement the Frontend

After defining the basic data structure, let's consider the frontend functions. The frontend needs to determine the frame's pose based on the binocular image, but there are many alternative implementation choices. For example, how should we use the right eye's image? Do we track the feature in both left and right eyes or just one of them? When computing triangulation, should we consider the left and right images in the same frame or chronological order? In fact, any two images can be triangulated (for example, the left image of the previous frame vs. the right image of the next frame), so everyone's realization will be different.

For simplicity, we first determine the frontend processing logic:

1. The frontend has three states: *initialization*, *normal tracking*, and *tracking lost*.
2. In the initialization state, we do the triangulation according to the optical flow matching between the left and right eyes. We will establish the initial map when successful.
3. In the tracking phase, the front end calculates the optical flow from the previous frame to the current frame and estimates the image pose based on the optical flow result. This optical flow is used only for the left eye image to save the computation resource.
4. If the tracked features are fewer than a threshold, we set the current frame as a keyframe. For keyframes, do the following things:

   - Extract new feature points;
   - Find the corresponding points of these points on the right, and use triangulation to create new landmarks;
   - Add new keyframes and landmarks to the map and trigger a backend optimization.
   - If the tracking is lost, reset the frontend system and reinitialize it.

According to this logic, the frontend processing flow is written as follows:

Listing 12.5 slambook2/ch13/src/frontend.cpp

```
bool Frontend::AddFrame(myslam::Frame::Ptr frame) {
  current_frame_ = frame;
  switch (status_) {
    case FrontendStatus::INITING:
```

```
      StereoInit();
      break;
    case FrontendStatus::TRACKING_GOOD:
    case FrontendStatus::TRACKING_BAD:
      Track();
      break;
    case FrontendStatus::LOST:
      Reset();
      break;
  }

  last_frame_ = current_frame_;
  return true;
}
```

The *Track()* is implemented as:

Listing 12.6 slambook2/ch13/src/frontend.cpp

```
bool Frontend::Track() {
  if (last_frame_) {
    current_frame_->SetPose(relative_motion_ * last_frame_->Pose());
  }

  int num_track_last = TrackLastFrame();
  tracking_inliers_ = EstimateCurrentPose();

  if (tracking_inliers_ > num_features_tracking_) {
    // tracking good
    status_ = FrontendStatus::TRACKING_GOOD;
  } else if (tracking_inliers_ > num_features_tracking_bad_) {
    // tracking bad
    status_ = FrontendStatus::TRACKING_BAD;
  } else {
    // lost
    status_ = FrontendStatus::LOST;
  }

  InsertKeyframe();
  relative_motion_ = current_frame_->Pose() * last_frame_->Pose().inverse()
      ;

  if (viewer_) viewer_->AddCurrentFrame(current_frame_);
  return true;
}
```

In the *TrackLastFrame()*, we call the optical flow of OpenCV to track the feature points

Listing 12.7 slambook2/ch13/src/frontend.cpp

```
int Frontend::TrackLastFrame() {
  // use LK flow to estimate points in the right image
  std::vector<cv::Point2f> kps_last, kps_current;
  for (auto &kp : last_frame_->features_left_) {
    if (kp->map_point_.lock()) {
      // use project point
      auto mp = kp->map_point_.lock();
      auto px =
      camera_left_->world2pixel(mp->pos_, current_frame_->Pose());
      kps_last.push_back(kp->position_.pt);
      kps_current.push_back(cv::Point2f(px[0], px[1]));
    } else {
      kps_last.push_back(kp->position_.pt);
      kps_current.push_back(kp->position_.pt);
```

```
    }
  }

  std::vector<uchar> status;
  Mat error;
  cv::calcOpticalFlowPyrLK(
  last_frame_->left_img_, current_frame_->left_img_, kps_last,
  kps_current, status, error, cv::Size(21, 21), 3,
  cv::TermCriteria(cv::TermCriteria::COUNT + cv::TermCriteria::EPS, 30,
       0.01),
  cv::OPTFLOW_USE_INITIAL_FLOW);

  int num_good_pts = 0;

  for (size_t i = 0; i < status.size(); ++i) {
    if (status[i]) {
      cv::KeyPoint kp(kps_current[i], 7);
      Feature::Ptr feature(new Feature(current_frame_, kp));
      feature->map_point_ = last_frame_->features_left_[i]->map_point_;
      current_frame_->features_left_.push_back(feature);
      num_good_pts++;
    }
  }

  LOG(INFO) << "Find " << num_good_pts << " in the last image.";
  return num_good_pts;
}
```

In the implementation, we often split the complex functions into some short functions until the underlying functions call OpenCV or *g2o* to achieve specific calculations. The readability and reusability of the program can be improved in this way. For example, the same feature extraction function can be used for both the initialization phase and the keyframe's new feature detection part. We recommend that readers go through this code by themselves (the frontend is less than 400 lines).

12.3.3 Implement the Backend

Compared with the frontend, the logic of the backend implementation will be more complicated. The overall backend implementation is as follows:

Listing 12.8 slambook2/ch13/include/myslam/backend.h

```
class Backend {
public:
  EIGEN_MAKE_ALIGNED_OPERATOR_NEW;
  typedef std::shared_ptr<Backend> Ptr;

  /// Start the backend thread in the constructor
  Backend();

  // Set cameras and fetch the params
  void SetCameras(Camera::Ptr left, Camera::Ptr right) {
    cam_left_ = left;
    cam_right_ = right;
  }

  void SetMap(std::shared_ptr<Map> map) { map_ = map; }

```

```
  /// optimize and update the map
  void UpdateMap();

  /// stop the backend
  void Stop();

private:
  /// backend thread
  void BackendLoop();

  /// optimize the sliding window
  void Optimize(Map::KeyframesType& keyframes, Map::LandmarksType&
      landmarks);

  std::shared_ptr<Map> map_;
  std::thread backend_thread_;
  std::mutex data_mutex_;

  std::condition_variable map_update_;
  std::atomic<bool> backend_running_;

  Camera::Ptr cam_left_ = nullptr, cam_right_ = nullptr;
};
```

After the backend is started, it will wait for the condition variable of map_update_. When the map update is triggered, take the activated keyframes and map points from the map and perform optimization:

Listing 12.9 slambook2/ch13/src/backend.cpp

```
void Backend::BackendLoop() {
  while (backend_running_.load()) {
    std::unique_lock<std::mutex> lock(data_mutex_);
    map_update_.wait(lock);

    /// optimize the active frames and landmarks
    Map::KeyframesType active_kfs = map_->GetActiveKeyFrames();
    Map::LandmarksType active_landmarks = map_->GetActiveMapPoints();
    Optimize(active_kfs, active_landmarks);
  }
}
```

The optimization function is similar to the BA we used before, except that the data needs to be obtained from the frame and map point objects:

Listing 12.10 slambook2/ch13/src/backend.cpp

```
void Backend::Optimize(Map::KeyframesType &keyframes,
Map::LandmarksType &landmarks) {
  // setup g2o
  typedef g2o::BlockSolver_6_3 BlockSolverType;
  typedef g2o::LinearSolverCSparse<BlockSolverType::PoseMatrixType>
  LinearSolverType;
  auto solver = new g2o::OptimizationAlgorithmLevenberg(
  g2o::make_unique<BlockSolverType>(
  g2o::make_unique<LinearSolverType>()));
  g2o::SparseOptimizer optimizer;
  optimizer.setAlgorithm(solver);

  // pose vertices, use keyframe id
  std::map<unsigned long, VertexPose *> vertices;
  unsigned long max_kf_id = 0;
  for (auto &keyframe : keyframes) {
    auto kf = keyframe.second;
```

```
    VertexPose *vertex_pose = new VertexPose(); // camera vertex_pose
    vertex_pose->setId(kf->keyframe_id_);
    vertex_pose->setEstimate(kf->Pose());
    optimizer.addVertex(vertex_pose);
    if (kf->keyframe_id_ > max_kf_id) {
      max_kf_id = kf->keyframe_id_;
    }

    vertices.insert({kf->keyframe_id_, vertex_pose});
  }

  // landmark vertices, use landmark id
  std::map<unsigned long, VertexXYZ *> vertices_landmarks;

  // K and extrinsics
  Mat33 K = cam_left_->K();
  SE3 left_ext = cam_left_->pose();
  SE3 right_ext = cam_right_->pose();

  // edges
  int index = 1;
  double chi2_th = 5.991; // robust kernel th
  std::map<EdgeProjection *, Feature::Ptr> edges_and_features;

  for (auto &landmark : landmarks) {
    if (landmark.second->is_outlier_) continue;
    unsigned long landmark_id = landmark.second->id_;
    auto observations = landmark.second->GetObs();
    for (auto &obs : observations) {
      if (obs.lock() == nullptr) continue;
      auto feat = obs.lock();
      if (feat->is_outlier_ || feat->frame_.lock() == nullptr) continue;

      auto frame = feat->frame_.lock();
      EdgeProjection *edge = nullptr;
      if (feat->is_on_left_image_) {
        edge = new EdgeProjection(K, left_ext);
      } else {
        edge = new EdgeProjection(K, right_ext);
      }

      // add landmark verices if not contained
      if (vertices_landmarks.find(landmark_id) ==
        vertices_landmarks.end()) {
        VertexXYZ *v = new VertexXYZ;
        v->setEstimate(landmark.second->Pos());
        v->setId(landmark_id + max_kf_id + 1);
        v->setMarginalized(true);
        vertices_landmarks.insert({landmark_id, v});
        optimizer.addVertex(v);
      }

      edge->setId(index);
      edge->setVertex(0, vertices.at(frame->keyframe_id_)); // pose
      edge->setVertex(1, vertices_landmarks.at(landmark_id)); // landmark
      edge->setMeasurement(toVec2(feat->position_.pt));
      edge->setInformation(Mat22::Identity());
      auto rk = new g2o::RobustKernelHuber();
      rk->setDelta(chi2_th);
      edge->setRobustKernel(rk);
      edges_and_features.insert({edge, feat});

      optimizer.addEdge(edge);

      index++;
    }
  }
```

```

  // do optimization and eliminate the outliers
  optimizer.initializeOptimization();
  optimizer.optimize(10);

  int cnt_outlier = 0, cnt_inlier = 0;
  int iteration = 0;
  while (iteration < 5) {
    cnt_outlier = 0;
    cnt_inlier = 0;
    // determine if we want to adjust the outlier threshold
    for (auto &ef : edges_and_features) {
      if (ef.first->chi2() > chi2_th) {
        cnt_outlier++;
      } else {
        cnt_inlier++;
      }
    }
    double inlier_ratio = cnt_inlier / double(cnt_inlier + cnt_outlier);
    if (inlier_ratio > 0.5) {
      break;
    } else {
      chi2_th *= 2;
      iteration++;
    }
  }

  for (auto &ef : edges_and_features) {
    if (ef.first->chi2() > chi2_th) {
      ef.second->is_outlier_ = true;
      // remove the observation
      ef.second->map_point_.lock()->RemoveObservation(ef.second);
    } else {
      ef.second->is_outlier_ = false;
    }
  }

  LOG(INFO) << "Outlier/Inlier in optimization: " << cnt_outlier << "/"
    << cnt_inlier;

  // Set pose and lanrmark position
  for (auto &v : vertices) {
    keyframes.at(v.first)->SetPose(v.second->estimate());
  }
  for (auto &v : vertices_landmarks) {
    landmarks.at(v.first)->SetPos(v.second->estimate());
  }
}
```

The back end is shorter than the front end, with less than 200 lines.

12.4 Experiment Results

Finally, we look at the running effect of this visual odometry. First, we need to download the Kitti dataset from http://www.cvlibs.net/datasets/kitti/eval_odometry.php. Its odometry data is about 22GB. After downloading, we decompress it and get several video segments. Let's take segment 0 as an example. After compiling the program in this section, fill in the path corresponding to the configuration file config/default.yaml. In my machine, it looks like this:

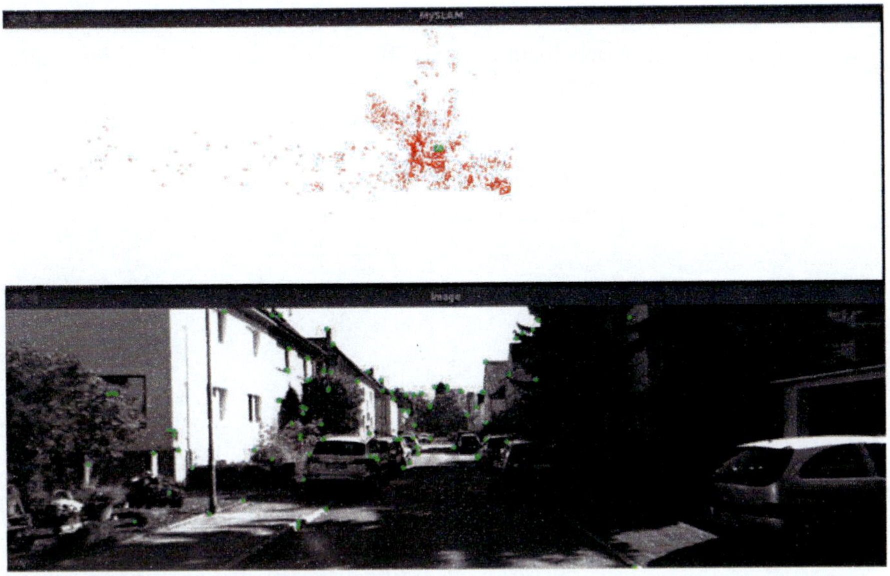

Fig. 12.4 Snapshot of the visual odometry

dataset_dir: /media/xiang/Data/Dataset/Kitti/dataset/sequences/00.

Then we run it with:

Listing 12.11 Terminal input:

```
bin/run_kitti_stereo
```

Then you can see the SLAM output, as shown in Fig. 12.4. During operation, the program will display the activated keyframes and maps, which should continue to grow and disappear as the camera moves.

We print the run time it takes for a single frame. When dealing with non-keyframes, the time is about 16 ms. When processing keyframes, the time-consuming will increase due to the additional steps of extracting features and the right eye matching. Moreover, since our map currently stores all keyframes and map points, it will cause memory growth after running for a while. If readers do not need all maps, they can keep only the active part.

Exercises

1. Do you understand all the C++ techniques used in this book? If there is something you are not familiar with, use search engines to learn the related knowledge, including range-based for loops, smart pointers, singleton patterns in design patterns, threads and locks, and so on.

2. Consider optimizing the system introduced in this lecture. For example, use a faster method of extracting feature points (this demo uses GFTT, which is not fast), use a 1-d search instead of 2d optical flow when matching left and right images, and use the direct method to estimate the pose and features at the same time, etc.
3. *Add a loop detection thread to the demo, and use a pose graph to optimize when a loop is detected to eliminate accumulated errors.

Chapter 13
Discussions and Outlook

> **Goal of Study**
> 1. Understand the classic SLAM implementation scheme.
> 2. Through experiments, compare the similarities and differences of various SLAM solutions.
> 3. Explore the future development direction of SLAM.

The previous chapters introduce the widely used SLAM pipelines, which are the conclusion of decades of researches. At present, in addition to the theoretical frameworks, we can also find many excellent open-source SLAM solutions. However, since most of their implementations are more complicated and not suitable for beginners' hands-on materials, we put them at the end of the book to introduce them. I believe that readers should be able to understand the basic principles by reading the previous content.

13.1 Open-Source Implementations

This lecture is a summary of the book. We will take readers to see how far the existing SLAM solutions have gone. In particular, we focus on solutions that provide open-source implementations. In the early days, it is not easy to see open-source solutions. Usually, the theories introduced in the research paper only account for 20% of the content, and the other 80% are in the code, which is not mentioned. These researchers' selfless dedication has promoted the entire SLAM community's rapid advancement and enabled subsequent researchers to have a higher starting point. Before we start to do SLAM, we should have an in-depth understanding of similar solutions and then conduct our own research.

X. Gao and T. Zhang, *Introduction to Visual SLAM*,
https://doi.org/10.1007/978-981-16-4939-4_13

Table 13.1 Open source SLAM systems

Name	Sensors*	URL
MonoSLAM	M	https://github.com/hanmekim/SceneLib2
PTAM	M	http://www.robots.ox.ac.uk/~gk/PTAM/
ORB-SLAM	M/S/R	http://webdiis.unizar.es/~raulmur/orbslam/
LSD-SLAM	M	http://vision.in.tum.de/research/vslam/lsdslam
SVO	M	https://github.com/uzh-rpg/rpg_svo
DTAM	R	https://github.com/anuranbaka/OpenDTAM
DVO	R	https://github.com/tum-vision/dvo_slam
DSO	M	https://github.com/JakobEngel/dso
VINS	M+IMU	https://github.com/HKUST-Aerial-Robotics/VINS-Mono
RTAB-MAP	S/R	https://github.com/introlab/rtabmap
RGBD-SLAM-V2	R	https://github.com/felixendres/rgbdslam_v2
Elastic fusion	R	https://github.com/mp3guy/ElasticFusion
Hector SLAM	L	http://wiki.ros.org/hector_slam
GMapping	L	http://wiki.ros.org/gmapping
OKVIS	M/S+IMU	https://github.com/ethz-asl/okvis
ROVIO	M+IMU	https://github.com/ethz-asl/rovio

*M=Monocular, S=Stereo, R=RGB-D, L=Lidar

The first half of this lecture will lead the readers to take a tour of the current visual SLAM program and comment on its historical status, advantages and disadvantages. Table 13.1 lists some popular open-source SLAM solutions. Readers can choose the interesting solutions for research and experiments.[1] We only selected a part of the representative ones due to space limitations, which is certainly not complete. We will explore some possible future development directions in the second half and provide some current research results.

13.1.1 MonoSLAM

When talking about visual SLAM, the first thing that many researchers think of is A. J. Davison's monocular SLAM [1, 55]. Professor Davison is a pioneer in this area. He proposed MonoSLAM in 2007 as the first real-time monocular visual SLAM system [1], which is considered the birthplace of many works. MonoSLAM uses the extended Kalman filter as the backend to track very sparse feature points on the frontend. Since EKF occupies a prominent dominant position in early SLAM, MonoSLAM is also based on EKF, using the current state of the camera and all landmark points as the state quantity, and updating the mean and covariance of the states.

[1] The book was basically written in 2016, so I did not list the newer systems after 2016.

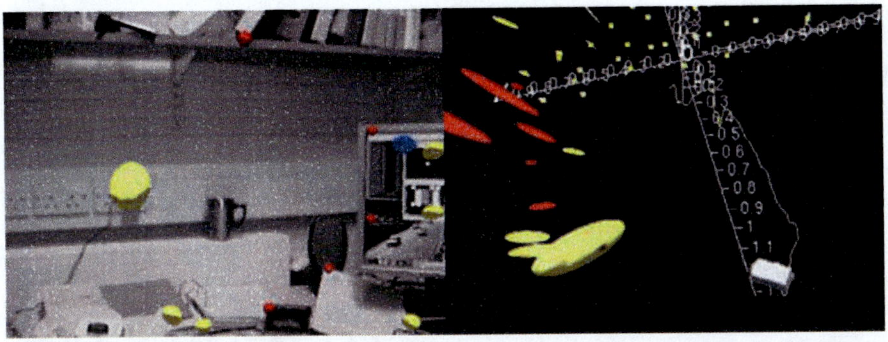

Fig. 13.1 Snapshot of MonoSLAM. Left: the tracked features. Right: the map point in 3D space

Figure 13.1 shows the snapshot of MonoSLAM. The monocular camera tracks very sparse feature points in an image using active tracking technology. In EKF, each feature's position obeys the Gaussian distribution, so we can express its mean value and uncertainty as an ellipsoid. In the right half of the picture, we can find some small ellipsoids distributed in 3D space. If the ellipsoid looks long in a specific direction, the landmark corresponding to it is more uncertain in that direction. We can imagine that if a feature point converges, we should see it change from a very long ellipsoid (initially very uncertain in the Z axis of the camera system) to a small point.

This approach seems to have many drawbacks today. Still, it was already a milestone work at that time because most of the previous visual SLAM systems could not run online. The advancement of CPU performance and sparse bundle adjustment combined to make a SLAM system run online. From a modern point of view, MonoSLAM has drawbacks such as the small scenarios, limited number of landmarks, and sparse feature points that are very easy to lose. Its maintaining work has also been stopped and replaced by more advanced theories and programming tools. But this does not prevent us from understanding and respecting the work of our predecessors.

13.1.2 PTAM

In 2007, Klein's team proposed PTAM (Parallel Tracking and Mapping) [97], which is also an important event in the development of visual SLAM. The important significance of PTAM lies in the following two points:

1. PTAM proposed and realized the parallelization of the tracking and mapping process. Most people think it is clear that the tracking part needs to respond to image data in real-time, and the optimization does not need to be calculated in real-time. Backend optimization can be done slowly in the background, and then thread synchronization can be performed when necessary. This is the first time the

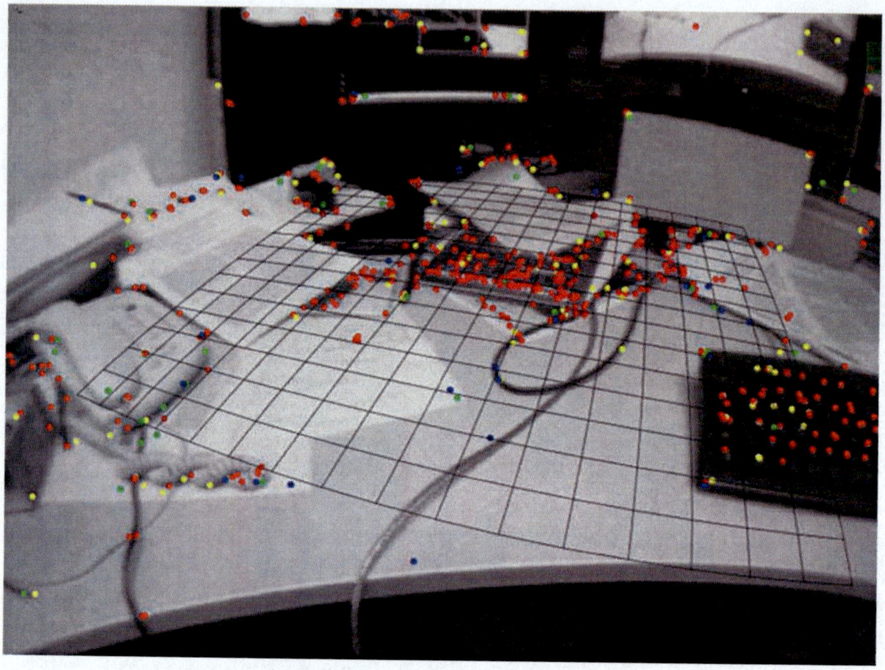

Fig. 13.2 Screenshot of PTAM demo. It can provide not only real-time SLAM functions but also superimpose virtual objects on a virtual plane

concept of front and backends have been distinguished in visual SLAM, leading to the structure of many later visual SLAM systems (most of the SLAMs we see now are divided into front and backends).

2. PTAM is the first real-time solution that uses nonlinear optimization instead of traditional filters in the backend. It introduces the keyframe mechanism: we don't need to process each image carefully but string together several key images and optimize its trajectory and map. Early SLAM mostly used EKF filters or their variants. After PTAM, visual SLAM research gradually turned to the backend dominated by nonlinear optimization.
   PTAM is also an augmented reality software that demonstrates promising AR effects (as shown in Fig. 13.2). According to the camera pose estimated by PTAM, we can place a virtual object on a virtual plane, which looks like it is in a real scene.

However, from a modern perspective, PTAM can be regarded as one of the early SLAM work combined with AR. Like many previous works, there are apparent defects: the scene is small, tracking is easy to lose, and so on. These were improved in the follow-up systems.

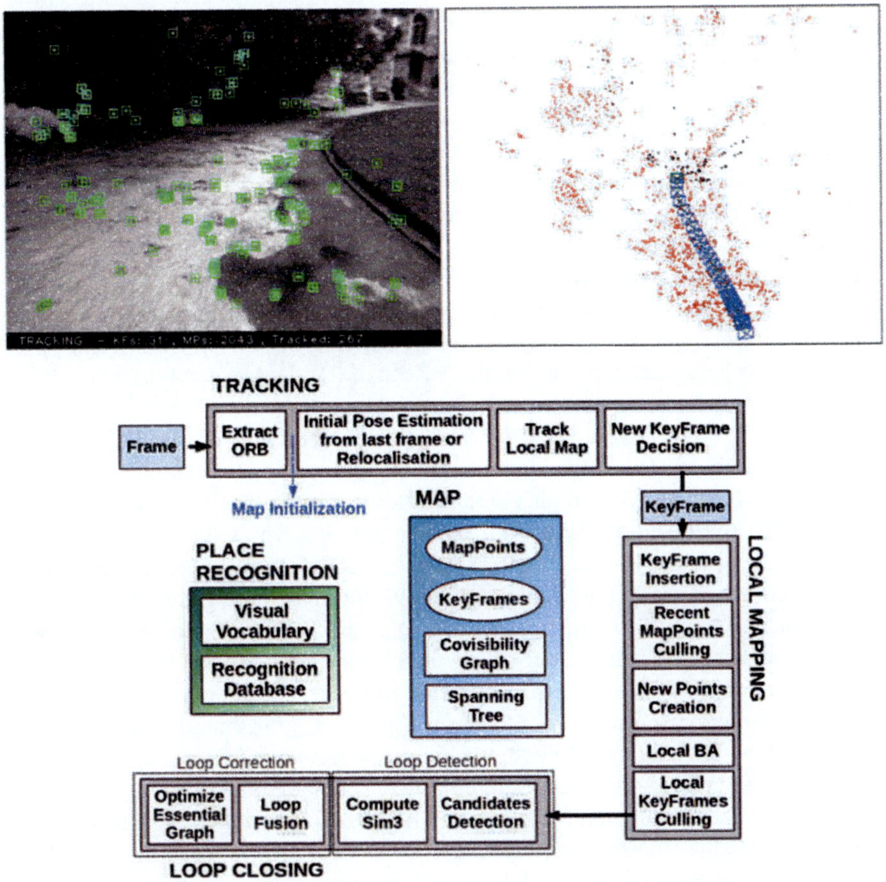

Fig. 13.3 ORB-SLAM running screenshot. The left side is the image and the tracked feature points, and the right side is the camera trajectory and the modeled feature point map. Below is the three-thread framework

13.1.3 ORB-SLAM Series

After introducing several historical solutions, let's look at some modern SLAM systems. ORB-SLAM is a very famous [88] among the successors of PTAM (see Fig. 13.3). It was proposed in 2015 and is one of the most complete and easy-to-use systems in modern SLAM systems (if not the most complete and easy-to-use). ORB-SLAM is a peak of the mainstream feature-based open-source SLAM. Compared with previous work, ORB-SLAM has the following obvious advantages:

1. It supports various sensor settings: monocular, binocular, and RGB-D. For most of the sensors, it is almost possible to test it on ORB-SLAM. It has good versatility.

2. The entire system is calculated with ORB features, including the ORB dictionary for visual odometry and loop detection. It reflects that the ORB feature is an excellent compromise between the computing platform's efficiency and accuracy at this stage. ORB is not as time-consuming as SIFT or SURF and can be calculated in real-time on the CPU; compared to simple corner features such as Harris corners, it has good rotation and scaling invariance. Also, ORB provides descriptors that enable us to perform loop detection and relocation in a large range of motion.
3. ORB's loop detection is its highlight. The excellent loop closure detection algorithm ensures that ORB-SLAM effectively prevents accumulated errors and can be quickly retrieved after being lost. Many existing SLAM systems are not perfect in the relocalization point. For this reason, ORB-SLAM needs to load a large ORB dictionary file.
4. ORB-SLAM innovatively uses the three threads structure in SLAM: Tracking thread for real-time tracking of feature points, optimization thread for local bundle adjustment (co-visibility Graph, commonly known as the *small graph*), and global pose graph optimization thread (essential graph, known as the *big graph*). The tracking thread is responsible for extracting ORB feature points for each new image, comparing it with the most recent keyframe, calculating the feature points' location, and roughly estimating the camera pose. The small graph solves a local bundle adjustment problem, including the feature points in the local space and the camera pose. This thread is responsible for refining camera poses and spatial locations of feature points. The tracking and local mapping threads construct good visual odometry. The third thread, the big graph thread, performs loop detection on the global map and keyframes to eliminate accumulated errors.
   Following the two-thread structure of PTAM, the three-thread design of ORB-SLAM has achieved excellent tracking and mapping effects, ensuring the global consistency of the trajectory and the map. This three-thread structure will also be accepted and adopted by subsequent researchers.
5. ORB-SLAM has carried out many optimizations around feature points. For example, based on OpenCV feature extraction, ORB-SLAM uniforms the distribution of image features. When optimizing the pose, a four-iteration optimization routine is used to obtain better matches. A more relaxed keyframe selection strategy than PTAM is also used. These small improvements make ORB-SLAM far more robust than other solutions: even for poor scenarios and poor calibration parameters, ORB-SLAM can work smoothly.

The advantages mentioned above make ORB-SLAM reach the state-of-the-art open-source visual SLAM system. Many research works use ORB-SLAM as the standard or follow-up development. Its code is known for being clear and easy to read with sufficient annotations, being friendly to later researchers.

Of course, ORB-SLAM also has some shortcomings. Since the entire SLAM system uses feature points for calculation, we must detect the ORB feature for each image, which is very time-consuming. ORB-SLAM's three-thread structure also brings a heavier burden to the CPU, making it only able to operate in real-time on the current PC's CPU. It is not easy to transform into embedded devices. Secondly, the ORB-SLAM map consists of sparse feature points. But it does not implement the storage and relocation functions after closing the SLAM software (although it is not difficult in terms of implementation). According to Chap. 11, the sparse feature point map can only meet our localization needs but cannot satisfy navigation, obstacle avoidance, interaction, or other requirements. However, if we only use ORB-SLAM to deal with the localization problem, it seems a bit too heavyweight. In contrast, some other programs provide a more lightweight localization, allowing us to run SLAM on low-level processors.

13.1.4 LSD-SLAM

LSD-SLAM (Large Scale Direct monocular SLAM) is a SLAM work [69, 75] proposed by J. Engel et al. in 2014. Analogous to ORB-SLAM to feature points, LSD-SLAM marks the direct method's successful application in monocular SLAM. The core contribution of LSD-SLAM is to apply the direct method to semi-dense monocular SLAM. It does not need to calculate the features but can also construct a semi-dense map. The semi-dense means estimating the pixel position with an obvious gradient. Its main advantages are as follows:

1. The direct method of LSD-SLAM is for pixels. The authors creatively proposed the relationship between pixel gradient and the direct method. And also, they note the angular relationship between pixel gradient and epipolar direction in dense reconstruction. These are discussed in Chaps. 7 and 11 in this book. However, LSD-SLAM performs semi-dense tracking on monocular images, and the implementation detail is more complicated than the routines in this book.
2. LSD-SLAM realizes the semi-dense reconstruction on CPU, which is rarely seen in previous schemes. The method based on feature points can only be sparse, and most of the dense reconstruction schemes use RGB-D sensors or use GPU to build dense maps [143].
3. The semi-dense tracking of LSD-SLAM uses some subtle tricks to ensure the tracking's real-time efficiency and stability. For example, LSD-SLAM does not use a single-pixel or image block for epipolar searching but takes five points on the epipolar line at equal distances to measure its SSD. In the depth estimation, LSD-SLAM first initializes the depth with random numbers and then normalizes the depth estimation after converges. When measuring the depth uncertainty, it considers the triangle's geometric relationship and the angle between the epipolar line and the image gradient. The loop closure thread uses the similarity transformation group SE(3) to express the scale explicitly, which can be used to reduce the scale drift in monocular SLAM.

Figure 13.4 shows a snapshot of the LSD-SLAM. We can observe how this semi-dense map is a form between the sparse map and the dense map. The semi-dense map

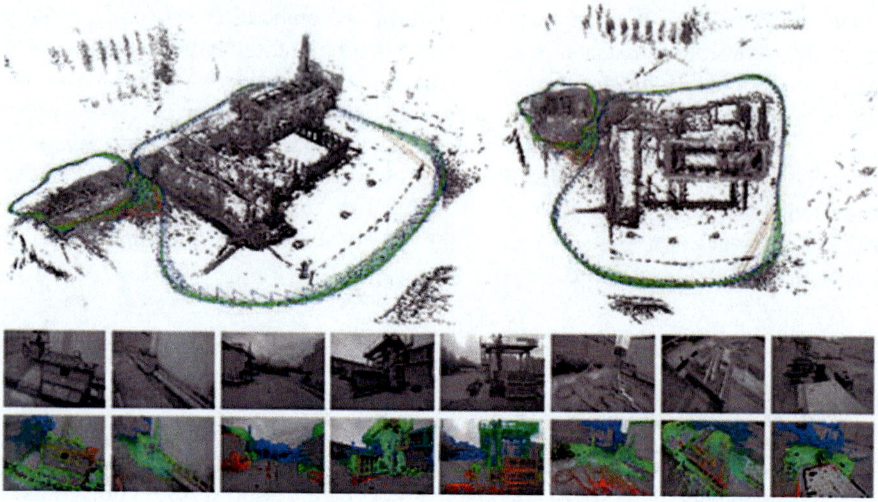

Fig. 13.4 Snapshots of LSD-SLAM. The upper part is the estimated trajectory and map, and the lower part is the reconstructed part of the image, that is, the part with better pixel gradients

models the parts with apparent gradients in the grayscale image. A large part of the map is displayed on the edges or textured parts of the object. LSD-SLAM tracks them and establishes the keyframes, and finally optimizes to get such a map. It seems that it has more information than a sparse map, but it does not have complete surfaces like a dense map (dense maps are generally considered to be unable to achieve real-time performance with CPU alone).

Since LSD-SLAM uses the direct tracking method, it has both the direct method's advantages and disadvantages. For example, LSD-SLAM is very sensitive to camera intrinsics and exposure time and is easily lost when the camera moves quickly. Also, in the loop detection part, LSD-SLAM still relies on the feature point method for loop detection and has not entirely eliminated the calculation of feature points.

13.1.5 SVO

SVO is the abbreviation of Semi-direct Visual Odometry [68]. It is visual odometry based on the sparse direct method proposed by Forster et al. in 2014. According to the author's name, it should be called the "semi-direct" method, but according to the conceptual framework of this book, it may be better to call it the "sparse direct method". The meaning of semi-direct in the original text refers to the mixed-use of feature points and direct methods: SVO tracks some key points (corner points, no descriptors), and then, like the direct method, uses the information of these key points to estimate camera movement and the map point's position (as shown in Fig. 13.5).

Fig. 13.5 Features tracked by SVO

In the implementation, SVO uses small blocks of 4×4 around the key points for block matching to estimate the camera's motion.

Compared with other programs, SVO's most significant advantage is high-speed. Due to the use of the sparse direct method, it does not have to laboriously calculate the descriptor, nor does it need to process as much information as dense and semi-dense approaches. So, it can achieve real-time performance even on low-level computing platforms. On PC platforms, it can reach a speed of more than 100 frames per second. In the subsequent SVO 2.0, the speed has reached an astonishing 400 frames per second. This makes SVO very suitable for scenarios with limited computing resources, such as the localization of drones and handheld AR/VR devices. UAV is also the target application platform for the author to develop SVO.

Another SVO innovation is that it puts forward the concept of depth filter and derives a depth filter based on uniform-Gaussian mixture distribution. This is mentioned in Chap. 11 of this book, but since the principle is more complicated, we did not explain it in detail. SVO uses this filter to estimate the key points' depth and uses the inverse depth parameterization to better calculate feature points' position.

SVO's code is clear and easy to read, which is very suitable for readers to analyze as the first SLAM example. However, the open-source version of SVO also has some problems:

1. Since the target application platform is the drone's top-down camera, the object in its field of view is mainly the ground. The camera's movement is mostly horizontal rotation and up-and-down movement. Many SVO details are designed around this application, making its performance not very good for front-seeing cameras. For example, SVO uses the decomposition of the **H** matrix instead of the traditional

F or **E** matrix during monocular initialization, which requires the assumption that the feature points are on the plane. This assumption is correct for a top-down camera, but it is usually not valid for a head-up camera. When SVO selects a keyframe, it uses the translation amount to determine a new keyframe without considering the rotation amount. This is also effective in the drone's top view configuration, but it is easy to lose in the head-up camera. Therefore, if readers want to use SVO in a head-up camera, some modification is needed.

2. SVO discards the backend optimization and loop closure detection part for speed and lightweight. And it basically has no mapping functions. This means that SVO pose estimation must have accumulative errors, and it is not easy to relocate after loss (because there is no descriptor for loop detection). Therefore, we call it a VO, not a complete SLAM.

13.1.6 RTAB-MAP

After introducing several monocular SLAM solutions, let's look at some SLAM solutions on RGB-D sensors. Compared with monocular and binocular cameras, the RGB-D SLAM principle is much more straightforward (though not necessarily in implementation). It can build dense maps in real-time on the CPU.

RTAB-MAP (Real-Time Appearance-Based Mapping) [122] is a classic scheme in RGB-D SLAM. It implements everything that should be included in RGB-D SLAM: feature-based visual odometry, bag-of-words-based loop detection, backend pose map optimization, and point cloud and triangular mesh maps. Therefore, RTAB-MAP provides a complete (but somewhat huge) RGB-D SLAM solution. At present, we can obtain the binary program directly from ROS. In addition, the App can also be used on Google Project Tango (as shown in Fig. 13.6).

RTAB-MAP supports some common RGB-D and binocular sensors, such as Kinect, Xtion, etc., and provides real-time localization and mapping functions. However, due to the high integration level, it becomes difficult for other developers to carry out secondary development. RTAB-MAP is more suitable for SLAM applications than research use.

13.1.7 Others

In addition to these open source solutions, readers can also find many other studies on websites such as `openslam.org`, for example, DVO-SLAM [144], RGBD-SLAM-V2 [100], DSO [70], and some Kinect Fusion related work, etc. With the development of the times, newer and better open-source SLAM works will also appear. Due to space limitations, we will not introduce them one by one here.

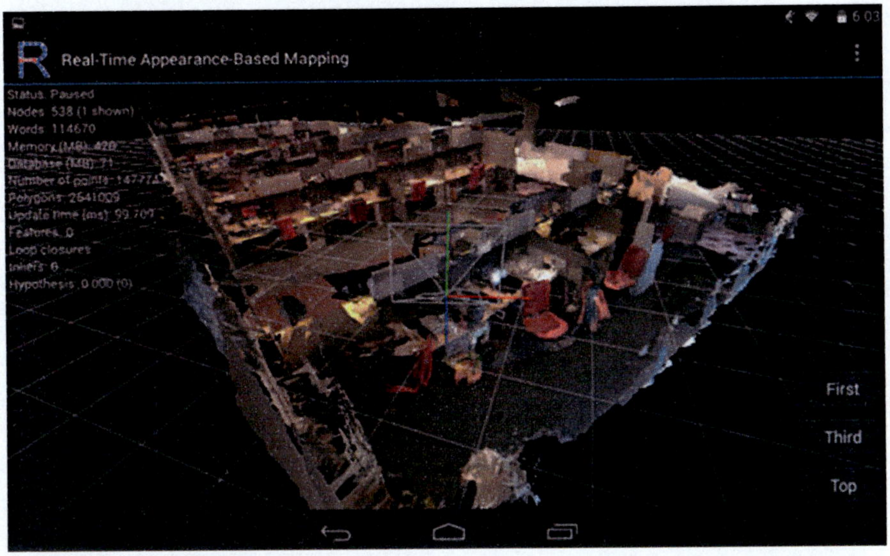

Fig. 13.6 RTAB-MAP in Google Project Tango

13.2 SLAM in Future

After studying the existing systems, let's discuss some future development directions.[2] Generally speaking, SLAM's future development trend is divided into two categories: one is to develop towards lightweight and miniaturization so that SLAM can run well on small devices such as embedded or mobile phones, and then consider high-level applications that use it as the underlying function. After all, in most cases, our real goal is to realize the functions of robots and AR/VR devices, such as sports, navigation, teaching, and entertainment. SLAM is to provide its own pose estimation for upper-level applications. In these applications, we do not want SLAM to occupy many computing resources, so there is a solid demand for SLAM's miniaturization and lightweight. The second is to use high-performance computing equipment to realize precise 3D reconstruction and scene understanding. In these applications, our goal is to reconstruct the scene perfectly, and there are no restrictions on the portability of computing resources and equipment. Since GPU can be used, this direction and deep learning also have combination points.

[2] Well, this part is my personal understanding, which may not be completely correct.

13.2.1 IMU Integrated VSLAM

First, we want to talk about a direction with a strong application background: the visual-inertial navigation scheme. Whether in robots or hardware devices, we do not rely on one sensor but often have multiple sensors. Researchers in academia love the big clean problems, such as implementing visual SLAM with a single camera. But friends in the industry pay more attention to making algorithms more practical and face complex and corner-case scenarios. In this application context, vision and inertial integrated SLAM have become a hot spot.

The inertial sensor (IMU) can measure the angular velocity and acceleration of the sensor body, which is considered to be evidently complementary with the camera sensor and has the potential to obtain a more complete SLAM system after fusion [145]. Why do we say that?

1. IMU can measure angular velocity and acceleration, but these quantities have a noticeable drift, making the pose data obtained by the twice integration unreliable. For example, if we put the IMU on the table without moving, the post obtained by integrating its readings will drift far away in a short time. However, for fast movements, IMU can provide some better estimates than the camera.
   When the movement is too fast, the rolling shutter camera will have a motion blur. The overlapping area between two frames is too small to perform any feature matching, so pure visual SLAM is very weak against the fast movement. With IMU, we can still maintain the pose estimation even when the camera images are invalid, which is not possible with pure visual SLAM.
2. Compared with raw IMU integration, the camera data will basically not drift. If the camera is placed in place and fixed in a static scene, visual SLAM's pose estimation is also fixed. Therefore, the camera data can effectively estimate and correct the drift in the IMU reading so that the pose estimation after slow motion is still valid.
3. When the image changes, we practically cannot know whether the camera has moved or the environmental objects have changed. In pure visual SLAM, it is difficult to deal with a large number of dynamic obstacles. But the IMU can feel its own motion. To some extent, VIO (Visual Inertial Odometry) can reduce the impact of dynamic objects.

In summary, we see that IMU provides a better solution for fast motion, and the camera can solve the drift problem of IMU under slow motion. In this sense, the two are complementary.

Although it sounds perfect, VIO is quite complicated, both in theory or practice. Its complexity mainly comes from the fact that IMU measures acceleration and angular velocity, so kinematic calculations have to be involved in the computation. At present, the VIO framework has been finalized into two categories: loosely coupled and tightly coupled [146]. Loosely coupled means that the IMU and the camera perform their own motion estimation separately, and we merge their pose estimation results. Tightly coupled refers to combining the IMU state with the camera, jointly

Fig. 13.7 More and more devices start to integrate IMU with cameras

constructing the motion equation and observation equation, and then performing state estimation. This is very similar to the theory we introduced before. We can foresee that tightly coupled theory will also be divided into two directions: filtering-based and optimization-based. In terms of filtering, the traditional EKF [147] and the improved MSCKF (Multi-State Constraint KF) [148] have achieved certain results, and the researchers have also carried out the EKF in-depth discussion (such as observability [149]). There are also corresponding solutions of optimization [89, 150]. It is worth mentioning that although optimization methods in pure visual SLAM have taken the mainstream, in VIO, because the frequency of IMU is very high, the amount of calculation required to optimize the state is greater, so it is still in filtering and optimizing the coexistence stage [151, 152]. Due to its complexity and limited space, we can only briefly introduce this direction here.

VIO provides a very effective direction for the miniaturization and cost reduction of SLAM in the future. And combined with the sparse direct method, we are expected to achieve good SLAM or VO effects on low-level hardware, which is very promising (Fig. 13.7).

13.2.2 Semantic SLAM

Another general direction of SLAM is to combine with deep learning technology. So far, SLAM solutions are still at the feature level or pixel level. We don't know what exactly these feature points or pixels come from. This makes the SLAM in computer vision not very similar to what we humans do. At least we never see the feature points, nor do we judge our own movement direction based on those features or pixels. We see the objects, judge their distance through the left and right eyes, and then guess the camera's movement based on their movement in the image.

A long time ago, researchers tried to incorporate object information into SLAM. For example, [153–156] once combined object recognition and visual SLAM to construct a map with object labels. On the other hand, by introducing the label information into the objective function and constraints of the BA or optimization end,

Fig. 13.8 Semantic SLAM results from [156, 158]

we can combine the feature point's position with the label information to optimize [157]. These tasks can be called semantic SLAM. In summary, the combination of SLAM and semantics mainly has two aspects [8]:

1. Semantics help SLAM. Traditional object recognition and segmentation algorithms often only consider one image, while in SLAM, we have a moving camera. If we put object labels on all the pictures during the movement, we can get a labeled map. Also, object information can bring more conditions for loop detection and BA optimization.
2. SLAM helps semantics. Both object recognition and segmentation require a lot of training data. For the classifier to recognize objects from various angles, it is necessary to collect the object's data from different perspectives and then perform manual calibration, which is very difficult. In SLAM, since we can estimate the camera movement, we can automatically get the object images from different angles, saving the cost of manual calibration. If there is automatically generated sample data with high-quality pose annotations, it can significantly accelerate the classifier's training process (Fig. 13.8).

Before deep learning is widely used, we can only use traditional tools such as support vector machines (SVM) and conditional random fields (CRF) to segment and recognize objects or scenes or directly compare observation data with samples in the database [123, 158]. Semantic maps are built by those technologies in early researches [156, 159–161]. Because these tools themselves have limitations on the classification accuracy, the effect is often not satisfactory. With the development of deep learning, we began to use the network to recognize, detect and segment images more and more accurately [162–167]. This lays a better foundation for constructing accurate semantic maps [168]. We see that some scholars are gradually introducing neural network methods into the object recognition and segmentation in SLAM, and even the pose estimation and loop detection of SLAM [120, 169, 170]. Although these methods have not yet become mainstream, the combination of SLAM and deep learning to process images is also a promising research direction.

In addition, SLAM based on line/surface features [171–173], SLAM in dynamic scenes [174–176], and multi-robot SLAM [82, 177, 178], etc., are all places where researchers are interested in concurrency. According to the opinion of the [8], visual SLAM has gone through three major eras: asking questions, searching for algorithms, and improving the algorithms. And we are currently in the third era, facing how to further improve the existing framework so that the visual SLAM system can operate stably under various interference conditions. This step requires the ongoing efforts of many researchers.

Of course, no one can predict the future, and we are not sure if one day the entire framework will be overturned and rewritten by new technologies. But even then, our contribution today will still be meaningful. Without today's research, there will be no future development. Finally, we hope that readers will fully understand the entire existing SLAM system after reading this book. We also look forward to your contribution to SLAM research!

Exercises

1. Choose one of the open-source SLAM systems mentioned in this lecture, compile and run it on your machine, and experience the process intuitively.
2. You should be able to understand most SLAM-related papers. Pick up paper and pen and start your research!

Appendix A
Gaussian Distribution

A.1 Gaussian Distribution

If way say a random variable x satisfies a Gaussian distribution $N(\mu, \sigma)$, then its *pdf* is:

$$p(x) = \frac{1}{\sqrt{2\pi}\sigma} \exp\left(-\frac{1}{2}\frac{(x-\mu)^2}{\sigma^2}\right). \tag{A.1}$$

The matrix form is:

$$p(\mathbf{x}) = \frac{1}{\sqrt{(2\pi)^N \det(\mathbf{\Sigma})}} \exp\left(-\frac{1}{2}(\mathbf{x}-\boldsymbol{\mu})^T \mathbf{\Sigma}^{-1}(\mathbf{x}-\boldsymbol{\mu})\right). \tag{A.2}$$

A.2 Transform of Gaussian Variables

A.2.1 Linear Transform

Suppose we have two independent Gaussian variables:

$$\mathbf{x} \sim N(\boldsymbol{\mu}_x, \mathbf{\Sigma}_{xx}), \quad \mathbf{y} \sim N(\boldsymbol{\mu}_y, \mathbf{\Sigma}_{yy}),$$

the sum of them are still Gaussian:

$$\mathbf{x} + \mathbf{y} \sim N(\boldsymbol{\mu}_x + \boldsymbol{\mu}_y, \mathbf{\Sigma}_{xx} + \mathbf{\Sigma}_{yy}). \tag{A.3}$$

If we multiply $\mathbf{x}$ with a constant factor a, then $a\mathbf{x}$ satisfies:

$$a\mathbf{x} \sim N(a\boldsymbol{\mu}_x, a^2\mathbf{\Sigma}_{xx}). \tag{A.4}$$

X. Gao and T. Zhang, *Introduction to Visual SLAM*,
https://doi.org/10.1007/978-981-16-4939-4

If we take a linear transform of $\mathbf{x}$ with $\mathbf{y} = \mathbf{A}\mathbf{x}$, then $\mathbf{y}$ satisfies:

$$\mathbf{y} \sim N(\mathbf{A}\boldsymbol{\mu}_x, \mathbf{A}\boldsymbol{\Sigma}_{xx}\mathbf{A}^T). \tag{A.5}$$

A.2.2 *Normalized Product*

If we want to fuse two Gaussian estimations like $\mathbf{x}$ and $\mathbf{y}$, assume the fused mean and covariance are $\boldsymbol{\mu}$ and $\boldsymbol{\Sigma}$, then we have:

$$\begin{aligned}\boldsymbol{\Sigma}^{-1} &= \boldsymbol{\Sigma}_{xx}^{-1} + \boldsymbol{\Sigma}_{yy}^{-1} \\ \boldsymbol{\Sigma}^{-1}\boldsymbol{\mu} &= \boldsymbol{\Sigma}_{xx}^{-1}\boldsymbol{\mu}_x + \boldsymbol{\Sigma}_{yy}^{-1}\boldsymbol{\mu}_y.\end{aligned} \tag{A.6}$$

This formula can be extended to the product of any number of Gaussian distributions.

A.2.3 *Joint and Conditional Distribution*

If $\mathbf{x}$ and $\mathbf{y}$ are not independent, then their joint distribution is:

$$p(\mathbf{x}, \mathbf{y}) = N\left(\begin{bmatrix}\boldsymbol{\mu}_x \\ \boldsymbol{\mu}_y\end{bmatrix}, \begin{bmatrix}\boldsymbol{\Sigma}_{xx} & \boldsymbol{\Sigma}_{xy} \\ \boldsymbol{\Sigma}_{yx} & \boldsymbol{\Sigma}_{yy}\end{bmatrix}\right). \tag{A.7}$$

Since $p(\mathbf{x}, \mathbf{y}) = p(\mathbf{x}|\mathbf{y})\, p(\mathbf{y})$, we can find that the conditional distribution $p(\mathbf{x}|\mathbf{y})$ satisfies:

$$p(\mathbf{x}|\mathbf{y}) = N\left(\boldsymbol{\mu}_x + \boldsymbol{\Sigma}_{xy}\boldsymbol{\Sigma}_{yy}^{-1}\left(\mathbf{y} - \boldsymbol{\mu}_y\right), \boldsymbol{\Sigma}_{xx} - \boldsymbol{\Sigma}_{xy}\boldsymbol{\Sigma}_{yy}^{-1}\boldsymbol{\Sigma}_{yx}\right). \tag{A.8}$$

A.3 Example of Joint Distribution

Let's take an example of the Kalman filter. Consider a random variable $\mathbf{x} \sim N(\boldsymbol{\mu}_x, \boldsymbol{\Sigma}_{xx})$, and another variable $\mathbf{y}$ has:

$$\mathbf{y} = \mathbf{A}\mathbf{x} + \mathbf{b} + \mathbf{w}, \tag{A.9}$$

where $\mathbf{A}, \mathbf{b}$ is the linear coefficient matrix and bias, $\mathbf{w}$ is the white noise that satisfies: $\mathbf{w} \sim N(\mathbf{0}, \mathbf{R})$.

Then we can calculate $\mathbf{y}$'s distribution:

$$p(\mathbf{y}) = N\left(\mathbf{A}\boldsymbol{\mu}_x + \mathbf{b}, \mathbf{R} + \mathbf{A}\boldsymbol{\Sigma}_{xx}\mathbf{A}^T\right). \tag{A.10}$$

This provides the theoretical basis for the prediction part of the Kalman filter.

Appendix B
Matrix Derivatives

First, the derivation of a scalar function to a scalar variable is obvious. Suppose a function $f(x)$ takes the derivative of x, then a derivative like $\frac{\mathrm{d}f}{\mathrm{d}x}$ will be obtained, which is obviously still a scalar. Below we discuss the cases when x is a vector and f is a vector function.

B.1 Scalar Function with Vector Variable

Suppose $\mathbf{x} \in \mathbb{R}^m$ is a column vector, then we have:

$$\frac{\mathrm{d}f}{\mathrm{d}\mathbf{x}} = \left[\frac{\mathrm{d}f}{\mathrm{d}x_1}, \cdots, \frac{\mathrm{d}f}{\mathrm{d}x_m}\right]^T \in \mathbb{R}^m. \tag{B.1}$$

The result is a $m \times 1$ vector. In some documents we may write the denominator as $\mathbf{x}^T$:

$$\frac{\mathrm{d}f}{\mathrm{d}\mathbf{x}^T} = \left[\frac{\mathrm{d}f}{\mathrm{d}x_1}, \cdots, \frac{\mathrm{d}f}{\mathrm{d}x_m}\right]. \tag{B.2}$$

The result is a row vector. We usually call $\frac{\mathrm{d}f}{\mathrm{d}\mathbf{x}}$ as gradient or Jacobian. But it should be noted that the name or manner is not exactly same in different fields.

B.2 Vector Function with Vector Variable

We may also take the derivative with a vector function to a vector variable. Consider a vector function $\mathbf{F}(\mathbf{x})$ like:

$$\mathbf{F}(\mathbf{x}) = [f_1(\mathbf{x}), \cdots, f_n(\mathbf{x})]^T,$$

X. Gao and T. Zhang, *Introduction to Visual SLAM*,
https://doi.org/10.1007/978-981-16-4939-4

where each f_k is a scalar function of $\mathbf{x}$. When taking derivative of $\mathbf{x}$, we usually write it as:

$$\frac{\partial \mathbf{F}}{\partial \mathbf{x}^T} = \begin{bmatrix} \frac{\partial f_1}{\partial \mathbf{x}^T} \\ \vdots \\ \frac{\partial f_n}{\partial \mathbf{x}^T} \end{bmatrix} = \begin{bmatrix} \frac{\partial f_1}{\partial x_1} & \frac{\partial f_1}{\partial x_2} & \cdots & \frac{\partial f_1}{\partial x_m} \\ \frac{\partial f_2}{\partial x_1} & \frac{\partial f_2}{\partial x_2} & \cdots & \frac{\partial f_2}{\partial x_m} \\ \vdots & \vdots & \ddots & \vdots \\ \frac{\partial f_n}{\partial x_1} & \frac{\partial f_n}{\partial x_2} & \cdots & \frac{\partial f_n}{\partial x_m} \end{bmatrix} \in \mathbb{R}^{n \times m} \tag{B.3}$$

which is a column vector function to a row vector variable. Then we get a $n \times m$ Jacobian matrix. Such writing style is standardized. We may write the derivative as:

$$\frac{\partial \mathbf{A}\mathbf{x}}{\partial \mathbf{x}^T} = \mathbf{A}, \tag{B.4}$$

or a row vector function to a column vector variable, and the result is transposed:

$$\frac{\partial \mathbf{F}^T}{\partial \mathbf{x}} = \left(\frac{\partial \mathbf{F}}{\partial \mathbf{x}^T} \right)^{\mathrm{T}}. \tag{B.5}$$

In this book we use the former definition by default. However, this writing style requires adding an extra transpose operator in the denominator of each equation, which is inconvenient if we have to write a lot of derivatives. So without ambiguity, we relax the notation by omitting the transpose operator like this:

$$\frac{\partial \mathbf{A}\mathbf{x}}{\partial \mathbf{x}} = \mathbf{A}, \tag{B.6}$$

which makes the notation more natural.

References

1. A. Davison, I. Reid, N. Molton, O. Stasse, MonoSLAM: real-time single camera SLAM. IEEE Trans. Pattern Anal. Mach. Intell. **29**(6), 1052–1067 (2007)
2. R. Hartley, A. Zisserman, *Multiple View Geometry in Computer Vision* (Cambridge University Press, Cambridge, 2003)
3. R.C. Smith, P. Cheeseman, On the representation and estimation of spatial uncertainty. Int. J. Robot. Res. **5**(4), 56–68 (1986)
4. S. Thrun, W. Burgard, D. Fox, *Probabilistic Robotics* (MIT Press, Cambridge, 2005)
5. T. Barfoot, State estimation for robotics: a matrix lie group approach (2016)
6. A. Pretto, E. Menegatti, E. Pagello, Omnidirectional dense large-scale mapping and navigation based on meaningful triangulation, in *2011 IEEE International Conference on Robotics and Automation (ICRA 2011)* (2011), pp. 3289–3296
7. B. Rueckauer, T. Delbruck, Evaluation of event-based algorithms for optical flow with ground-truth from inertial measurement sensor. Front. Neurosci. **10** (2016)
8. C. Cadena, L. Carlone, H. Carrillo, Y. Latif, D. Scaramuzza, J. Neira, I.D. Reid, J.J. Leonard, Past, present, and future of simultaneous localization and mapping: towards the robust-perception age (2016), arXiv:1606.05830
9. N. Silberman, D. Hoiem, P. Kohli, R. Fergus, Indoor segmentation and support inference from RGBD images, in *Computer Vision–ECCV 2012* (Springer, 2012), pp. 746–760
10. K. Ho, P. Newman, Detecting loop closure with scene sequences. Int. J. Comput. Vis. **74**(3), 261–286 (2007)
11. R. Smith, M. Self, P. Cheeseman, Estimating uncertain spatial relationships in robotics, *Autonomous Robot Vehicles* (Springer, Berlin, 1990), pp. 167–193
12. P. Beeson, J. Modayil, B. Kuipers, Factoring the mapping problem: mobile robot map-building in the hybrid spatial semantic hierarchy. Int. J. Robot. Res. **29**(4), 428–459 (2010)
13. H. Strasdat, J.M. Montiel, A.J. Davison, Visual SLAM: why filter? Image Vis. Comput. **30**(2), 65–77 (2012)
14. J. Fuentes-Pacheco, J. Ruiz-Ascencio, J.M. Rendón-Mancha, Visual simultaneous localization and mapping: a survey. Artif. Intell. Rev. **43**(1), 55–81 (2015)
15. J. Boal, Á. Sánchez-Miralles, Á. Arranz, Topological simultaneous localization and mapping: a survey. Robotica **32**, 803–821 (2014)
16. S.Y. Chen, Kalman filter for robot vision: a survey. IEEE Trans. Ind. Electron. **59**(11), 4409–4420 (2012)
17. Z. Chen, J. Samarabandu, R. Rodrigo, Recent advances in simultaneous localization and map-building using computer vision. Adv. Robot. **21**(3–4), 233–265 (2007)

X. Gao and T. Zhang, *Introduction to Visual SLAM*,
https://doi.org/10.1007/978-981-16-4939-4

18. J. Stuelpnagel, On the parametrization of the three-dimensional rotation group. SIAM Rev. **6**(4), 422–430 (1964)
19. T. Barfoot, J.R. Forbes, P.T. Furgale, Pose estimation using linearized rotations and quaternion algebra. Acta Astronaut. **68**(1–2), 101–112 (2011)
20. V.S. Varadarajan, *Lie Groups, Lie Algebras, and Their Representations*, vol. 102 (Springer Science & Business Media, New York, 2013)
21. Y. Ma, S. Soatto, J. Kosecka, S.S. Sastry, *An Invitation to 3-d vision: from images to geometric models*, vol. 26 (Springer Science & Business Media, New York, 2012)
22. J. Sola, Quaternion kinematics for the error-state KF, Laboratoire dAnalyse et dArchitecture des Systemes-Centre national de la recherche scientifique (LAAS-CNRS), Toulouse, France, Technical report (2012)
23. J. Sturm, N. Engelhard, F. Endres, W. Burgard, D. Cremers, A benchmark for the evaluation of RGB-D SLAM systems, in *2012 IEEE/RSJ International Conference on Intelligent Robots and Systems (IROS)* (IEEE, 2012), pp. 573–580
24. H. Strasdat, Local accuracy and global consistency for efficient visual SLAM. Ph.D. thesis, Citeseer (2012)
25. Z. Zhang, Flexible camera calibration by viewing a plane from unknown orientations, in *The Proceedings of the Seventh IEEE International Conference on Computer Vision, 1999*, vol. 1 (IEEE, 1999), pp. 666–673
26. H. Hirschmuller, Stereo processing by semiglobal matching and mutual information. IEEE Trans. Pattern Anal. Mach. Intell. **30**(2), 328–341 (2008)
27. D. Scharstein, R. Szeliski, A taxonomy and evaluation of dense two-frame stereo correspondence algorithms. Int. J. Comput. Vis. **47**(1–3), 7–42 (2002)
28. S.M. Seitz, B. Curless, J. Diebel, D. Scharstein, R. Szeliski, A comparison and evaluation of multi-view stereo reconstruction algorithms, in *Null* (IEEE, 2006), pp. 519–528
29. S. Agarwal, N. Snavely, I. Simon, S.M. Seitz, R. Szeliski, Building Rome in a day, in *2009 IEEE 12th International Conference on Computer Vision* (IEEE, 2009), pp. 72–79
30. P. Wolfe, Convergence conditions for ascent methods. SIAM Rev. **11**(2), 226–235 (1969)
31. J. Nocedal, S. Wright, *Numerical Optimization* (Springer Science & Business Media, New York, 2006)
32. M.I. Lourakis, A.A. Argyros, SBA: a software package for generic sparse bundle adjustment. ACM Trans. Math. Softw. (TOMS) **36**(1), 2 (2009)
33. G. Sibley, Relative bundle adjustment, Department of Engineering Science, Oxford University, Technical report, vol. 2307(09) (2009)
34. B. Triggs, P.F. McLauchlan, R.I. Hartley, A.W. Fitzgibbon, Bundle adjustment: a modern synthesis, in *Vision Algorithms: Theory and Practice* (Springer, 2000), pp. 298–372
35. S. Agarwal, K. Mierle, and Others, Ceres solver, http://ceres-solver.org
36. R. Kummerle, G. Grisetti, H. Strasdat, K. Konolige, W. Burgard, G2o: a general framework for graph optimization, in *IEEE International Conference on Robotics and Automation (ICRA)* (IEEE, 2011), pp. 3607–3613
37. Wikipedia, Feature (computer vision) (2016), https://en.wikipedia.org/wiki/Feature_(computer_vision). Accessed 09 July 2016
38. C. Harris, M. Stephens, A combined corner and edge detector, in *Alvey Vision Conference, Citeseer*, vol. 15 (1988), pp. 10–5244
39. E. Rosten, T. Drummond, Machine learning for high-speed corner detection, in *Computer Vision–ECCV 2006* (Springer, 2006), pp. 430–443
40. J. Shi, C. Tomasi, Good features to track, in *1994 IEEE Computer Society Conference on Computer Vision and Pattern Recognition, 1994. Proceedings CVPR'94* (IEEE, 1994), pp. 593–600
41. D.G. Lowe, Distinctive image features from scale-invariant keypoints. Int. J. Comput. Vis. **60**(2), 91–110 (2004)
42. H. Bay, T. Tuytelaars, L. Van Gool, SURF: speeded up robust features, in *Computer Vision–ECCV 2006* (Springer, 2006), pp. 404–417
43. E. Rublee, V. Rabaud, K. Konolige, G. Bradski, ORB: an efficient alternative to SIFT or SURF, in *2011 IEEE International Conference on Computer Vision (ICCV)* (IEEE, 2011), pp. 2564–2571

44. M. Calonder, V. Lepetit, C. Strecha, P. Fua, BRIEF: binary robust independent elementary features, in *European Conference on Computer Vision* (Springer, 2010), pp. 778–792
45. M. Nixon, A.S. Aguado, *Feature Extraction and Image Processing for Computer Vision* (Academic, New York, 2012)
46. P.L. Rosin, Measuring corner properties. Comput. Vis. Image Underst. **73**(2), 291–307 (1999)
47. M. Muja, D.G. Lowe, Fast approximate nearest neighbors with automatic algorithm configuration, in *VISAPP (1)* (2009), pp. 331–340
48. R.I. Hartley, In defense of the eight-point algorithm. IEEE Trans. Pattern Anal. Mach. Intell. **19**(6), 580–593 (1997)
49. H.C. Longuet-Higgins, A computer algorithm for reconstructing a scene from two projections, in *Readings in Computer Vision: Issues, Problems, Principles, and Paradigms*, ed. by M.A. Fischler, O. Firschein (1987), pp. 61–62
50. H. Li, R. Hartley, Five-point motion estimation made easy, in *18th International Conference on Pattern Recognition (ICPR'06)*, vol. 1 (IEEE, 2006), pp. 630–633
51. D. Nistér, An efficient solution to the five-point relative pose problem. IEEE Trans. Pattern Anal. Mach. Intell. **26**(6), 756–770 (2004)
52. O.D. Faugeras, F. Lustman, Motion and structure from motion in a piecewise planar environment. Int. J. Pattern Recognit. Artif. Intell. **2**(03), 485–508 (1988)
53. Z. Zhang, A.R. Hanson, 3d reconstruction based on homography mapping, *ARPA Image Understanding Workshop* (1996), pp. 1007–1012
54. E. Malis, M. Vargas, Deeper understanding of the homography decomposition for vision-based control. Ph.D. thesis, INRIA (2007)
55. A.J. Davison, Real-time simultaneous localisation and mapping with a single camera, in *Ninth IEEE International Conference on Computer Vision, 2003. Proceedings* (IEEE, 2003), pp. 1403–1410
56. X.-S. Gao, X.-R. Hou, J. Tang, H.-F. Cheng, Complete solution classification for the perspective-three-point problem. IEEE Trans. Pattern Anal. Mach. Intell. **25**, 930–943 (2003)
57. V. Lepetit, F. Moreno-Noguer, P. Fua, EPnP: an accurate O(n) solution to the PnP problem. Int. J. Comput. Vis. **81**(2), 155–166 (2008)
58. A. Penate-Sanchez, J. Andrade-Cetto, F. Moreno-Noguer, Exhaustive linearization for robust camera pose and focal length estimation. IEEE Trans. Pattern Anal. Mach. Intell. **35**(10), 2387–2400 (2013)
59. L. Chen, C.W. Armstrong, D.D. Raftopoulos, An investigation on the accuracy of three-dimensional space reconstruction using the direct linear transformation technique. J. Biomech. **27**(4), 493–500 (1994)
60. B.F. Green, The orthogonal approximation of an oblique structure in factor analysis. Psychometrika **17**(4), 429–440 (1952)
61. iplimage, P3p(blog) (2016), http://iplimage.com/blog/p3p-perspective-point-overview/
62. K.S. Arun, T.S. Huang, S.D. Blostein, Least-squares fitting of two 3-d point sets. IEEE Trans. Pattern Anal. Mach. Intell. **5**, 698–700 (1987)
63. F. Pomerleau, F. Colas, R. Siegwart, A review of point cloud registration algorithms for mobile robotics. Found. Trends in Robot. (FnTROB) **4**(1), 1–104 (2015)
64. O.D. Faugeras, M. Hebert, The representation, recognition, and locating of 3-d objects. Int. J. Robot. Res. **5**(3), 27–52 (1986)
65. B.K. Horn, Closed-form solution of absolute orientation using unit quaternions. JOSA A **4**(4), 629–642 (1987)
66. G.C. Sharp, S.W. Lee, D.K. Wehe, ICP registration using invariant features. IEEE Trans. Pattern Anal. Mach. Intell. **24**(1), 90–102 (2002)
67. G. Silveira, E. Malis, P. Rives, An efficient direct approach to visual SLAM. IEEE Trans. Robot. **24**(5), 969–979 (2008)
68. C. Forster, M. Pizzoli, D. Scaramuzza, SVO: fast semi-direct monocular visual odometry, in *2014 IEEE International Conference on Robotics and Automation (ICRA)* (rs, ed.) (IEEE, 2014), pp. 15–22

69. J. Engel, T. Schöps, D. Cremers, LSD-SLAM: large-scale direct monocular SLAM, in *Computer Vision–ECCV 2014* (Springer, 2014), pp. 834–849
70. J. Engel, V. Koltun, D. Cremers, Direct sparse odometry (2016), arXiv:1607.02565
71. B.D. Lucas, T. Kanade et al., An iterative image registration technique with an application to stereo vision (1981)
72. B.K. Horn, B.G. Schunck, Determining optical flow. Artif. Intell. **17**(1–3), 185–203 (1981)
73. S. Baker, I. Matthews, Lucas-Kanade 20 years on: a unifying framework. Int. J. Comput. Vis. **56**(3), 221–255 (2004)
74. A. Geiger, P. Lenz, C. Stiller, R. Urtasun, Vision meets robotics: the KITTI dataset. Int. J. Robot. Res. (2013)
75. J. Engel, J. Sturm, D. Cremers, Semi-dense visual odometry for a monocular camera, in *Proceedings of the IEEE International Conference on Computer Vision* (2013), pp. 1449–1456
76. J. Sherman, W.J. Morrison, Adjustment of an inverse matrix corresponding to a change in one element of a given matrix. Ann. Math. Stat. **21**(1), 124–127 (1950)
77. V. Sujan, S. Dubowsky, Efficient information-based visual robotic mapping in unstructured environments. Int. J. Robot. Res. **24**(4), 275–293 (2005)
78. F. Janabi-Sharifi, M. Marey, A Kalman-filter-based method for pose estimation in visual servoing. IEEE Trans. Robot. **26**(5), 939–947 (2010)
79. S. Li, P. Ni, Square-root unscented Kalman filter based simultaneous localization and mapping, in *2010 IEEE International Conference on Information and Automation (ICIA)* (IEEE, 2010), pp. 2384–2388
80. R. Sim, P. Elinas, J. Little, A study of the Rao-Blackwellised particle filter for efficient and accurate vision-based SLAM. Int. J. Comput. Vis. **74**(3), 303–318 (2007)
81. J.S. Lee, S.Y. Nam, W.K. Chung, Robust RBPF-SLAM for indoor mobile robots using sonar sensors in non-static environments. Adv. Robot. **25**(9–10), 1227–1248 (2011)
82. A. Gil, O. Reinoso, M. Ballesta, M. Julia, Multi-robot visual SLAM using a Rao-Blackwellized particle filter. Robot. Auton. Syst. **58**(1), 68–80 (2010)
83. G. Sibley, L. Matthies, G. Sukhatme, Sliding window filter with application to planetary landing. J. Field Robot. **27**(5), 587–608 (2010)
84. L.M. Paz, J.D. Tardós, J. Neira, Divide and conquer: EKF SLAM in O(n). IEEE Trans. Robot. **24**(5), 1107–1120 (2008)
85. O.G. Grasa, J. Civera, J. Montiel, EKF monocular SLAM with relocalization for laparoscopic sequences, in *2011 IEEE International Conference on Robotics and Automation (ICRA)* (IEEE, 2011), pp. 4816–4821
86. E. Süli, D.F. Mayers, *An Introduction to Numerical Analysis* (Cambridge University Press, Cambridge, 2003)
87. L. Polok, V. Ila, M. Solony, P. Smrz, P. Zemcik, Incremental block Cholesky factorization for nonlinear least squares in robotics, in *Robotics: Science and Systems* (2013)
88. R. Mur-Artal, J. Montiel, J.D. Tardos, ORB-SLAM: a versatile and accurate monocular SLAM system (2015), arXiv:1502.00956
89. S. Leutenegger, S. Lynen, M. Bosse, R. Siegwart, P. Furgale, Keyframe-based visual-inertial odometry using nonlinear optimization. Int. J. Robot. Res. **34**(3), 314–334 (2015)
90. Bundle adjustment in the large, http://grail.cs.washington.edu/projects/bal/
91. G. Sibley, L. Matthies, G. Sukhatme, A sliding window filter for incremental SLAM, in *Unifying Perspectives in Computational and Robot Vision* (Springer, Berlin, 2008), pp. 103–112
92. T. Qin, P. Li, S. Shen, VINS-Mono: a robust and versatile monocular visual-inertial state estimator. IEEE Trans. Robot. **34**(4), 1004–1020 (2018)
93. H. Strasdat, A.J. Davison, J.M.M. Montiel, K. Konolige, Double window optimisation for constant time visual SLAM, in *2011 IEEE International Conference on Computer Vision (ICCV)* (2011), pp. 2352–2359
94. G. Dubbelman, B. Browning, COP-SLAM: closed-form online pose-chain optimization for visual SLAM. IEEE Trans. Robot. **31**, 1194–1213 (2015)

95. D. Lee, H. Myung, Solution to the SLAM problem in low dynamic environments using a pose graph and an RGB-D sensor. Sensors **14**(7), 12467–12496 (2014)
96. Y. Latif, C. Cadena, J. Neira, Robust loop closing over time for pose graph SLAM. Int. J. Robot. Res. **32**(14), 1611–1626 (2013)
97. G. Klein, D. Murray, Parallel tracking and mapping for small AR workspaces, in *6th IEEE and ACM International Symposium on Mixed and Augmented Reality, 2007. ISMAR 2007* (IEEE, 2007), pp. 225–234
98. M. Kaess, A. Ranganathan, F. Dellaert, ISAM: incremental smoothing and mapping. IEEE Trans. Robot. **24**(6), 1365–1378 (2008)
99. M. Kaess, H. Johannsson, R. Roberts, V. Ila, J.J. Leonard, F. Dellaert, ISAM2: incremental smoothing and mapping using the Bayes tree. Int. J. Robot. Res. 0278364911430419 (2011)
100. F. Endres, J. Hess, J. Sturm, D. Cremers, W. Burgard, 3-d mapping with an RGB-D camera. IEEE Trans. Robot. **30**(1), 177–187 (2014)
101. D. Hahnel, W. Burgard, D. Fox, S. Thrun, An efficient fastSLAM algorithm for generating maps of large-scale cyclic environments from raw laser range measurements, in *2003 IEEE/RSJ International Conference on Intelligent Robots and Systems, 2003. (IROS 2003). Proceedings*, vol. 1 (IEEE, 2003), pp. 206–211
102. I. Ulrich, I. Nourbakhsh, Appearance-based place recognition for topological localization, in *IEEE International Conference on Robotics and Automation, 2000. Proceedings. ICRA'00*, vol. 2 (IEEE, 2000), pp. 1023–1029
103. X. Gao, T. Zhang, Unsupervised learning to detect loops using deep neural networks for visual SLAM system. Auton. Robot. 1–18 (2015)
104. S. Lloyd, Least squares quantization in PCM. IEEE Trans. Inf. Theory **28**(2), 129–137 (1982)
105. D. Arthur, S. Vassilvitskii, K-means++: the advantages of careful seeding, in *Proceedings of the Eighteenth Annual ACM-SIAM Symposium on Discrete Algorithms* (Society for Industrial and Applied Mathematics, 2007), pp. 1027–1035
106. C. Chow, C. Liu, Approximating discrete probability distributions with dependence trees. IEEE Trans. Inf. Theory **14**(3), 462–467 (1968)
107. M. Cummins, P. Newman, FAB-MAP: probabilistic localization and mapping in the space of appearance. Int. J. Robot. Res. **27**(6), 647–665 (2008)
108. M. Cummins, P. Newman, Accelerating FAB-MAP with concentration inequalities. IEEE Trans. Robot. **26**(6), 1042–1050 (2010)
109. M. Cummins, P. Newman, Appearance-only SLAM at large scale with FAB-MAP 2.0. Int. J. Robot. Res. **30**(9), 1100–1123 (2011)
110. D. Galvez-Lopez, J.D. Tardos, Bags of binary words for fast place recognition in image sequences. IEEE Trans. Robot. **28**(5), 1188–1197 (2012)
111. J.L. Bentley, Multidimensional binary search trees used for associative searching. Commun. ACM **18**(9), 509–517 (1975)
112. J. Sivic, A. Zisserman, Video google: a text retrieval approach to object matching in videos, in *Ninth IEEE International Conference on Computer Vision, 2003. Proceedings* (IEEE, 2003), pp. 1470–1477
113. S. Robertson, Understanding inverse document frequency: on theoretical arguments for IDF. J. Doc. **60**(5), 503–520 (2004)
114. D. Nister, H. Stewenius, Scalable recognition with a vocabulary tree, in *2006 IEEE Computer Society Conference on Computer Vision and Pattern Recognition (CVPR'06)*, vol. 2 (IEEE, 2006), pp. 2161–2168
115. C. Cadena, D. Galvez-Lopez, J.D. Tardos, J. Neira, Robust place recognition with stereo sequences. IEEE Trans. Robot. **28**(4), 871–885 (2012)
116. Y. Hou, H. Zhang, S. Zhou, Convolutional neural network-based image representation for visual loop closure detection, in *2015 IEEE International Conference on Information and Automation* (IEEE, 2015), pp. 2238–2245
117. S. Cascianelli, G. Costante, E. Bellocchio, P. Valigi, M.L. Fravolini, T.A. Ciarfuglia, Robust visual semi-semantic loop closure detection by a covisibility graph and CNN features. Robot. Auton. Syst. **92**, 53–65 (2017)

118. R. Arandjelovic, P. Gronat, A. Torii, T. Pajdla, J. Sivic, NetVLAD: CNN architecture for weakly supervised place recognition, in *Proceedings of the IEEE Conference on Computer Vision and Pattern Recognition* (2016), pp. 5297–5307
119. M. Angelina Uy, G. Hee Lee, PointNetVLAD: deep point cloud based retrieval for large-scale place recognition, in *Proceedings of the IEEE Conference on Computer Vision and Pattern Recognition* (2018), pp. 4470–4479
120. A. Kendall, M. Grimes, R. Cipolla, PoseNet: a convolutional network for real-time 6-DOF camera relocalization, in *Proceedings of the IEEE International Conference on Computer Vision* (2015), pp. 2938–2946
121. B. Williams, M. Cummins, J. Neira, P. Newman, I. Reid, J. Tardós, A comparison of loop closing techniques in monocular SLAM. Robot. Auton. Syst. **57**(12), 1188–1197 (2009)
122. M. Labbé, F. Michaud, Online global loop closure detection for large-scale multi-session graph-based SLAM, in *2014 IEEE/RSJ International Conference on Intelligent Robots and Systems* (IEEE, 2014), pp. 2661–2666
123. R.F. Salas-Moreno, R.A. Newcombe, H. Strasdat, P.H.J. Kelly, A.J. Davison, SLAM++: simultaneous localisation and mapping at the level of objects, in *2013 IEEE Conference on Computer Vision and Pattern Recognition (CVPR)* (2013), pp. 1352–1359
124. Correlation based similarity measure-summary, https://siddhantahuja.wordpress.com/tag/stereo-matching/
125. H. Hirschmuller, D. Scharstein, Evaluation of cost functions for stereo matching, in *2007 IEEE Conference on Computer Vision and Pattern Recognition* (IEEE, 2007), pp. 1–8
126. G. Vogiatzis, C. Hernández, Video-based, real-time multi-view stereo. Image Vis. Comput. **29**(7), 434–441 (2011)
127. A. Handa, R.A. Newcombe, A. Angeli, A.J. Davison, Real-time camera tracking: when is high frame-rate best?, in *European Conference on Computer Vision* (Springer, 2012), pp. 222–235
128. M. Pizzoli, C. Forster, D. Scaramuzza, REMODE: probabilistic, monocular dense reconstruction in real time, in *2014 IEEE International Conference on Robotics and Automation (ICRA)* (IEEE, 2014), pp. 2609–2616
129. J. Montiel, J. Civera, A.J. Davison, Unified inverse depth parametrization for monocular SLAM. Analysis **9**, 1 (2006)
130. J. Civera, A.J. Davison, J.M. Montiel, Inverse depth parametrization for monocular SLAM. IEEE Trans. Robot. **24**(5), 932–945 (2008)
131. A. Handa, T. Whelan, J. McDonald, A. Davison, A benchmark for RGB-D visual odometry, 3D reconstruction and SLAM, in *IEEE International Conference on Robotics and Automation, ICRA, Hong Kong, China* (2014)
132. M. Kazhdan, M. Bolitho, H. Hoppe, Poisson surface reconstruction, in *Proceedings of the Fourth Eurographics Symposium on Geometry Processing*, vol. 7 (2006)
133. J. Stuckler, S. Behnke, Multi-resolution surfel maps for efficient dense 3d modeling and tracking. J. Vis. Commun. Image Represent. **25**(1), 137–147 (2014)
134. M. Alexa, J. Behr, D. Cohen-Or, S. Fleishman, D. Levin, C.T. Silva, Computing and rendering point set surfaces. IEEE Trans. Vis. Comput. Graph. **9**(1), 3–15 (2003)
135. Z.C. Marton, R.B. Rusu, M. Beetz, On fast surface reconstruction methods for large and noisy datasets, in *Proceedings of the IEEE International Conference on Robotics and Automation (ICRA), Kobe, Japan, 12–17 May 2009*
136. A. Hornung, K.M. Wurm, M. Bennewitz, C. Stachniss, W. Burgard, OctoMap: an efficient probabilistic 3d mapping framework based on octrees. Auton. Robot. **34**(3), 189–206 (2013)
137. M. Burri, H. Oleynikova, M.W. Achtelik, R. Siegwart, Real-time visual-inertial mapping, re-localization and planning onboard MAVs in unknown environments, in *2015 IEEE/RSJ International Conference on Intelligent Robots and Systems (IROS)* (IEEE, 2015), pp. 1872–1878
138. R.A. Newcombe, A.J. Davison, S. Izadi, P. Kohli, O. Hilliges, J. Shotton, D. Molyneaux, S. Hodges, D. Kim, A. Fitzgibbon, KinectFusion: real-time dense surface mapping and tracking, in *2011 10th IEEE International Symposium on Mixed and Augmented Reality (ISMAR)* (IEEE, 2011), pp. 127–136

139. R.A. Newcombe, D. Fox, S.M. Seitz, DynamicFusion: reconstruction and tracking of non-rigid scenes in real-time, in *Proceedings of the IEEE Conference on Computer Vision and Pattern Recognition* (2015), pp. 343–352
140. T. Whelan, S. Leutenegger, R.F. Salas-Moreno, B. Glocker, A.J. Davison, ElasticFusion: dense SLAM without a pose graph, in *Proceedings of Robotics: Science and Systems, Rome, Italy* (2015)
141. M. Dou, S. Khamis, Y. Degtyarev, P. Davidson, S.R. Fanello, A. Kowdle, S.O. Escolano, C. Rhemann, D. Kim, J. Taylor et al., Fusion4d: real-time performance capture of challenging scenes. ACM Trans. Graph. (TOG) **35**(4), 114 (2016)
142. M. Innmann, M. Zollhöfer, M. Nießner, C. Theobalt, M. Stamminger, VolumeDeform: real-time volumetric non-rigid reconstruction (2016), arXiv:1603.08161
143. C. Kerl, J. Sturm, D. Cremers, Robust odometry estimation for RGB-D cameras, in *2013 IEEE International Conference on Robotics and Automation (ICRA)* (IEEE, 2013), pp. 3748–3754
144. C. Kerl, J. Sturm, D. Cremers, Dense visual SLAM for RGB-D cameras, in *2013 IEEE/RSJ International Conference on Intelligent Robots and Systems* (IEEE, 2013), pp. 2100–2106
145. J. Gui, D. Gu, S. Wang, H. Hu, A review of visual inertial odometry from filtering and optimisation perspectives. Adv. Robot. **29**, 1289–1301 (2015)
146. A. Martinelli, Closed-form solution of visual-inertial structure from motion. Int. J. Comput. Vis. **106**(2), 138–152 (2014)
147. M. Bloesch, S. Omari, M. Hutter, R. Siegwart, Robust visual inertial odometry using a direct EKF-based approach, in *2015 IEEE/RSJ International Conference on Intelligent Robots and Systems (IROS)* (IEEE, 2015), pp. 298–304
148. M. Li, A.I. Mourikis, High-precision, consistent EKF-based visual-inertial odometry. Int. J. Robot. Res. **32**, 690–711 (2013)
149. G. Huang, M. Kaess, J.J. Leonard, Towards consistent visual-inertial navigation, in *2014 IEEE International Conference on Robotics and Automation ICRA, Hong Kong, Peoples R China, May 31–June 07* (2014), pp. 4926–4933
150. C. Forster, L. Carlone, F. Dellaert, D. Scaramuzza, IMU preintegration on manifold for efficient visual-inertial maximum-a-posteriori estimation, in *Robotics: Science and Systems XI*, no. EPFL-CONF-214687 (2015)
151. M. Tkocz, K. Janschek, Towards consistent state and covariance initialization for monocular SLAM filters. J. Intell. Robot. Syst. **80**, 475–489 (2015)
152. V. Usenko, J. Engel, J. Stueckler, D. Cremers, Direct visual-inertial odometry with stereo cameras, in *IEEE International Conference on Robotics and Automation (ICRA)* (2016)
153. A. Nüchter, J. Hertzberg, Towards semantic maps for mobile robots. Robot. Auton. Syst. **56**(11), 915–926 (2008)
154. J. Civera, D. Gálvez-López, L. Riazuelo, J.D. Tardós, J. Montiel, Towards semantic SLAM using a monocular camera, in *2011 IEEE/RSJ International Conference on Intelligent Robots and Systems (IROS)* (IEEE, 2011), pp. 1277–1284
155. H.S. Koppula, A. Anand, T. Joachims, A. Saxena, Semantic labeling of 3d point clouds for indoor scenes, in *Advances in Neural Information Processing Systems* (2011), pp. 244–252
156. A. Anand, H.S. Koppula, T. Joachims, A. Saxena, Contextually guided semantic labeling and search for three-dimensional point clouds. Int. J. Robot. Res. 0278364912461538 (2012)
157. N. Fioraio, L. Di Stefano, Joint detection, tracking and mapping by semantic bundle adjustment, in *2013 IEEE Conference on Computer Vision and Pattern Recognition (CVPR)* (2013), pp. 1538–1545
158. R.F. Salas-Moreno, B. Glocken, P.H. Kelly, A.J. Davison, Dense planar SLAM, in *2014 IEEE International Symposium on Mixed and Augmented Reality (ISMAR)* (IEEE, 2014), pp. 157–164
159. J. Stückler, N. Biresev, S. Behnke, Semantic mapping using object-class segmentation of RGB-D images, in *2012 IEEE/RSJ International Conference on Intelligent Robots and Systems* (IEEE, 2012), pp. 3005–3010
160. I. Kostavelis, A. Gasteratos, Learning spatially semantic representations for cognitive robot navigation. Robot. Auton. Syst. **61**(12), 1460–1475 (2013)

161. C. Couprie, C. Farabet, L. Najman, Y. LeCun, Indoor semantic segmentation using depth information (2013), arXiv:1301.3572
162. J. Deng, W. Dong, R. Socher, L.-J. Li, K. Li, L. Fei-Fei, ImageNet: a large-scale hierarchical image database, in *CVPR09* (2009)
163. A. Krizhevsky, I. Sutskever, G.E. Hinton, ImageNet classification with deep convolutional neural networks, in *Advances in Neural Information Processing Systems* (2012), pp. 1097–1105
164. K. He, X. Zhang, S. Ren, J. Sun, Deep residual learning for image recognition (2015), arXiv:1512.03385
165. S. Ren, K. He, R. Girshick, J. Sun, Faster R-CNN: towards real-time object detection with region proposal networks, in *Advances in Neural Information Processing Systems* (2015), pp. 91–99
166. J. Long, E. Shelhamer, T. Darrell, Fully convolutional networks for semantic segmentation (2014), arXiv:1411.4038
167. S. Zheng, S. Jayasumana, B. Romera-Paredes, V. Vineet, Z. Su, D. Du, C. Huang, P. Torr, Conditional random fields as recurrent neural networks, in *International Conference on Computer Vision (ICCV)* (2015)
168. S. Gupta, P. Arbeláez, R. Girshick, J. Malik, Indoor scene understanding with RGB-D images: bottom-up segmentation, object detection and semantic segmentation. Int. J. Comput. Vis. 1–17 (2014)
169. K. Konda, R. Memisevic, Learning visual odometry with a convolutional network, in *International Conference on Computer Vision Theory and Applications* (2015)
170. Y. Hou, H. Zhang, S. Zhou, Convolutional neural network-based image representation for visual loop closure detection (2015), arXiv:1504.05241
171. S.Y. An, J.G. Kang, L.K. Lee, S.Y. Oh, Line segment-based indoor mapping with salient line feature extraction. Adv. Robot. **26**(5–6), 437–460 (2012)
172. H. Zhou, D. Zou, L. Pei, R. Ying, P. Liu, W. Yu, StructSLAM: visual slam with building structure lines. IEEE Trans. Veh. Technol. **64**, 1364–1375 (2015)
173. D. Benedettelli, A. Garulli, A. Giannitrapani, Cooperative SLAM using M-space representation of linear features. Robot. Auton. Syst. **60**(10), 1267–1278 (2012)
174. J.P. Saarinen, H. Andreasson, T. Stoyanov, A.J. Lilienthal, 3d normal distributions transform occupancy maps: an efficient representation for mapping in dynamic environments. Int. J. Robot. Res. **32**(14), 1627–1644 (2013)
175. W. Maddern, M. Milford, G. Wyeth, CAT-SLAM: probabilistic localisation and mapping using a continuous appearance-based trajectory. Int. J. Robot. Res. **31**(4SI), 429–451 (2012)
176. H. Wang, Z.-G. Hou, L. Cheng, M. Tan, Online mapping with a mobile robot in dynamic and unknown environments. Int. J. Model. Identif. Control **4**(4), 415–423 (2008)
177. D. Zou, P. Tan, CoSLAM: collaborative visual SLAM in dynamic environments. IEEE Trans. Pattern Anal. Mach. Intell. **35**(2), 354–366 (2013)
178. T.A. Vidal-Calleja, C. Berger, J. Sola, S. Lacroix, Large scale multiple robot visual mapping with heterogeneous landmarks in semi-structured terrain. Robot. Auton. Syst. **59**(9), 654–674 (2011)